Modular forms and functions

Modular forms and functions

ROBERT A. RANKIN

Professor of Mathematics in the University of Glasgow

CAMBRIDGE UNIVERSITY PRESS

CAMBRIDGE

LONDON · NEW YORK · MELBOURNE

CAMBRIDGE UNIVERSITY PRESS
Cambridge, New York, Melbourne, Madrid, Cape Town, Singapore, São Paulo, Delhi

Cambridge University Press
The Edinburgh Building, Cambridge CB2 8RU, UK

Published in the United States of America by Cambridge University Press, New York

www.cambridge.org
Information on this title: www.cambridge.org/9780521212120

First published 1977
This digitally printed version 2008

A catalogue record for this publication is available from the British Library

Library of Congress Cataloguing in Publication data

Rankin, Robert Alexander, 1915–
Modular forms and functions

Bibliography: p. 361 Includes indexes
1. Forms, modular 2. Functions, modular I. Title
QA243.R36 512.9′44 76–11089

ISBN 978-0-521-21212-0 hardback
ISBN 978-0-521-09168-8 paperback

TO MY WIFE

Contents

	Preface	*page* xi
1	**Groups of matrices and bilinear mappings**	
1.1	Notation	1
1.2	The modular group	7
1.3	The subgroups Γ^2, Γ^3, Γ^4 and $\Gamma'(1)$	15
1.4	The level of a subgroup; congruence subgroups	19
1.5	Groups of level 2	28
1.6	Groups of level 3	30
1.7	Further results	32
2	**Mapping properties**	
2.1	Conformal mappings	39
2.2	Fixed points	42
2.3	Fundamental regions	47
2.4	Construction of fundamental regions for $\hat{\Gamma}(1)$ and its subgroups	51
2.5	Further results	66
3	**Automorphic factors and multiplier systems**	
3.1	Introduction	70
3.2	The functions $\sigma(S, T)$ and $w(S, T)$	73
3.3	Automorphic factors on subgroups of the modular group	77
3.4	Multiplier systems on $\Gamma(1)$, Γ^2 and Γ^3	82
3.5	Further results	86
4	**General properties of modular forms**	
4.1	Definitions and general theorems	88
4.2	Dimensions of spaces of modular forms	102
4.3	Relations between modular forms	108
4.4	Modular forms of weight 2	122
4.5	Orders of magnitude	128
4.6	Further results	130

5 Construction of modular forms

5.1 Poincaré series 134
5.2 The Hilbert space of cusp forms 144
5.3 The Fourier coefficients of Poincaré series 155
5.4 Kloosterman sums 164
5.5 Poincaré series belonging to $\Gamma(N)$ 172
5.6 Poincaré series on $\Gamma(N^2)$ 179
5.7 Modular forms on $\bar{\Gamma}(N)$ of weight 2 183
5.8 Further results 191

6 Functions belonging to the full modular group

6.1 Modular forms of even weight with constant multiplier system 194
6.2 Poincaré series 201
6.3 The case $\delta_k = 1$ 202
6.4 Modular forms of any real weight 205
6.5 Modular equations 209
6.6 Further remarks 212

7 Groups of level 2 and sums of squares

7.1 Hermite and theta functions 215
7.2 Modular functions of level 2 226
7.3 Eisenstein series belonging to $\Gamma(2)$ 232
7.4 Functions belonging to $\Gamma_V(2)$; sums of squares 238
7.5 Further remarks 243

8 Modular forms of level *N*

8.1 Forms of fixed character or divisor 245
8.2 The interaction of the operators R^χ and D_t 250
8.3 The operator J_n 254
8.4 The operator H_q 260
8.5 The operator L_q 264
8.6 The conjugate linear map K 269
8.7 Historical remarks 271

9 Hecke operators and congruence groups

9.1 Double coset modules 273
9.2 Definition and properties of Hecke operators 289
9.3 The effect of Hecke operators on Poincaré series 307

9.4 Eigenforms 317
9.5 Historical and other remarks 333

10 Applications
10.1 Dirichlet series 337
10.2 Eigenforms for the full modular group 342
10.3 Eigenforms for $\Gamma_V(2)$ and $\Gamma(2)$ 343
10.4 Eigenforms of level 4 350
10.5 Final remarks 359

Bibliography 361
Index of special symbols 369
Index of authors 380
Subject index 382

Preface

This book has grown out of lectures given in 1963–64 in Indiana University and before that in the University of Cambridge. It aims to provide a reasonably elementary introduction to the theory of elliptic modular functions and forms.

Chapter 1 is concerned with the study of the modular group $SL(2, \mathbb{Z})$ and its more important subgroups. These are studied mainly as matrix groups but also as groups of mappings, although it is in the second chapter that the mapping properties are most closely investigated and it is there that fundamental regions are constructed. Since our concern is not with the more general theory of automorphic functions on Fuchsian groups, it is possible to give a relatively simple account.

In another respect our treatment of the theory is fuller than is customary in a textbook on automorphic or modular forms, since it is not confined solely to modular forms of integral weight (dimension). Multiplier systems of arbitrary real weight possess a complicated structure. We have therefore presented the elementary theory in chapter 3 in a more general form than is necessary for our purposes, so that it can be applied to $SL(2, \mathbb{R})$ and its subgroups if desired; it is only in §3.3 that the theory is restricted to subgroups of the modular group.

Chapter 4 contains the basic definitions and properties of modular forms of arbitrary real weight with corresponding multiplier systems. It is important to know the dimensions of vector spaces of modular forms belonging to various subgroups. Where possible we have, in this chapter and later, determined these dimensions by elementary arguments and without recourse to the Riemann–Roch theorem. This is, in fact, always possible when the genus is zero, and in certain other cases also.

There are many ways of constructing modular forms. In chapter 5 we have followed Petersson in using as our building blocks a special family of Poincaré series. This method is particularly convenient for forms of real weight $k > 2$ and arbitrary multiplier system. The Fourier coefficients of these series, although complicated in form, are of interest and involve Kloosterman sums of various types. In particular cases these sums are considered in detail and are estimated with sufficient accuracy to yield results in the limiting case when the weight $k=2$. The Petersson inner product is introduced in this chapter. This product is of importance, not only because it allows us to regard the space of cusp forms as a Hilbert space, but also because it can be used to provide explicit formulae for the Fourier coefficients of an arbitrary cusp form.

Chapter 6 is concerned with the simplest situation when the group considered is the full modular group. The standard functions, such as the modular discriminant Δ, Klein's invariant J and the Eisenstein series E_k $(k \geq 4)$ are introduced and studied. In the last section the six different multiplier systems of real weight are found by considering real powers of Δ.

Modular forms of level 2 are discussed in chapter 7. The most important of these are the three simple theta functions ϑ_2, ϑ_3, ϑ_4 and certain polynomials in the graded ring they generate. The corresponding multiplier systems are described and dimensions of various subspaces found. Modular equations are obtained in a few very simple cases. The Eisenstein series of level 2 and integral weight $k \geq 3$ are studied and their Fourier coefficients are evaluated. The theory is then applied to the problem of finding the number of representations of a number as the sum of $2k$ squares.

In chapter 8 we return to a more general situation and discuss forms of level N and integral weight k with simple multiplier systems. The vector space of forms of this type can be expressed, as was done by Hecke, as a direct sum of subspaces each corresponding to a different character χ modulo N and divisor t of N. The action on these subspaces of various linear operators is studied in preparation for the following chapter, which is concerned with the theory of Hecke operators. These operators and their effect on Fourier coefficients and on Poincaré series are examined in detail. The Hecke–Petersson theory of eigenforms and the more recent Atkin–Lehner theory of newforms are developed and extended.

The final chapter begins by describing the connexion between these theories and Dirichlet series having Euler products, and concludes with a number of applications to spaces of levels 1 and 2, particular attention being paid to the eigenforms that arise in the study of the representations of a number as a sum of squares. When the weight is odd, these eigenforms have coefficients with rather curious multiplicative properties.

The material in the book is taken in the main from articles in the research journals and relies heavily on the fundamental work of Hecke and Petersson, but includes also some new results, particularly in chapters 3, 9 and 10. Modular function theory is now so vast a subject that a book of reasonable length cannot include every aspect and so necessarily reflects the particular current interests of the author. A quarter of a century ago, when the idea of writing a book on the subject was first suggested to me by my friend Mr R. J. L. Kingsford, who was then Secretary to the Syndics of the Cambridge University Press, I had visions of writing a comprehensive treatise that would give a complete account of every part of the subject. Such a task was difficult then and would be virtually impossible today. Thus in the present work there will be found only a brief account of order results concerning the Fourier coefficients of modular forms (an earlier interest of the author) and even briefer reference to the important recent work of Deligne and Serre in that area. Moreover, there is no mention of such subjects as the Selberg trace formula, congruence properties of coefficients, transcendentality of values of coefficients, complex multiplication and elliptic curves.

In conclusion, it is a pleasure to acknowledge my indebtedness to the individuals and institutions that have aided me in the writing of this book and of its earlier versions. In particular, my thanks are due to the Universities of Cambridge, Glasgow and Indiana, and to Clare Hall, Cambridge, where the major part of chapters 5–7 was completed during my three months' tenure of a Visiting Fellowship there. I am grateful to Dr Bruce C. Berndt and Dr W. Wilson Stothers for their detailed comments and criticisms of the manuscript; it is scarcely necessary to state that the imperfections and inaccuracies that remain are the sole responsibility of the author. Finally, I wish to express my gratitude to Miss Doris M. Caldwell, who typed the greater part of the manuscript, for her skill and patience.

R.A.R.

Glasgow
March 1976

1: Groups of matrices and bilinear mappings

1.1. Notation. Modular functions and forms will first be defined in chapter 4. In this chapter we study the groups on which these functions are defined. We write:

$\mathbb{C}$ for the set of all (finite) complex numbers, with the usual topology.

$\mathbb{R}$ for the set of all (finite) real numbers.

$\mathbb{Q}$ for the set of all rational numbers.

$\mathbb{Z}$ for the set of all (rational) integers.

$\mathbb{Z}^+$ for the set of all positive integers.

$\mathbb{H}=\{z: z\in\mathbb{C}, \operatorname{Im} z>0\}$, the *upper half-plane.*

$\bar{\mathbb{C}}=\mathbb{C}\cup\{\infty\}$, *the extended complex plane.* This is the one-point compactification of $\mathbb{C}$. In the topology on $\bar{\mathbb{C}}$, a set A is open if either (i) A is an open subset of $\mathbb{C}$, or (ii) $\infty\in A$ and $\bar{\mathbb{C}}-A$ is compact in $\mathbb{C}$. With this topology, $\bar{\mathbb{C}}$ is homeomorphic to the two-sphere, i.e. to the Riemann sphere $\{(x, y, z)\in\mathbb{R}^3: x^2+y^2+z^2=1\}$.

$\bar{\mathbb{R}}=\mathbb{R}\cup\{\infty\}$, *the one-point compactification of* $\mathbb{R}$.

$\bar{\mathbb{H}}=\mathbb{H}\cup\bar{\mathbb{R}}$.

$\mathbb{P}=\mathbb{Q}\cup\{\infty\}$.

$\mathbb{H}'=\mathbb{H}\cup\mathbb{P}$.

Other subsets of $\bar{\mathbb{C}}$ will be defined later.

We write throughout

$$T=\begin{bmatrix} a & b \\ c & d \end{bmatrix}, \quad S=\begin{bmatrix} \alpha & \beta \\ \gamma & \delta \end{bmatrix} \tag{1.1.1}$$

for $a, b, c, d, \alpha, \beta, \gamma, \delta\in\mathbb{C}$ and put

$$|T|=\det T=ad-bc.$$

Let

$$\Theta=\{T: a, b, c, d\in\mathbb{C}, |T|=1\}, \tag{1.1.2}$$

and

$$\Omega=\{T: a, b, c, d\in\mathbb{R}, |T|=1\}. \tag{1.1.3}$$

Then Θ is a group under matrix multiplication with the identity element

$$I=\begin{bmatrix}1 & 0\\ 0 & 1\end{bmatrix}, \tag{1.1.4}$$

and Θ contains Ω as a subgroup. In group theory the groups Θ and Ω are referred to as the *special linear* groups $\mathrm{SL}(2,\mathbb{C})$ and $\mathrm{SL}(2,\mathbb{R})$, respectively. It is easily verified that, for each $T\in\Theta$,

$$T^{-1}=\begin{bmatrix}d & -b\\ -c & a\end{bmatrix}. \tag{1.1.5}$$

The subgroup consisting of I and

$$-I=\begin{bmatrix}-1 & 0\\ 0 & -1\end{bmatrix} \tag{1.1.6}$$

is denoted by Λ; it is the centre of Θ and Ω.

For each $T\in\Theta$ we write

$$\operatorname{tr} T=a+d=2\cos\theta_T,$$

where θ_T is any complex number for which this equation is valid. Then it is easily shown by induction that, for any $q\in\mathbb{Z}^+$,

$$T^q=F_qT-F_{q-1}I, \tag{1.1.7}$$

where

$$F_q=\frac{\sin q\theta_T}{\sin\theta_T}. \tag{1.1.8}$$

This is a polynomial of degree $q-1$ in $\cos\theta_T$ and so is defined even when $\sin\theta_T=0$. We note also that

$$\operatorname{tr}(T^q)=2\cos q\theta_T. \tag{1.1.9}$$

Further, if $\theta_T=\pi k/q$ for some integer k, then

$$T^q=(-1)^kI. \tag{1.1.10}$$

With each $T\in\Theta$ we associate a bilinear† mapping, which we also call T, defined on $\bar{\mathbb{C}}$ by

$$w=T(z)=\frac{az+b}{cz+d}\quad(z\in\bar{\mathbb{C}}),$$

† Other terms used are *linear, linear fractional* and *Möbius*. No confusion should arise with the use of the word *bilinear* in multilinear algebra.

and we also, for brevity, write Tz in place of $T(z)$. For example,

$$T(-d/c)=\infty \quad \text{and} \quad T\infty=a/c.$$

These relations hold even when $c=0$, for then a and d are non-zero, so that $-d/c$ and a/c both mean ∞; this is a consequence of the rules

$$z+\infty=\infty+z=\infty, \quad \frac{z}{\infty}=0 \quad (z\in\mathbb{C}),$$

$$z\infty=\infty z=\infty, \quad \frac{z}{0}=\infty \quad (z\in\bar{\mathbb{C}}-\{0\}).$$

The mapping T is a bijective mapping of $\bar{\mathbb{C}}$ onto itself, the inverse mapping being given by

$$z=T^{-1}(w)=\frac{dw-b}{-cw+a}.$$

if Γ is any subgroup of Θ, the mappings T defined by matrices $T\in\Gamma$ form a group $\hat{\Gamma}$ under composition as group operation; i.e., if $S\in\Gamma$, $T\in\Gamma$, then $(ST)(z)$ means $S\{T(z)\}$. This is easily checked. For, by the definitions of S and T,

$$ST=\begin{bmatrix}\alpha a+\beta c & \alpha b+\beta d\\ \gamma a+\delta c & \gamma b+\delta d\end{bmatrix}, \tag{1.1.11}$$

while

$$S\{T(z)\}=\frac{\alpha\dfrac{az+b}{cz+d}+\beta}{\gamma\dfrac{az+b}{cz+d}+\delta}=\frac{(\alpha a+\beta c)z+(\alpha b+\beta d)}{(\gamma a+\delta c)z+(\gamma b+\delta d)}.$$

The group $\hat{\Gamma}$ is called the *inhomogeneous group* associated with Γ, which is called a *homogeneous group.*

Let ϕ denote the mapping:

$$\phi\colon \text{matrix } T\mapsto \text{bilinear mapping } T.$$

Then the above remarks show that ϕ is a homomorphism of Θ onto $\hat{\Theta}$. Let Γ be a subgroup of Θ, so that ϕ is a homomorphism of Γ onto $\hat{\Gamma}$. The subgroups $\hat{\Gamma}$ that we consider will usually act not on the whole of $\bar{\mathbb{C}}$, but on some subset $\mathbb{D}$. We suppose that $\mathbb{D}$ is a subset of $\bar{\mathbb{C}}$ such that $\hat{\Gamma}\mathbb{D}=\mathbb{D}$, i.e. $T\mathbb{D}=\mathbb{D}$ for all $T\in\hat{\Gamma}$. We suppose further that $\mathbb{D}$ contains more than two points; usually $\mathbb{D}$ will be $\bar{\mathbb{C}}$, $\mathbb{C}$, $\mathbb{H}$ or $\mathbb{H}'$.

The identity mapping in $\hat{\Gamma}$ is $w = z$ $(z \in \mathbb{D})$, and we have

$$T(z) = \frac{az+b}{cz+d} = z \quad \text{for all } z \in \mathbb{D}$$

if and only if $az + b = cz^2 + dz$ for all $z \in \mathbb{D}$. Since $\mathbb{D}$ contains more than two points, this gives

$$b = c = a - d = 0;$$

i.e. $a = d = \pm 1$, $b = c = 0$, so that $T = \pm I$. Thus the kernel of ϕ is Λ if $-I \in \Gamma$ and is I if $-I \notin \Gamma$. Hence we have

$$\hat{\Gamma} \cong \Gamma/\Lambda \quad (\text{if } -I \in \Gamma), \quad \hat{\Gamma} \cong \Gamma \quad (\text{if } -I \notin \Gamma). \quad (1.1.12)$$

In particular, $\hat{\Theta} \cong \Theta/\Lambda$ and $\hat{\Omega} \cong \Omega/\Lambda$. The groups $\hat{\Theta}$ and $\hat{\Omega}$ are referred to in group theory as the *linear fractional* groups LF(2, $\mathbb{C}$) and LF(2, $\mathbb{R}$), respectively.

When $-I \notin \Gamma$ we can adjoin $-I$ to Γ, obtaining an overgroup $\bar{\Gamma}(=\Gamma\Lambda = \Lambda\Gamma)$ of Γ having Γ as a subgroup of index 2, and then

$$\hat{\Gamma} \cong \Gamma \cong \bar{\Gamma}/\Lambda. \quad (1.1.13)$$

By writing Γ for the homogeneous group and $\hat{\Gamma}$ for the associated inhomogeneous group we indicate that we regard the latter as being determined by the former. This point of view is especially convenient when we are concerned with algebraic properties of groups and, in particular, with multiplier systems. On the other hand, a different point of view can be taken when analytic properties are under discussion. For we shall be concerned with classes of functions f defined on $\mathbb{H}$ for which the quotient $f(Tz)/f(z)$ is the same for each member of the class, when T belongs to a certain given group of bilinear transformations. Since this quotient takes the same value for $-T$ as for T, it is often convenient to assume that $-I$ belongs to the associated matrix group. Accordingly, if we start with a group $\hat{\Gamma}$ of bilinear transformations, we may define the associated homogeneous group Γ to consist of all matrices T such that the associated bilinear transformation belongs to $\hat{\Gamma}$; it then follows that $-T \in \Gamma$ whenever $T \in \Gamma$. It is easily checked that Γ is in fact a group. (Authors who adopt this analytic point of view commonly write Γ for the inhomogeneous group and $\bar{\Gamma}$ for the homogeneous group.)

We now introduce a notation that we shall find useful when dealing with coset representatives. Let S be a set closed under an operation, which we call multiplication, and let A and B be subsets of S. Then we write, as is customary,

$$AB=\{x\in S: x=ab, a\in A, b\in B\}.$$

An element $x\in AB$ may be expressible in more than one way as a product ab for $a\in A, b\in B$. If, however, each $x\in AB$ is expressible in exactly one way as a product ab for $a\in A$, $b\in B$, we write

$$AB=A\cdot B.$$

It is easily verified that $A\cdot(B\cdot C)=(A\cdot B)\cdot C$, provided that one side is defined, in which case the other is also.

If A is a finite set, we write $|A|$ for the number of its elements.

Now suppose that $\Gamma_2\subseteq\Gamma_1\subseteq\Theta$, Γ_1 and Γ_2 being subgroups of Θ. Then the statements

$$\Gamma_1=\Gamma_2\cdot\mathcal{R},\quad \Gamma_1=\mathcal{L}\cdot\Gamma_2$$

are equivalent to the statements that $\mathcal{R}$ is a set of right coset representatives of Γ_1 modulo Γ_2, and that $\mathcal{L}$ is a set of left coset representatives, respectively. We call $\mathcal{R}$ a *right transversal*, and $\mathcal{L}$ a *left transversal*, of Γ_2 in Γ_1. If $|\mathcal{R}|$ is finite, so is $|\mathcal{L}|$ and

$$|\mathcal{R}|=|\mathcal{L}|=[\Gamma_1:\Gamma_2],$$

the index of Γ_2 in Γ_1. Note that, if $T\in\mathcal{R}$ and $-I\in\Gamma_2$, then $-T\notin\mathcal{R}$.

Similar notations can be used with inhomogeneous groups. We note also that, if $-I\in\Gamma_2\subseteq\Gamma_1\subseteq\Theta$ and $\Gamma_1=\Gamma_2\cdot\mathcal{R}$, then $\hat{\Gamma}_1=\hat{\Gamma}_2\cdot\hat{\mathcal{R}}$, where there is a one-to-one correspondence between matrices in $\mathcal{R}$ and transformations in $\hat{\mathcal{R}}$; for this reason we shall usually write not only $\Gamma_1=\Gamma_2\cdot\mathcal{R}$ but also $\hat{\Gamma}_1=\hat{\Gamma}_2\cdot\mathcal{R}$.

Theorem 1.1.1. *Let Γ_2 and Γ_1 be subgroups of Θ with $\Gamma_2\subseteq\Gamma_1$. Then, if $\Gamma_1=\Gamma_2\cdot\mathcal{R}$ and S is any member of Γ_1, $\Gamma_1=\Gamma_2\cdot(\mathcal{R}S)$. A similar result holds in the inhomogeneous case.*

Proof. For $\Gamma_1=\Gamma_1 S=(\Gamma_2\cdot\mathcal{R})S=\Gamma_2\cdot(\mathcal{R}S)$.

Theorem 1.1.2. *Let Γ_2 be a subgroup of finite index μ in a group Γ_1, and let S be a fixed member of Γ_1. Then there exist a finite number of* *elements $L_1, L_2, \ldots, L_m$, say, in Γ_1 and m disjoint sets*

$$\mathscr{S}_i = \bigcup\{L_i S^k : 0 \le k < \sigma_i\} \quad (1 \le i \le m),$$

where

$$\sigma_i = \min\{k : S^k \in L_i^{-1}\Gamma_2 L_i, k \in \mathbb{Z}^+\}, \tag{1.1.14}$$

such that

$$\mu = \sigma_1 + \sigma_2 + \cdots + \sigma_m \tag{1.1.15}$$

and

$$\Gamma_1 = \Gamma_2 \cdot \bigcup_{i=1}^{m} \mathscr{S}_i.$$

Moreover, if S has finite order σ, then σ_i divides σ for $1 \le i \le m$. Also, if Γ_2 is normal in Γ_1, then $\sigma_i = \sigma_0$, say, for $1 \le i \le m$, and so

$$\mu = m\sigma_0. \tag{1.1.16}$$

Proof. Take any $L_1 \in \Gamma_1$ and define σ_1 by (1.1.14); since $L_1^{-1}\Gamma_2 L_1$ has finite index μ in Γ_1, σ_1 is a finite positive number and the members of $\mathscr{S}_1$ belong to σ_1 different right cosets of Γ_2 in Γ_1. If $\mu = \sigma_1$, this completes the proof and $m = 1$ in this case. If $\mu > \sigma_1$ we take any L_2 not belonging to $\Gamma_2\mathscr{S}_1$ and define σ_2 by (1.1.14). As before, the σ_2 elements L_2S^k $(0 \le k < \sigma_2)$ belong to different right cosets of Γ_2. Moreover $L_2S^k \notin \Gamma_2\mathscr{S}_1$; for if $L_2S^k \in \Gamma_2\mathscr{S}_1$ then $L_2 \in \Gamma_2\mathscr{S}_1S^{-k} = \Gamma_2\mathscr{S}_1$, which is false. If $\mu = \sigma_1 + \sigma_2$ the theorem follows; if $\mu > \sigma_1 + \sigma_2$, we take an $L_3 \notin \Gamma_2(\mathscr{S}_1 \cup \mathscr{S}_2)$ and proceed similarly. Since μ is finite and $\sigma_i > 0$ for each i, there exists a positive integer m such that (1.1.15) holds and the process then terminates, giving the required result. The two final sentences in the enunciation are immediate consequences.

Note that, when $L_1, L_2, \ldots, L_r$ $(r < m)$ have been chosen, L_{r+1} may be any member of Γ_1 not in $\Gamma_2\bigcup_{i=1}^{r}\mathscr{S}_i$. Also, although the elements $L_1, \ldots, L_m$ are not uniquely determined, the integer m is the same and so are the numbers $\sigma_1, \sigma_2, \ldots, \sigma_m$ (in some order) for all choices of $L_1, \ldots, L_m$.

Theorem 1.1.3. *Let Γ_2 be a normal subgroup of finite index μ in a group Γ_1, and let S be a fixed member of Γ_1. Let σ be the least positive* *integer such that $S^\sigma \in \Gamma_2$ and write*

$$\mathscr{S} = \bigcup\{S^k : 0 \le k < \sigma\}.$$

Then there exist $m = \mu/\sigma$ distinct elements $L_1, L_2, \ldots, L_m$ of Γ_1 such that

$$\Gamma_1 = \mathscr{S} \cdot \Gamma_2 \cdot \mathscr{L}, \tag{1.1.17}$$

where $\mathscr{L} = \bigcup\{L_i : 1 \le i \le m\}$. Also, if Γ_1^* and Γ_2^* are the subgroups of Γ_1 generated by S and S^σ, respectively, and $\Gamma_2 = \Gamma_2^* \cdot \mathscr{R}$, then

$$\Gamma_1 = \Gamma_1^* \cdot \mathscr{R} \cdot \mathscr{L}. \tag{1.1.18}$$

Proof. The proof of (1.1.17) is similar to that of theorem 1.1.2. The normality of Γ_2 in Γ_1 comes in when we infer from

$$S^k T_2 L_i = S^l T_2' L_j \quad (0 \le l \le k < \sigma;\ T_2, T_2' \in \Gamma_2),$$

that $L_j \in \mathscr{S}\Gamma_2 L_i$, and so $L_j = L_i$, $l = k$ and $T_2' = T_2$. We then have, since $\Gamma_1^* = \mathscr{S} \cdot \Gamma_2^*$,

$$\Gamma_1 = \mathscr{S} \cdot \Gamma_2 \cdot \mathscr{L} = \mathscr{S} \cdot \Gamma_2^* \cdot \mathscr{R} \cdot \mathscr{L} = \Gamma_1^* \cdot \mathscr{R} \cdot \mathscr{L},$$

which is (1.1.18).

We note that, by (1.1.11),

$$\operatorname{tr} ST = \operatorname{tr} TS \tag{1.1.19}$$

whenever S and T belong to Θ. In particular, we deduce that

$$\operatorname{tr} L^{-1}TL = \operatorname{tr} T \tag{1.1.20}$$

whenever L and T belong to Θ; i.e. conjugate elements have the same trace.

If A and B are elements of a group, we write $\langle A, B\rangle$ for the subgroup generated by them, and use similar notations for any number of generators.

We conclude this subsection by introducing another convenient notation. We shall write

$$x := y \quad \text{or} \quad y =: x$$

to denote that x is a new symbol, which is defined to be equal to y; y will often be a rather complicated expression.

1.2. The modular group. We write

$$\Gamma(1) := \{T \in \Omega : a, b, c, d \in \mathbb{Z}\}. \tag{1.2.1}$$

It follows from (1.1.5, 11) that this set of matrices is a group, and it is known as the (*homogeneous*) *modular group*.† The corresponding group of mappings is denoted by $\hat{\Gamma}(1)$ and is called the (*inhomogeneous*) *modular group*. By (1.1.12), $\hat{\Gamma}(1) \cong \Gamma(1)/\Lambda$. Alternative notations are

$$\Gamma(1) = \mathrm{SL}(2, \mathbb{Z}), \quad \hat{\Gamma}(1) = \mathrm{LF}(2, \mathbb{Z}).$$

The following special matrices belonging to $\Gamma(1)$ occur with great frequency:

$$U := \begin{bmatrix} 1 & 1 \\ 0 & 1 \end{bmatrix}, \quad V := \begin{bmatrix} 0 & -1 \\ 1 & 0 \end{bmatrix}, \quad W := \begin{bmatrix} 1 & 0 \\ 1 & 1 \end{bmatrix}. \tag{1.2.2}$$

The corresponding mappings are given by

$$Uz = z + 1, \quad Vz = -1/z, \quad Wz = \frac{z}{z+1}.$$

We note that, for any $k \in \mathbb{Z}$,

$$U^k = \begin{bmatrix} 1 & k \\ 0 & 1 \end{bmatrix}. \tag{1.2.3}$$

Further,

$$V^2 = -I, \quad P^3 = -I, \tag{1.2.4}$$

where

$$P := VU = \begin{bmatrix} 0 & -1 \\ 1 & 1 \end{bmatrix}. \tag{1.2.5}$$

Also

$$P^2 = \begin{bmatrix} -1 & -1 \\ 1 & 0 \end{bmatrix}, \quad W = UVU. \tag{1.2.6}$$

The mappings V and P therefore have periods 2 and 3 respectively.

We denote by Γ_U the subgroup of $\Gamma(1)$ generated by $\pm U$; it consists of all matrices $\pm U^n$ $(n \in \mathbb{Z})$. The corresponding mappings are the translations

$$w = z + n \quad (n \in \mathbb{Z}).$$

† German: *Modulgruppe*.

We write similarly, Γ_{U^k} ($k \in \mathbb{Z}^+$) for the group generated by $\pm U^k$; the corresponding mappings are the translations

$$w = x + k \quad (n \in \mathbb{Z}).$$

We now consider the group $\Gamma(1)$ more closely. Let c, d be any two coprime integers. Then we can find integers a, b such that $ad - bc = 1$; i.e. we can find a $T \in \Gamma(1)$ with second row $[c, d]$. Further if a', b' is any other pair of integers with $a'd - b'c = 1$, then, by elementary number theory,

$$a' = a + nc, \quad b' = b + nd$$

for some $n \in \mathbb{Z}$, and conversely. Thus the only matrices in $\Gamma(1)$ with second row $[c, d]$ are the matrices

$$T' = U^n T \quad (n \in \mathbb{Z}).$$

Hence, if S is any member of the right coset $\Gamma_U T$ of Γ_U in $\Gamma(1)$, then $[\gamma, \delta] = \pm[c, d]$, and conversely. We deduce

Theorem 1.2.1. *Let $\mathcal{R}$ be a set of matrices $T \in \Gamma(1)$ with the property that, for each $S \in \Gamma(1)$ there is exactly one $T \in \mathcal{R}$ such that $[c, d] = \pm[\gamma, \delta]$. Then*

$$\Gamma(1) = \Gamma_U \cdot \mathcal{R}. \tag{1.2.7}$$

Conversely, if (1.2.7) holds, then $\mathcal{R}$ has the property stated. A similar result holds for the inhomogeneous groups $\hat{\Gamma}(1)$, $\hat{\Gamma}_U$ and a set of mappings $\hat{\mathcal{R}}$ with corresponding properties.

The next theorem shows that if T is any element of $\Gamma(1)$ we can find a conjugate element $L^{-1}TL$ of a certain simple form.

Theorem 1.2.2. *If $T \in \Gamma(1)$ and* tr $T = t$, *then there exists an $L \in \Gamma(1)$ such that, if $S = L^{-1}TL$, then*

$$|\alpha - \tfrac{1}{2}t| \le \tfrac{1}{2}|\gamma|, \quad |\delta - \tfrac{1}{2}t| \le \tfrac{1}{2}|\gamma|, \quad |\gamma| \le |\beta|, \quad 3\gamma^2 \le |t^2 - 4|,$$

where S is given by (1.1.1). Further, L belongs to the subgroup generated by U and V.

Proof. For any $n \in \mathbb{Z}$

$$T_1 := U^{-n}TU^n = \begin{bmatrix} 1 & -n \\ 0 & 1 \end{bmatrix} \begin{bmatrix} a & b \\ c & d \end{bmatrix} \begin{bmatrix} 1 & n \\ 0 & 1 \end{bmatrix}$$

$$= \begin{bmatrix} a-nc & b+n(a-d)-n^2c \\ c & d+nc \end{bmatrix} =: \begin{bmatrix} a_1 & b_1 \\ c_1 & d_1 \end{bmatrix}.$$

Note that tr $T_1 = t$. We choose n to make

$$|a_1 - \tfrac{1}{2}t| = |\tfrac{1}{2}(a-d) - nc| = |d_1 - \tfrac{1}{2}t| \le \tfrac{1}{2}|c| = \tfrac{1}{2}|c_1|;$$

this is possible even when $c = 0$, since then $a = d$. If $|c_1| \le |b_1|$ we stop the process here; if $|c_1| > |b_1|$ we form

$$T_2 := V^{-1}T_1V = \begin{bmatrix} 0 & 1 \\ -1 & 0 \end{bmatrix} \begin{bmatrix} a_1 & b_1 \\ c_1 & d_1 \end{bmatrix} \begin{bmatrix} 0 & -1 \\ 1 & 0 \end{bmatrix} = \begin{bmatrix} d_1 & -c_1 \\ -b_1 & a_1 \end{bmatrix}$$

$$=: \begin{bmatrix} a_2 & b_2 \\ c_2 & d_2 \end{bmatrix}$$

and note that we then have $|c_2| < |c_1|$ and tr $T_2 = t$. We now form $T_3 := U^{-m}T_2U^m$, choosing m as above to make

$$|a_3 - \tfrac{1}{2}t| = |d_3 - \tfrac{1}{2}t| \le \tfrac{1}{2}|c_3| = \tfrac{1}{2}|c_2| = \tfrac{1}{2}|b_1|,$$

and stop the process here if $|c_3| \le |b_3|$; if not, then $|c_3| > |b_3|$ and we proceed as above, obtaining a matrix T_4 with $|c_4| < |c_3| < |c_1|$. The process must stop after a finite number k of steps when we reach a matrix $S = T_{2k+1}$ with

$$|\alpha - \tfrac{1}{2}t| = |\delta - \tfrac{1}{2}t| \le \tfrac{1}{2}|\gamma| \quad \text{and} \quad |\gamma| \le |\beta|.$$

Hence

$$|t^2 - 4| = |(\alpha+\delta)^2 - 4 = |4\beta\gamma + (\alpha-\delta)^2|$$
$$\ge 4|\beta||\gamma| - |\alpha-\delta|^2 \ge 4\gamma^2 - \gamma^2 = 3\gamma^2.$$

Theorem 1.2.3. *Let $T \in \Gamma(1)$. If T is of finite order then T is conjugate to one of the matrices*

$$\pm I, \pm V, \pm P, \pm P^2$$

and $|\text{tr } T| \le 2$. Conversely, if $|\text{tr } T| \le 2$, then T is conjugate to one of these matrices or else T is a conjugate of $\pm U^k$ for some $k \in \mathbb{Z}$.

Proof. From the inequalities in theorem 1.2.2 we get the possibilities shown in table 1 for S when $|t| \le 2$. If $t > 2$, we can put

Table 1

$t=\alpha+\delta$	α	β	γ	δ	S
0	0	∓ 1	± 1	0	$\pm V$
1	0 1	∓ 1	± 1	1 0	$P, -V^{-1}P^2V$ $-P^2, V^{-1}PV$
-1	0 -1	∓ 1	± 1	-1 0	$-P, V^{-1}P^2V$ $P^2, -V^{-1}PV$
2	1	k	0	1	U^k
-2	-1	$-k$	0	-1	$-U^k$

$t=2\cosh\theta$, for some $\theta>0$, so that, by (1.1.9),

$$\operatorname{tr}(S^q)=2\cosh q\theta>2$$

for all $q\in\mathbb{Z}^+$. It follows that, if $|t|>2$, then $S^q\neq\pm I$ for all $q\in\mathbb{Z}^+$.

Theorem 1.2.4. *$\Gamma(1)$ is generated by U and V; every element $T\in\Gamma(1)$ can be written in the form*

$$T=U^{q_0}VU^{q_1}V\cdots VU^{q_n} \tag{1.2.8}$$

where $q_i\in\mathbb{Z}$ $(0\le i\le n)$; this representation is not unique.

Proof. Let $T\in\Gamma(1)$ and take $S=L^{-1}TL$ as in theorem 1.2.2. By theorem 1.2.3, we may suppose that $|t|>2$, where $t=\operatorname{tr}T$. Then $S\notin\Gamma_U$, and so we can choose $q\in\mathbb{Z}$ so that, if $t_1=\operatorname{tr}U^qS$,

$$|t_1|=|t+q\gamma|\le\tfrac{1}{2}|\gamma|;$$

this is possible since $\gamma\neq 0$. Then

$$|t_1|\le\tfrac{1}{2}|\gamma|\le\tfrac{1}{2}\{\tfrac{1}{3}|t^2-4|\}^{\frac{1}{2}}<|t|,$$

and some transform of U^qS will satisfy the inequalities in theorem 1.2.2. In this way we can derive an element S' of trace t', where $|t'|\le 2$, and where S' is derived from T by multiplication on left and right by powers of U and V. Since, by table 1, S' is also a product of matrices U and V, the required result follows. The representation is not unique, since

$$P^6=(VU)^6=I.$$

Theorem 1.2.5. *$\Gamma(1)$ is generated by V and P^2; every element $T\in\Gamma(1)$ can be written uniquely in the form*

$$T=(-1)^r P^{2p_0} V P^{2p_1} V \cdots V P^{2p_n} \tag{1.2.9}$$

where

$$0\le r\le 1, \quad 0\le p_i \le 2 \quad (0\le i\le n), \quad p_i>0 \quad (0<i<n). \tag{1.2.10}$$

(When $n=0$, the expression on the right-hand side of (1.2.9) is $(-1)^r P^{2p_0}$.)

Proof. It is clear from (1.2.4, 5, 8) that T can be written in the form (1.2.9) subject to the conditions (1.2.10). The representation is unique if we can have

$$I=(-1)^r P^{2p_0} V P^{2p_1} \cdots V P^{2p_n} \tag{1.2.11}$$

only when $n=r=p_0=0$. We call the right-hand side of (1.2.11) a *word of length n* and suppose that (1.2.11) holds for some $n>0$ and that n is the least positive integer for which this holds. Then

$$I=(-1)^r V P^{2p_1} V P^{2p_2} \cdots V P^{2p_n+2p_0} \tag{1.2.12}$$

and $p_n+p_0 \not\equiv 0 \pmod 3$. For if $p_n+p_0\equiv 0 \pmod 3$, then $n\ge 2$ and

$$I=(-1)^{r-1} P^{2p_1} V \cdots V P^{2p_{n-1}},$$

which is a word of shorter length $n-2$; hence $n=2$ and so $p_1=0$, which is false. Since $P^6=I$, we may assume that $p_n+p_0=1$ or 2, so that we deduce from (1.2.12) that $\pm I$ can be represented as a product of n factors each of which is either

$$VP^4=U=\begin{bmatrix}1 & 1\\ 0 & 1\end{bmatrix}, \quad \text{or} \quad -VP^2=W=\begin{bmatrix}1 & 0\\ 1 & 1\end{bmatrix}. \tag{1.2.13}$$

Since the entries in U and W are all non-negative the same is true of the product, which must be $+I$. But, if either

$$I=US \quad \text{or} \quad I=WS,$$

then S has one negative entry, and this gives the desired contradiction.

It follows from theorem 1.2.5 that

$$\Gamma(1)=\langle P^2, V\rangle=\langle P, V\rangle, \tag{1.2.14}$$

with the notation introduced at the end of §1.1. Further, since $P^4 = -P$, it is clear that theorem 1.2.5 remains true, possibly with a different value of r, if each exponent $2p_i$ $(0 \le i \le n)$ in (1.2.9) is replaced by p_i $(0 \le i \le n)$.

Since the matrices T and $-T$ give rise to the same bilinear mapping, it is clear that we have

Theorem 1.2.6. *$\hat{\Gamma}(1)$ is generated by the mappings V and P, which have orders* 2 *and* 3, *respectively; i.e.*

$$\hat{\Gamma}(1) = \langle P, V \rangle. \tag{1.2.15}$$

Further, every mapping T of $\hat{\Gamma}(1)$ can be written uniquely in the form

$$T = P^{p_0} V P^{p_1} \cdots V P^{p_n} \tag{1.2.16}$$

as a composition of mappings, where

$$0 \le p_i \le 2 \quad (0 \le i \le n), \quad p_i > 0 \quad (0 < i < n). \tag{1.2.17}$$

Theorem 1.2.6 can be expressed by stating that $\hat{\Gamma}(1)$ is the free product of the cyclic groups $\langle V \rangle$ and $\langle P \rangle$. We use an asterisk (*) to denote a free product, so that we have

$$\hat{\Gamma}(1) = \langle V \rangle * \langle P \rangle. \tag{1.2.18}$$

The integer n occurring in (1.2.9) is called the *length* of the group element or *word* T and we write $l(T) := n$. Thus $l(T) = 0$ if and only if $T = P^q$ for some integer q.

We conclude this section by investigating the automorphism groups of $\Gamma(1)$ and $\hat{\Gamma}(1)$; see Hua and Reiner (1951). We write Aut G and Inn G for the groups of all automorphisms and all inner automorphisms, respectively, of a group G. We also recall that Klein's four-group is the direct product of two cyclic groups of order 2.

Theorem 1.2.7. *Let ψ be an automorphism of $\Gamma(1)$. Then ψ is determined uniquely by its action on the generators V and P, and we must have, for some $L \in \Gamma(1)$,*

$$\psi(V) = L^{-1} V^{\mu} L, \quad \psi(P) = L^{-1} P^{\nu} L, \tag{1.2.19}$$

where $\mu = \pm 1$, $\nu = \pm 1$. Accordingly, Aut $\Gamma(1)$/Inn $\Gamma(1)$ *is isomorphic to the four-group.*

Proof. Clearly, if $\psi \in \operatorname{Aut} \Gamma(1)$, then, for some $L_1, L_2 \in \Gamma(1)$,

$$\psi(V) = L_1^{-1} V^{\mu} L_1, \quad \psi(P) = L_2^{-1} P^{\nu} L_2,$$

by theorem 1.2.3, where $\mu = \pm 1$, $\nu = \pm 1$. By applying an inner automorphism to ψ we obtain an automorphism $\varphi \in \operatorname{Aut} \Gamma(1)$ such that

$$\varphi(V) = L^{-1} V^{\mu} L, \quad \varphi(P) = P^{\nu} \tag{1.2.20}$$

for some $L \in \Gamma(1)$. Further, we may assume either that $L = I$, or else that

$$m := l(L) \geq 1,$$

in which case L has the canonical form

$$L = P^{2q_0} V P^{2q_1} V \cdots V P^{2q_m}, \tag{1.2.21}$$

where $0 < q_0 \leq 2$ and $q_m = 0$; for V^{μ} and P^{ν} are unaltered when conjugated by V and P, respectively. Note that $l(L^{-1} V L) = 2m + 1$.

Now take any $T \in \Gamma(1)$ as in (1.2.9), so that $n = l(T)$. From (1.2.9, 20, 21) we can work out $\varphi(T)$ and express it in canonical form. We find that

$$l(\varphi(T)) = n(2m + 1).$$

From this it is clear that φ is a surjection if and only if $m = 0$, i.e. $L = I$. The theorem follows.

Note that, when $\mu = -1$, $\nu = 1$, we deduce from (1.2.19) that, for all $T \in \Gamma(1)$,

$$\psi(T) = (-1)^{l(T)} L^{-1} T L \quad \text{for some } L \in \Gamma(1). \tag{1.2.22}$$

Write

$$J = \begin{bmatrix} 1 & 0 \\ 0 & -1 \end{bmatrix}, \tag{1.2.23}$$

so that J is its own inverse and does not belong to Θ. Note that

$$J^{-1} T J = \begin{bmatrix} a & -b \\ -c & d \end{bmatrix}. \tag{1.2.24}$$

and that

$$J^{-1} V J = -V, \quad J^{-1} P J = V^{-1} P^{-1} V. \tag{1.2.25}$$

We now deduce immediately the following theorem from theorem 1.2.7.

Theorem 1.2.8. Aut $\hat{\Gamma}(1) = \langle J \rangle$ Inn $\hat{\Gamma}(1)$. *Thus* Aut $\hat{\Gamma}(1)$/Inn $\hat{\Gamma}(1)$ *is a cyclic group of order* 2.

In fact, if $\psi \in$ Aut $\hat{\Gamma}(1)$, then, for each $T \in \hat{\Gamma}(1)$,

$$\psi(T) = L^{-1}TL \quad \text{or} \quad \psi(T) = J^{-1}L^{-1}TLJ$$

for some fixed $L \in \hat{\Gamma}(1)$.

In conclusion, we note that the automorphism

$$T \mapsto T' = \begin{bmatrix} d & c \\ b & a \end{bmatrix}$$

is an outer automorphism, since

$$T' = (VJ)^{-1}T(VJ).$$

Further, the outer automorphism $T \mapsto J^{-1}TJ$ is closely associated with the non-analytic map

$$z \mapsto J^*(z) := -\bar{z}, \tag{1.2.26}$$

where the bar denotes the complex conjugate. For

$$(J^*)^{-1}TJ^*(z) = J^{-1}TJ(z) = \frac{az-b}{-cz+d}.$$

1.3. The subgroups $\Gamma^2, \Gamma^3, \Gamma^4$ and $\Gamma'(1)$. Let $T \in \Gamma(1)$, and suppose that T is expressed as in theorem 1.2.5. We define

$$h(T) = n + 2r, \quad p(T) = p_0 + p_1 + \cdots + p_n. \tag{1.3.1}$$

It then follows from (1.2.4) that, for any $T_1, T_2 \in \Gamma(1)$,

$$h(T_1T_2) \equiv h(T_1) + h(T_2) \pmod 4 \tag{1.3.2}$$

and

$$p(T_1T_2) \equiv p(T_1) + p(T_2) \pmod 3. \tag{1.3.3}$$

Thus h and p are homomorphisms of $\Gamma(1)$ onto the additive groups of residue classes modulis 4 and 3, respectively. The kernels of these homomorphisms, namely

$$\Gamma^4 := \{T : T \in \Gamma(1), h(T) \equiv 0 \pmod 4\} \tag{1.3.4}$$

and

$$\Gamma^3 := \{T : T \in \Gamma(1), p(T) \equiv 0 \pmod 3\}, \tag{1.3.5}$$

are therefore normal subgroups of $\Gamma(1)$ of indices 4 and 3, respectively. We also write

$$\Gamma^2 := \{T: T \in \Gamma(1), h(T) \equiv 0 \pmod 2\} \qquad (1.3.6)$$

so that $\Gamma^4 \subseteq \Gamma^2 \subseteq \Gamma(1)$, and Γ^2 is a normal subgroup of $\Gamma(1)$ of index 2.

From (1.2.9) and the definitions of the three subgroups it is easily seen that Γ^2 is generated by P and

$$P_1 := V^{-1}PV = \begin{bmatrix} 1 & -1 \\ 1 & 0 \end{bmatrix}; \qquad (1.3.7)$$

thus

$$\Gamma^2 = \langle P, P_1 \rangle.$$

Similarly,

$$\Gamma^4 = \langle P^2, P_1^2 \rangle, \qquad (1.3.8)$$

and

$$\Gamma^3 = \langle V, V_1, V_2 \rangle, \qquad (1.3.9)$$

where

$$V_1 := P^{-1}VP = P^2VP^4 = \begin{bmatrix} -1 & -2 \\ 1 & 1 \end{bmatrix},$$
$$V_2 := P^{-2}VP^2 = P^4VP^2 = \begin{bmatrix} -1 & -1 \\ 2 & 1 \end{bmatrix}. \qquad (1.3.10)$$

Note that $-I$ belongs to Γ^2 and Γ^3 but not to Γ^4. Thus

$$\Gamma^4 \cong \Gamma^2/\Lambda.$$

It is clear from theorem 1.2.5 that

$$\Gamma^4 = \langle P^2 \rangle * \langle P_1^2 \rangle. \qquad (1.3.11)$$

We now consider the commutator group $\Gamma'(1)$ of $\Gamma(1)$. This group is generated by the commutators

$$[S, T] := STS^{-1}T^{-1}, \qquad (1.3.12)$$

for $S, T \in \Gamma(1)$. We prove

Theorem 1.3.1. *The commutator group $\Gamma'(1)$ is a free group of index 12 in $\Gamma(1)$. $\Gamma'(1)$ has rank 2 and is generated by UW and WU.*

Further $\Gamma'(1) = \Gamma^3 \cap \Gamma^4$, so that $T \in \Gamma'(1)$ if and only if

$$h(T) \equiv 0 \pmod 4 \quad \textit{and} \quad p(T) \equiv 0 \pmod 3. \qquad (1.3.13)$$

The factor group $\Gamma(1)/\Gamma'(1)$ is a cyclic group of order 12 and is generated by the coset $\Gamma'(1)U$. Also, writing Γ' for $\Gamma'(1)$, we have

$$-I \in \Gamma' U^6, \quad V \in \Gamma' U^9, \quad P \in \Gamma' U^{10}, \quad W \in \Gamma' U^{-1}. \qquad (1.3.14)$$

In fact,

$$U^6 = -[UW, (WU)^{-1}].$$

Finally,

$$\Gamma(1) = \Gamma^3 \Gamma^4, \quad \textit{and} \quad \bar{\Gamma}'(1) = \bar{\Gamma}^2 \cap \bar{\Gamma}^3.$$

Proof. For any $S, T \in \Gamma(1)$ it follows from (1.3.2, 3, 12) that

$$h([S, T]) \equiv 0 \pmod 4, \quad p([S, T]) \equiv 0 \pmod 3.$$

It follows from this that every $T \in \Gamma'(1)$ satisfies (1.3.13).

Conversely, if $T \in \Gamma(1)$ and T satisfies (1.3.13), we can express T in the form (1.2.9), where (1.2.10) holds. Write

$$q_m = p_0 + p_1 + \cdots + p_m \quad (0 \le m \le n).$$

Then n is even and it is easily verified that

$$T = [P^{2q_0}, V][V, P^{2q_1}][P^{2q_2}, V][V, P^{2q_3}] \ldots [P^{2q_{n-2}}, V][V, P^{2q_{n-1}}],$$

so that $T \in \Gamma'(1)$. Now

$$[V, P] = [V, P^4] = [P^4, V]^{-1} = UW$$

and

$$[V, P^2] = [P^2, V]^{-1} = WU,$$

so that $\Gamma'(1)$ is generated by UW and WU.

From these facts it is clear that $\Gamma'(1)$ is a free group of rank 2 having UW and WU as generators; for otherwise we could express I as a product of the form (1.2.9) with $n > 0$, and this is impossible. The index of $\Gamma'(1)$ in $\Gamma(1)$ is 12 since each coset corresponds to a different pair (h, p) of residue classes and consists of those $T \in \Gamma(1)$ for which

$$h(T) \equiv h \pmod 4 \quad \text{and} \quad p(T) \equiv p \pmod 3.$$

In fact

$$\Gamma(1) = \Gamma'(1) \cdot \mathscr{R},$$

where $\mathscr{R}$ consists of the 12 matrices $P^{2p}V^h$ $(0 \le p < 3, 0 \le h < 4)$.

If we use $\sim$ to denote the equivalence relation of belonging to the same coset of $\Gamma'(1)$ in $\Gamma(1)$, we deduce from

$$V = -U^3[U^{-2}, V][V, U^{-1}]$$

that $-I \sim V^2 \sim U^6$, $V \sim U^9$, $P \sim VU \sim U^{10}$ and $W \sim WUU^{-1} \sim U^{-1}$. In fact we have

$$P^{2p}V^h \sim U^{8p+9h} \quad (0 \le p < 3, 0 \le h < 4).$$

Finally, it is obvious from the earlier part of the proof that $\Gamma'(1) = \Gamma^3 \cap \Gamma^4$, and so $\bar{\Gamma}'(1) = \Gamma^2 \cap \Gamma^3$. Also $\Gamma(1) = \Gamma^3\Gamma^4$ since $\Gamma^3\Gamma^4$ contains the generators V and P^2.

We now state the corresponding results for the associated inhomogeneous groups. For this purpose we use the results already obtained, together with (1.1.12, 13).

Theorems 1.3.2. *The associated normal inhomogeneous groups have the following properties:*

$$\hat{\Gamma}^2 = \hat{\Gamma}^4 = \langle P^2 \rangle * \langle P_1^2 \rangle, \quad \hat{\Gamma}^4 \cong \Gamma^4, \quad [\hat{\Gamma}(1) : \hat{\Gamma}^2] = 2, \tag{1.3.15}$$

$$\hat{\Gamma}^3 = \langle V \rangle * \langle V_1 \rangle * \langle V_2 \rangle, \quad \hat{\Gamma}^3 \cong \Gamma^3/\Lambda, \quad [\hat{\Gamma}(1) : \hat{\Gamma}^3] = 3, \tag{1.3.16}$$

$$\hat{\Gamma}'(1) = \hat{\Gamma}^2 \cap \hat{\Gamma}^3 \cong \Gamma'(1), \quad [\hat{\Gamma}(1) : \hat{\Gamma}'(1)] = 6, \tag{1.3.17}$$

$$\hat{\Gamma}(1) = \hat{\Gamma}^2\hat{\Gamma}^3. \tag{1.3.18}$$

The factor group $\hat{\Gamma}(1)/\hat{\Gamma}'(1)$ is a cyclic group of order 6 generated by the coset containing U.

We conclude by giving alternative definitions of Γ^2, Γ^3 and $\bar{\Gamma}'(1)$. For this purpose we define

$$Q := [W, U] = VU^3 = \begin{bmatrix} 0 & -1 \\ 1 & 3 \end{bmatrix}. \tag{1.3.19}$$

Theorem 1.3.3. *For $\nu = 2, 3$ and 6 define*

$$G^\nu := \{T \in \Gamma(1) : Q^{-1}TQ \equiv \pm T \pmod{\nu}\}. \tag{1.3.20}$$

Then $G^2 = \Gamma^2$, $\quad G^3 = \Gamma^3 \quad$ and $\quad G^6 = \bar{\Gamma}'(1)$.

Proof. G^2 and G^3 are certainly subgroups of $\Gamma(1)$ and are proper subgroups since they do not contain U. Further, G^2 contains the generators P and P_1 of Γ^2, while G contains the generators V, V_1 and V_2 of Γ^3. We deduce that $G^2 = \Gamma^2$ and $G^3 = \Gamma^3$. Accordingly

$$G^6 = G^2 \cap G^3 = \Gamma^2 \cap \Gamma^3 = \bar{\Gamma}'(1).$$

1.4. The level of a subgroup; congruence subgroups. Let Γ be any subgroup of $\Gamma(1)$, not necessarily of finite index. For each $L \in \Gamma(1)$, define n_L to be the least positive integer such that

$$U^{n_L} \in L\Gamma L^{-1}. \tag{1.4.1}$$

When Γ has infinite index in $\Gamma(1)$, no such finite positive n_L need exist and we put $n_L = \infty$ in this case; on the other hand, when Γ has finite index in $\Gamma(1)$, n_L always exists and is finite. Consider the set $\{n_L : L \in \Gamma(1)\}$. If the numbers in this set have a finite least common multiple n, we call n the *level*† of Γ and write

$$\operatorname{lev} \Gamma = n. \tag{1.4.2}$$

In all other cases we put $\operatorname{lev} \Gamma = \infty$.

In particular, $\operatorname{lev} \Gamma$ is always finite when Γ has finite index in $\Gamma(1)$, since then Γ has only a finite number of conjugate subgroups in $\Gamma(1)$. When (1.4.2) holds for finite n we have

$$U^n \in L\Gamma L^{-1} \quad \text{for all } L \in \Gamma(1). \tag{1.4.3}$$

We now write $\Delta(n)$ for the normal closure of the cyclic group $\langle U^n \rangle$; i.e.

$$\Delta(n) = \langle L^{-1}U^nL : L \in \Gamma(1) \rangle \tag{1.4.4}$$

and is the smallest normal subgroup containing $\langle U^n \rangle$. Clearly n is the smallest positive integer such that $\Delta(n) \subseteq \Gamma$.

We make exactly similar definitions for inhomogeneous groups. Thus, if $\hat{\Gamma}$ is any subgroup of $\hat{\Gamma}(1)$, $\operatorname{lev} \hat{\Gamma}$ is defined and the group $\hat{\Delta}(n)$ is the normal closure of the group $\langle U^n \rangle$ of mappings. By way of example we note that

$$\operatorname{lev} \Gamma(1) = \operatorname{lev} \hat{\Gamma}(1) = 1, \quad \operatorname{lev} \Delta(n) = \operatorname{lev} \hat{\Delta}(n) = n,$$

and $\operatorname{lev} \langle U^n \rangle = \infty$ for both homogeneous and inhomogeneous groups $\langle U^n \rangle$. Also, by theorems 1.3.1, 2,

$$\operatorname{lev} \Gamma'(1) = 12, \quad \operatorname{lev} \hat{\Gamma}'(1) = 6.$$

† German: *Stufe*.

An important class of subgroups of the modular group consists of what are called congruence subgroups. If S and $T \in \Gamma(1)$, we write

$$S \equiv T \pmod{n},$$

where $n \in \mathbb{Z}^+$, if and only if

$$\alpha \equiv a, \quad \beta \equiv b, \quad \gamma \equiv c \quad \text{and} \quad \delta \equiv d \pmod{n}.$$

It follows at once from (1.1.5) and (1.1.11) that if

$$S_1 \equiv S_2 \pmod{n} \quad \text{and} \quad T_1 \equiv T_2 \pmod{n},$$

then

$$T_1^{-1} \equiv T_2^{-1} \pmod{n} \quad \text{and} \quad S_1T_1 \equiv S_2T_2 \pmod{n}. \quad (1.4.5)$$

We write

$$\Gamma(n) := \{S \in \Gamma(1) : S \equiv I \pmod{n}\}; \quad (1.4.6)$$

this agrees with the previous definition when $n = 1$. Then $\Gamma(n)$ is a group; for if $S \equiv T \equiv I \pmod{n}$, then

$$ST^{-1} \equiv II^{-1} \equiv I \pmod{n}.$$

We also write

$$\bar{\Gamma}(n) := \Lambda\Gamma(n) = \{S \in \Gamma(1) : S \equiv \pm I \pmod{n}\}, \quad (1.4.7)$$

and can prove similarly that this is a group. The two homogeneous groups $\Gamma(n)$ and $\bar{\Gamma}(n)$ give rise to the same inhomogeneous group, which we denote by $\hat{\Gamma}(n)$. Each of the groups $\bar{\Gamma}(n)$ and $\hat{\Gamma}(n)$ is called a *principal congruence group*,† and the same title is sometimes conferred on $\Gamma(n)$.

Since $-I \in \Gamma(n)$ if and only if $n = 1$ or 2, we have

$$\hat{\Gamma}(n) \cong \Gamma(n)/\Lambda \cong \bar{\Gamma}(n)/\Lambda \quad (n = 1, 2), \quad (1.4.8)$$

$$\hat{\Gamma}(n) \cong \Gamma(n) \cong \bar{\Gamma}(n)/\Lambda \quad (n \geq 3). \quad (1.4.9)$$

Both $\Gamma(n)$ and $\bar{\Gamma}(n)$ are normal subgroups of $\Gamma(1)$, and $\hat{\Gamma}(n)$ is a normal subgroup of $\hat{\Gamma}(1)$. For, if $S \in \Gamma(n)$ and $T \in \Gamma(1)$, then

$$T^{-1}ST \equiv T^{-1}IT \equiv I \pmod{n},$$

and the proof is similar in the other cases.

† German: *Hauptkongruenzgruppe*.

Clearly

$$\operatorname{lev}\Gamma(n)=\operatorname{lev}\bar{\Gamma}(n)=\operatorname{lev}\hat{\Gamma}(n)=n,$$

and also

$$\Delta(n)\subseteq\Gamma(n),\quad \hat{\Delta}(n)\subseteq\hat{\Gamma}(n).$$

Further, since $J^{-1}U^nJ=U^{-n}$ and $l(U^n)=n$, it follows from (1.2.22) that $\Delta(n)$ and $\Gamma(n)$ are invariant under the automorphism $T\mapsto J^{-1}TJ$, but not under the automorphism (1.2.22) unless n is even. On the other hand, $\bar{\Gamma}(n)$ is invariant under all automorphisms of $\Gamma(1)$, and $\hat{\Delta}(n)$ and $\hat{\Gamma}(n)$ are invariant under all automorphisms of $\hat{\Gamma}(1)$.

The factor groups

$$G(n):=\Gamma(1)/\Gamma(n),\quad \bar{G}(n):=\Gamma(1)/\bar{\Gamma}(n) \tag{1.4.10}$$

and

$$\hat{G}(n):=\hat{\Gamma}(1)/\hat{\Gamma}(n) \tag{1.4.11}$$

are called *modular*† groups of level n. By (1.4.8–11) we have

$$\hat{G}(n)\cong\bar{G}(n)\cong G(n)\quad(n=1,2) \tag{1.4.12}$$

and

$$\hat{G}(n)\cong\bar{G}(n)\cong G(n)/\Lambda\quad(n\geq 3). \tag{1.4.13}$$

Since the number of incongruent matrices T modulo n is clearly less than or equal to n^4, the three modulary groups are clearly finite and their orders are denoted by $\mu(n)$, $\bar{\mu}(n)$ and $\hat{\mu}(n)$, respectively; $\mu(n)$ should not be confused with the Möbius function.

Theorem 1.4.1.

$$\mu(n)=n^3\prod_{p|n}\left(1-\frac{1}{p^2}\right),$$

where the product is taken over all primes p dividing n, and

$$\hat{\mu}(n)=\bar{\mu}(n)=\mu(n)\quad(n=1,2),\quad \hat{\mu}(n)=\bar{\mu}(n)=\tfrac{1}{2}\mu(n)\quad(n\geq 3).$$

Proof. By (1.4.12, 13), it is enough to find $\mu(n)$, i.e. the number of incongruent matrices S modulo n. Our proof is similar to that given by Gunning (1962). We say that a pair of integers c, d is a *primitive*

† German: *Modulargruppe.*

pair modulo n if and only if $(c, d, n) = 1$, and denote by $\lambda(n)$ the number of incongruent primitive pairs modulo n.

Lemma 1. *If c, d is a primitive pair modulo n there exists an $S \in \Gamma(1)$ such that $\gamma \equiv c$, $\delta \equiv d \pmod{n}$. Conversely, if $S \in \Gamma(1)$, then γ, δ is a primitive pair modulo n.*

Proof. The last part is obvious since $(\gamma, \delta) = 1$, so that $(\gamma, \delta, n) = 1$.

As usual we write $q|r$ to mean that q divides r. Suppose that c, d is a primitive pair modulo n, so that $(c, d, n) = 1$. Take $\gamma = c$ and write

$$c = c_1 c_2,$$

where c_1 is the largest divisor of c that is prime to n. Take $m \in \mathbb{Z}$ so that

$$\delta := d + mn \equiv 1 \pmod{c_1},$$

which is possible since $(n, c_1) = 1$. Then $(\gamma, \delta) = 1$. For if p is a prime divisor of γ and δ, then p divides both c and $d + mn$. If $p|c_1$ then, since $\delta \equiv 1 \pmod{c_1}$, $p \nmid \delta$, which is false. If $p|c_2$, then $p|n$ and so $p|d$; thus $p|(c, d, n)$, which is false. Hence we have found γ, δ with $\gamma \equiv c$, $\delta \equiv d \pmod{n}$ and $(\gamma, \delta) = 1$. We can now find $\alpha, \beta \in \mathbb{Z}$ such that $\alpha\delta - \beta\gamma = 1$ and so $S \in \Gamma(1)$.

Lemma 2. *For each primitive pair, c, d of integers modulo n there are exactly n matrices $S \in \Gamma(1)$ modulo n for which*

$$\gamma \equiv c, \quad \delta \equiv d \pmod{n}.$$

Proof. Let S_1 and S_2 be two members of $\Gamma(1)$ with second rows congruent to $[c, d]$ modulo n. Then

$$S_1 S_2^{-1} \equiv \begin{bmatrix} 1 & k \\ 0 & 1 \end{bmatrix} \pmod{n},$$

for some $k \in \mathbb{Z}$. As there are only n incongruent values of k modulo n, it follows that there are exactly n incongruent matrices $S \in \Gamma(1)$ with $[\gamma, \delta] \equiv [c, d] \pmod{n}$.

Lemma 3. *$\lambda(n)$ is a multiplicative function; i.e., if $(n_1, n_2) = 1$, then $\lambda(n_1 n_2) = \lambda(n_1)\lambda(n_2)$.*

Proof. Let γ_j, δ_j be a primitive pair modulo n_j ($j=1,2$). Then $\gamma_1 n_2+\gamma_2 n_1$, $\delta_1 n_2+\delta_2 n_1$ is a primitive pair modulo $n_1 n_2$, since $(n_1, n_2)=1$. Also, incongruent pairs for n_1 and n_2 lead to incongruent pairs for $n_1 n_2$; i.e., if

$$\gamma_1' n_2+\gamma_2' n_1 \equiv \gamma_1 n_2+\gamma_2 n_1 \pmod{n_1 n_2}$$

and

$$\delta_1' n_2+\delta_2' n_1 \equiv \delta_1 n_2+\delta_2 n_1 \pmod{n_1 n_2},$$

then $\gamma_1' \equiv \gamma_1 \pmod{n_1}$, $\delta_1' \equiv \delta_1 \pmod{n_1}$, $\gamma_2' \equiv \gamma_2 \pmod{n_2}$ and $\delta_2' \equiv \delta_2 \pmod{n_2}$. Thus $\lambda(n_1)\lambda(n_2) \le \lambda(n_1 n_2)$. Conversely, let γ, δ be a primitive pair modulo $n_1 n_2$; then γ, δ is a primitive pair modulo n_1 and modulo n_2. Also, since $(n_1, n_2)=1$, incongruent pairs modulo $n_1 n_2$ cannot give rise to congruent ones modulo n_1 and modulo n_2. Thus $\lambda(n_1 n_2) \le \lambda(n_1)\lambda(n_2)$.

Lemma 4. *If p is prime and $k \ge 1$, $\lambda(p^k)=p^{2k}(1-p^{-2})$.*

Proof. There are $p^k(1-1/p)$ incongruent integers c modulo p^k such that $(c, p)=1$. For any one of these, each of the p^k incongruent values of d will give a primitive pair. Since these pairs are all incongruent modulo p^k we have $p^{2k}(1-1/p)$ such pairs. There are p^{k-1} values of c such that $(c, p)=p$. To each of these correspond $p^k(1-1/p)$ values of d incongruent modulo p^k and such that $(d, p)=1$. This gives $p^{2k-1}(1-1/p)$ primitive pairs. Addition gives

$$\lambda(p^k)=p^{2k}\left(1-\frac{1}{p^2}\right).$$

From lemmas 3 and 4 we obtain

$$\lambda(n)=n^2\prod_{p|n}\left(1-\frac{1}{p^2}\right)$$

and, since $\mu(n)=n\lambda(n)$, by lemma 2, the theorem follows.

Theorem 1.4.2. *If m and n are positive integers,*

$$\Gamma(m)\cap\Gamma(n)=\Gamma(\{m, n\}) \tag{1.4.14}$$

and

$$\Gamma(m)\Gamma(n)=\Gamma((m, n)), \tag{1.4.15}$$

where $\{m, n\}$ is the least common multiple of m and n.

Proof. Write $h=(m, n), l=\{m, n\}$. We have $T\in\Gamma(m)\cap\Gamma(n)$ if and only if $T\equiv I \pmod m$ and $T\equiv I \pmod n$; this holds if and only if $T\equiv I \pmod l$. This proves (1.4.14).

If $T\in\Gamma(m)\Gamma(n)$ then $T=S_1S_2$ where $S_1\equiv I \pmod m$ and $S_2\equiv I \pmod n$. Hence $S_1\equiv I \pmod h$ and $S_2\equiv I \pmod h$, so that $T\equiv I \pmod h$. It follows that

$$\Gamma(m)\Gamma(n)\subseteq\Gamma(h). \tag{1.4.16}$$

Now, by (1.4.14) and one of the isomorphism theorems,

$$\{\Gamma(m)\Gamma(n)\}/\Gamma(n)\cong\Gamma(m)/\Gamma(l), \tag{1.4.17}$$

so that

$$\begin{aligned}\mu(n)=[\Gamma(1):\Gamma(n)]&=[\Gamma(1):\Gamma(m)\Gamma(n)][\Gamma(m)\Gamma(n):\Gamma(n)]\\&=[\Gamma(1):\Gamma(m)\Gamma(n)][\Gamma(m):\Gamma(l)]\\&=[\Gamma(1):\Gamma(m)\Gamma(n)]\mu(l)/\mu(m).\end{aligned}$$

Since $\mu(l)\mu(h)=\mu(m)\mu(n)$ by the formula in theorem 1.4.1 it follows that

$$[\Gamma(1):\Gamma(m)\Gamma(n)]=\mu(h)$$

and this combined with (1.4.16) gives (1.4.15).

Theorem 1.4.3. *If* $(m, n)=1$, *then*

$$G(mn)\cong G(m)\times G(n).$$

Proof. By (1.4.15, 17),

$$\Gamma(1)/\Gamma(n)\cong\Gamma(m)/\Gamma(mn) \tag{1.4.18}$$

so that we can identify $G(n)$ with $\Gamma(m)/\Gamma(mn)$ and $G(m)$ with $\Gamma(n)/\Gamma(mn)$. These are both subgroups of $\Gamma(1)/\Gamma(mn)=G(mn)$. We have, by theorem 1.4.2,

$$G(m)\cap G(n)=\{I\},\quad G(m)nG(n)=G(mn),$$

where I is the identity in $G(mn)$. From this the theorem follows; see theorem 2.5.1 of Hall (1959).

By repeated applications of theorem 1.4.3 we see that, if

$$n=p_1^{\alpha_1}p_2^{\alpha_2}\cdots p_k^{\alpha_k},$$

where the p_i are different primes and $\alpha_i > 0$ $(1 \le i \le k)$, then $G(n)$ is the direct product of the k groups $G(p_i^{\alpha_i})$. Thus the structure of $G(n)$ is known when we know the structure of $G(p^k)$ for each prime p and $k \in \mathbb{Z}^+$. It follows from (1.4.8, 9) that some knowledge of the structure of $\hat{G}(n)$ may be obtained when we know the structure of the groups $\hat{G}(p^k)$.

So far we have considered principal congruence groups only. If Γ is a subgroup of $\Gamma(1)$ and if, for some $n \in \mathbb{Z}^+$,

$$\Gamma(n) \subseteq \Gamma \subseteq \Gamma(1), \tag{1.4.19}$$

we say that Γ is a *congruence group*; Γ is then necessarily of finite index in $\Gamma(1)$. Further, since $U^n \in \Gamma$, it follows that the level of Γ is a divisor of n.

Suppose that Γ is a congruence group satisfying (1.4.19) and that

$$\Gamma(1) = \Gamma(n) \cdot \mathcal{R}.$$

It follows that $\Gamma = \Gamma(n) \cdot \mathcal{R}_0$, where $\mathcal{R}_0$ is a subset of $\mathcal{R}$; i.e. Γ is the set of all matrices T that are congruent to a member of $\mathcal{R}_0$ modulo n. Congruence groups are often defined in this way in terms of a finite set $\mathcal{R}_0$ of matrices.

As an example of a non-principal congruence group, let

$$\Gamma_0(n) = \{T \in \Gamma(1) : c \equiv 0 \pmod{n}\}. \tag{1.4.20}$$

That $\Gamma_0(n)$ is a group follows from (1.1.5, 11). Clearly

$$\hat{\Gamma}_0(n) \cong \Gamma_0(n)/\Lambda.$$

Similarly, we define

$$\Gamma^0(n) = \{T \in \Gamma(1) : b \equiv 0 \pmod{m}\} \tag{1.4.21}$$

and

$$\begin{aligned} \Gamma_0^0(n) &= \{T \in \Gamma(1) : b \equiv c \equiv 0 \pmod{n}\} \\ &= \Gamma_0(n) \cap \Gamma^0(n). \end{aligned} \tag{1.4.22}$$

Since

$$V^{-1}TV = \begin{bmatrix} d & -c \\ -b & a \end{bmatrix},$$

$\Gamma_0(n)$ and $\Gamma^0(n)$ are conjugate subgroups of $\Gamma(1)$ and

$$\Gamma^0(n) = V^{-1}\Gamma_0(n)V, \quad \hat{\Gamma}^0(n) = V^{-1}\hat{\Gamma}_0(n)V.$$

Since $\Gamma(n) \in \Gamma^0(n)$ and $U^r \in \Gamma^0(n)$ only when n divides r, it follows that

$$n = \operatorname{lev} \Gamma^0(n) = \operatorname{lev} \Gamma_0(n) = \operatorname{lev} \Gamma_0^0(n)$$

and similar results hold for the corresponding inhomogeneous groups.

We now calculate the index of $\Gamma_0(n)$ in $\Gamma(1)$.

If $c \equiv 0 \pmod n$ and $ad - bc = 1$, it follows that $(d, n) = 1$, and this holds for

$$\phi(n) := n \prod_{p|n} \left(1 - \frac{1}{p}\right)$$

incongruent values of d modulo n. Each of the $\phi(n)$ pairs c, d is a primitive pair modulo n, and lemmas 1 and 2 show that there are exactly $n\phi(n)$ incongruent matrices modulo $n\,\Gamma_0(n)$. It follows that

$$\psi(n) := [\Gamma(1) : \Gamma_0(n)] = n \prod_{p|n} \left(1 + \frac{1}{p}\right) \tag{1.4.23}$$

and that

$$\hat{\psi}(n) := [\hat{\Gamma}(1) : \hat{\Gamma}_0(n)] = \psi(n). \tag{1.4.24}$$

Values of $\hat{\mu}(n)$ and $\hat{\psi}(n)$ are given for $n \leq 16$ in table 2.

Table 2

n	$\hat{\mu}(n)$	$\hat{\psi}(n)$	n	$\hat{\mu}(n)$	$\hat{\psi}(n)$
1	1	1	9	324	12
2	6	3	10	360	18
3	12	4	11	660	12
4	24	6	12	576	24
5	60	6	13	1092	14
6	72	12	14	1008	24
7	168	8	15	1440	24
8	192	12	16	1536	24

Theorems 1.4.4. *For any positive integers n and N,*

$$\Gamma(N)\Gamma^0(n) = \Gamma^0((N, n)), \quad \Gamma(N)\Gamma_0(n) = \Gamma_0((N, n)). \tag{1.4.25}$$

Proof. It suffices to prove the first identity. Write $N_1=(N, n)$. Since $\Gamma(N)\subseteq\Gamma^0(N_1)$ and $\Gamma^0(n)\subseteq\Gamma^0(N_1)$ we need only prove that

$$\Gamma^0(N_1)\subseteq\Gamma(N)\Gamma^0(n).$$

Take any $S\in\Gamma^0(N_1)$, so that $\beta\equiv 0 \pmod{N_1}$ and therefore

$$\alpha\delta\equiv 1 \pmod{N_1}.$$

It follows that $(\delta, N_1)=1$. For any integer r write

$$S'=W^{rN}S=\begin{bmatrix}\alpha' & \beta'\\ \gamma' & \delta'\end{bmatrix},$$

say. Then

$$\delta'=\delta+\beta rN=(\delta, N)(A+rB),$$

say, where $(A, B)=1$. We can therefore choose r so that $A+rB$ is a prime greater than n. It follows that

$$(\delta', n)=((\delta, N), n)=(\delta, N, n)=(\delta, N_1)=1,$$

and therefore

$$(N\delta', n)=(N, n)=N_1.$$

Accordingly we can choose an integer s such that

$$\beta+sN\delta'\equiv 0 \pmod{n}.$$

It follows that, for this choice of r and s,

$$U^{sN}W^{rN}S\in\Gamma^0(n)$$

and so $S\in\Gamma(N)\Gamma^0(n)$.

Observe that we have in fact proved slightly more, namely that

$$\Delta(N)\Gamma^0(n)=\Gamma^0((N, n)). \qquad (1.4.26)$$

We also require to know the index in $\Gamma(1)$ of the group

$$\Gamma_0(m, n):=\Gamma_0(m)\cap\Gamma^0(n). \qquad (1.4.27)$$

In particular, we note that $\Gamma_0^0(n)=\Gamma_0(n, n)$ and that $[\Gamma_0^0(n):\Gamma(n)]$ is $\varphi(n)$, since it is the number of solutions modulo n of the congruence $ad\equiv 1 \pmod{n}$.

Now the m matrices U^{nr} $(0\leq r<m)$ are easily seen to form a left transversal of $\Gamma_0(m, mn)$ in $\Gamma_0(m, n)$. Hence

$$[\Gamma_0(m, n):\Gamma_0(m, mn)]=m,$$

and, similarly,

$$[\Gamma_0(m, mn):\Gamma_0(mn, mn)] = n.$$

Accordingly,

$$[\Gamma_0(m, n):\Gamma(mn)] = mn\varphi(mn),$$

from which we deduce that

$$[\hat{\Gamma}(1):\hat{\Gamma}_0(m, n)] = [\Gamma(1):\Gamma_0(m, n)] = \psi(mn). \qquad (1.4.28)$$

In particular, the index of $\Gamma_0^0(n)$ in $\Gamma(1)$ is

$$\psi(n^2) = n\psi(n).$$

Theorem 1.4.5. *Let m, n and q be positive integers. Then*

$$\Gamma_0(m, n) = \langle \Gamma_0(m, nq), U^n \rangle.$$

Proof. It is enough to prove that, if $S \in \Gamma_0(m, n)$, integers r and s can be found such that

$$S' := U^{rm} S U^{sn} \in \Gamma_0(m, nq).$$

We have, in an obvious notation,

$$\beta' = (\alpha + rm\gamma)sn + \beta + rm\delta = n\{(\alpha + rm\gamma)s + \beta_1 + r\delta\},$$

say. Now $(\alpha, n\gamma) = (\alpha, n) = 1$, since $\alpha\delta \equiv 1 \pmod{n}$, so that we can choose r to make $\alpha + rm\gamma$ a prime greater than q. Then $(\alpha + rm\gamma, q) = 1$, and therefore s can be chosen to make $\beta' \equiv 0 \pmod{nq}$, as required.

1.5. Groups of level 2. We shall need to study such groups when we introduce theta functions. We note that the following six matrices, defined in (1.1.4) and (1.2.2, 5, 6) form a set of coset representatives of $\Gamma(1)$ modulo $\Gamma(2)$.

$$I = \begin{bmatrix} 1 & 0 \\ 0 & 1 \end{bmatrix}, \quad U = \begin{bmatrix} 1 & 1 \\ 0 & 1 \end{bmatrix}, \quad V = \begin{bmatrix} 0 & -1 \\ 1 & 0 \end{bmatrix},$$

$$W = \begin{bmatrix} 1 & 0 \\ 1 & 1 \end{bmatrix}, \quad P = \begin{bmatrix} 0 & -1 \\ 1 & 1 \end{bmatrix}, \quad P^2 = \begin{bmatrix} -1 & -1 \\ 1 & 0 \end{bmatrix}.$$

The corresponding mappings form a transversal of $\hat{\Gamma}(2)$ in $\hat{\Gamma}(1)$.

We have

$$P^3 \equiv U^2 \equiv V^2 \equiv W^2 \equiv I \pmod 2,$$

$$P \equiv VU, \quad P^2 \equiv UV, \quad W = UVU \equiv VUV \pmod 2.$$

This illustrates the fact that $G(2)$ is isomorphic to the dihedral group of order 6, i.e. the symmetric group on three symbols. In fact we can set up a correspondence between

$$I, U, V, W, P, P^2$$

and the identical permutation e, (12), (13), (23), (123), (132), respectively; we use here the usual cycle notation.

Between $\Gamma(1)$ and $\Gamma(2)$ we have four groups

$$\Gamma_P(2), \Gamma_U(2), \Gamma_V(2), \Gamma_W(2)$$

as in fig. 1, where the index is marked and where normal subgroups

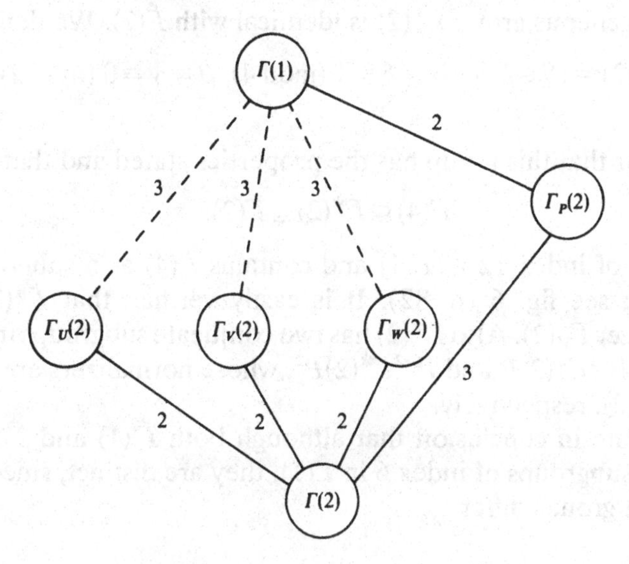

Fig. 1. Groups between $\Gamma(1)$ and $\Gamma(2)$.

are indicated with a continuous line. Here

$$\Gamma_U(2) := \{S \in \Gamma(1) : S \equiv I \text{ or } U \pmod 2\}, \tag{1.5.1}$$

and $\Gamma_V(2)$, $\Gamma_W(2)$ are defined similarly, while

$$\Gamma_P(2) := \{S \in \Gamma(1) : \; S = I, P \text{ or } P^2 \pmod 2\}. \tag{1.5.2}$$

Note that $\Gamma_U(2) = \Gamma_0(2)$, $\Gamma_W(2) = \Gamma^0(2)$. The three subgroups $\Gamma_U(2)$, $\Gamma_V(2)$ and $\Gamma_W(2)$ are conjugate; in fact

$$\Gamma_V(2) = P^{-1}\Gamma_U(2)P, \quad \Gamma_W(2) = P^{-2}\Gamma_U(2)P^2. \tag{1.5.3}$$

Each of the four groups just defined clearly has level 2.

The matrix $-I$ belongs to all six homogeneous groups mentioned above, so that an exactly similar figure can be drawn for the associated inhomogeneous groups.

The group $\Gamma_P(2)$ is, in fact, the group Γ^2 defined in (1.3.6). To prove this it is enough to show that $\Gamma_P(2) \subseteq \Gamma^2$, since both groups have index 2 in $\Gamma(1)$. This follows since Γ^2 is generated by P and P_1, where P_1 is defined by (1.3.7), and $P_1 \equiv P^2 \pmod 2$. We have also

$$\hat{\Gamma}_P(2) = \hat{\Gamma}^2.$$

For certain purposes it is convenient to define a homogeneous group $\Gamma^*(2)$, which does not contain $-I$ and whose associated inhomogeneous group $\hat{\Gamma}^*(2)$ is identical with $\hat{\Gamma}(2)$. We define

$$\Gamma^*(2) := \{S \in \Gamma(1) : \alpha \equiv \delta \equiv 1 \pmod 4, \beta \equiv \gamma \equiv 0 \pmod 2\}. \tag{1.5.4}$$

It is clear that this group has the properties stated and that

$$\Gamma(4) \subseteq \Gamma^*(2) \subseteq \Gamma(2).$$

$\Gamma^*(2)$ is of index 12 in $\Gamma(1)$ and contains $\Gamma(4)$ as a subgroup of index 4; see fig. 6 (p. 82). It is easily verified that $\Gamma^*(2)$ has normalizer $\Gamma_V(2)$. Also $\Gamma^*(2)$ has two conjugate subgroups in $\Gamma(1)$, namely $P^{-1}\Gamma^*(2)P$ and $P^{-2}\Gamma^*(2)P^2$, whose normalizers are $\Gamma_U(2)$ and $\Gamma_W(2)$, respectively.

We note in conclusion that although both $\hat{\Gamma}'(1)$ and $\hat{\Gamma}(2)$ are normal subgroups of index 6 in $\hat{\Gamma}(1)$, they are distinct, since their quotient groups differ.

1.6. Groups of level 3. It is easily verified that the transformations associated with the following twelve matrices constitute a transversal of $\hat{\Gamma}(3)$ in $\hat{\Gamma}(1)$.

$$\left.\begin{array}{llll} I, & V, & V_1 & V_2, \\ U, & U^2, & W, & W^2, \\ P, & P^2, & P_1, & P_1^2. \end{array}\right\} \tag{1.6.1}$$

Here P_1, V_1, V_2 are defined by (1.3.7, 10) and the remaining matrices are defined in the previous section. $\hat{G}(3)$ is isomorphic to the alternating group on four symbols and the transformations listed in (1.6.1) can be put into one-to-one correspondence with the permutations

$$\begin{array}{llll} e, & (12)(34), & (13)(24), & (14)(23), \\ (234), & (243), & (134), & (143), \\ (124), & (142), & (132), & (123), \end{array}$$

respectively. We write

$$\Gamma_U(3) = \{T \in \Gamma(1) : \pm T = I, U \text{ or } U^2 \pmod 3\} \qquad (1.6.2)$$

and conjugate subgroups $\Gamma_W(3)$, $\Gamma_P(3)$ and $\Gamma_{P_1}(3)$ are defined similarly, U being replaced by W, P, P_1 at each occurrence in (1.6.2). Then

$$\Gamma_W(3) = V^{-1}\Gamma_U(3)V, \quad \Gamma_P(3) = V_2^{-1}\Gamma_U(3)V_2,$$
$$\Gamma_{P_1} = V_1^{-1}\Gamma_U(3)V_1. \qquad (1.6.3)$$

Clearly $\Gamma_U(3) = \Gamma_0(3)$ and $\Gamma_W(3) = \Gamma^0(3)$.

Further,

$$\Gamma^3 = \{T \in \Gamma(1) : \pm T \equiv I, V, V_1 \text{ or } V_2 \pmod 3\}; \qquad (1.6.4)$$

for the set Γ on the right-hand side of (1.6.4) is clearly a subgroup of $\Gamma(1)$ containing $\bar{\Gamma}(3)$ as a subgroup of index 4, and so has index 3 in $\Gamma(1)$. Further, by (1.3.9), Γ^3 is contained in this subgroup Γ and, since $[\Gamma(1):\Gamma^3] = 3$, $\Gamma^3 = \Gamma$. The factor group $\Gamma^3/\bar{\Gamma}(3)$ is isomorphic to the four-group. There are three conjugate groups between $\bar{\Gamma}(3)$ and Γ^3, namely the groups

$$\Gamma_V(3), \quad \Gamma_{V_1}(3) \quad \text{and} \quad \Gamma_{V_2}(3)$$

where

$$\Gamma_V(3) := \{T \in \Gamma(1) : \pm T \equiv I \text{ or } V \pmod 3\}, \qquad (1.6.5)$$

and the other groups are defined similarly. We have

$$\Gamma_{V_1}(3) = P^{-1}\Gamma_V(3)P, \quad \Gamma_{V_2}(3) = P^{-2}\Gamma_V(3)P^2. \qquad (1.6.6)$$

It is easy to see that

$$\Gamma_V(3) = \{T \in \Gamma(1) : Q^{-1}TQ \equiv T \pmod 3\}, \qquad (1.6.7)$$

where Q is given by (1.3.19).

Corresponding results for the associated inhomogeneous groups and their relative structure is illustrated in fig. 2. With the exception of the modular group itself, all the groups considered in this section have level 3.

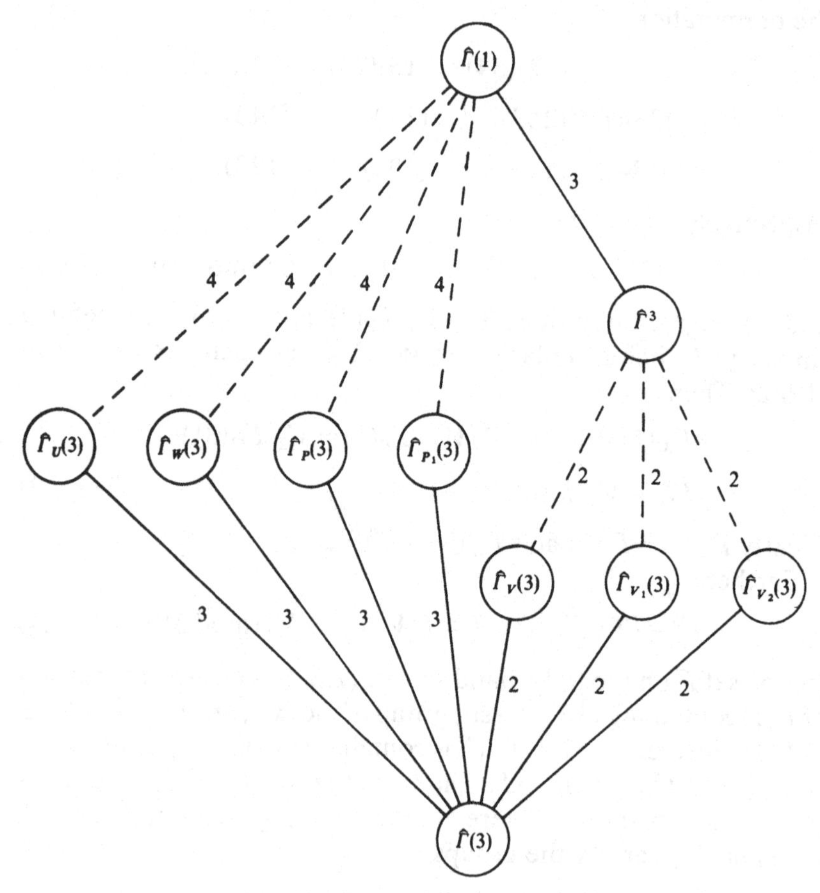

Fig. 2. Groups between $\hat{\Gamma}(1)$ and $\hat{\Gamma}(3)$.

1.7. Further results. We give here a brief sketch of further results of a similar kind, and some references for further reading.

From the monumental treatise by Klein and Fricke (1890, 1892) a great deal of information about particular subgroups of the modular group can be excavated by the persevering reader, and some of this information is given in a more easily assimilated form by Vivanti (1906, 1910). For example, the following alternative definitions can be given.

$$\hat{\Gamma}^2 = \{T: T \in \hat{\Gamma}(1),\ ab + bc + cd \equiv 0 \pmod 2\}, \tag{1.7.1}$$

$$\hat{\Gamma}^3 = \{T: T \in \hat{\Gamma}(1),\ ab + cd \equiv 0 \pmod 3\}, \tag{1.7.2}$$

$$\hat{\Gamma}'(1) = \hat{\Gamma}^2 \cap \hat{\Gamma}^3 = \{T:\ T \in \hat{\Gamma}(1),\ ab + 3bc + cd \equiv 0 \pmod 6\}. \tag{1.7.3}$$

See Klein and Fricke, vol. 1, p. 627; Weber (1909), §§54, 71; Petersson (1953); Wohlfahrt (1964). These definitions can be deduced from theorem 1.3.3 by writing the congruence defining G^ν in the form

$$T^{-1}QT \equiv \pm Q \pmod \nu.$$

We have seen in (1.4.20, 21) that each of the congruence subgroups $\Gamma_0(n)$ and $\Gamma^0(n)$ can be defined by a single linear congruence satisfied by entries of the matrices belonging to the group in question. The same holds for each group conjugate to $\Gamma_0(n)$. Thus, when $n = 2$ there are three conjugate subgroups, namely $\Gamma_0(2) = \Gamma_U(2)$, $\Gamma^0(2) = \Gamma_W(2)$ and

$$\Gamma_V(2) = P^{-1}\Gamma_0(2)P = \{T \in \Gamma(1):\ a + b - c - d \equiv 0 \pmod 2\}. \tag{1.7.4}$$

Similarly, when $n = 4$ we have again the three groups

$$\Gamma_0(4) =: \Gamma_U(4), \quad \Gamma^0(4) =: \Gamma_W(4)$$

and

$$\Gamma_V(4) := P^{-1}\Gamma_0(1)P = \{T \in \Gamma(1): a + b - c - d \equiv 0 \pmod 4\}. \tag{1.7.5}$$

Also, when $n = 3$, there are four conjugate subgroups listed in (1.6.2, 3), namely $\Gamma_U(3) = \Gamma_0(3)$, $\Gamma_W(3) = \Gamma^0(3)$ and also

$$\Gamma_P(3) = \{T \in \Gamma(1): a + b - c - d \equiv 0 \pmod 3\} \tag{1.7.6}$$

and

$$\Gamma_{P_1}(3) = \{T \in \Gamma(1): a - b + c - d \equiv 0 \pmod 3\}. \tag{1.7.7}$$

See Rankin (1973*b*).

Newman (1962) has studied a family of groups, which we denote by $\hat{\Gamma}^m$ ($m \in \mathbb{Z}^+$); here $\hat{\Gamma}^m$ denotes the subgroup of $\hat{\Gamma}(1)$ generated by the mth powers of elements of $\hat{\Gamma}(1)$. For $m = 2$, 3 it can be shown that the groups $\hat{\Gamma}^2$, $\hat{\Gamma}^3$ are the ones we have been studying.

For each $m \in \mathbb{Z}^+$, $\hat{\Gamma}^m$ is normal in $\hat{\Gamma}(1)$ and Newman has shown that

$$\hat{\Gamma}^m\hat{\Gamma}^n = \hat{\Gamma}^{(m,n)}, \quad \hat{\Gamma}^m = \hat{\Gamma}(1) \quad \text{if } (m, 6) = 1, \tag{1.7.8}$$

$$\hat{\Gamma}^{2m} = \hat{\Gamma}^2 \quad \text{if } (m, 3) = 1, \quad \hat{\Gamma}^{3m} = \hat{\Gamma}^3 \quad \text{if } (m, 2) = 1. \tag{1.7.9}$$

When $m \equiv 0 \pmod 6$, the structure of $\hat{\Gamma}^m$ is not known in all cases. Thus $\hat{\Gamma}^6$ is known to be a subgroup of index 216 in $\hat{\Gamma}(1)$ lying between $\hat{\Gamma}'(1)$ and $\hat{\Gamma}''(1)$, while, for $n > 1$, $[\hat{\Gamma}^6 : \hat{\Gamma}^{6n}] \geq n^{37}$ and may be infinite. See also Rankin (1969).

In the further algebraic study of the subgroups of the modular group the following three group-theoretic theorems are useful.

***Theorem 1.7.1* (Kurosh).** *Let the group G be a free product of subgroups A_α. We write this as $G = \prod_\alpha^* A_\alpha$. Then, if H is a subgroup of G, we have*

$$H = F * \prod_\beta^* B_\beta,$$

where F is a free group and, for each β, B_β is conjugate to one of the subgroups A_α.

We note that the free product $\prod_\beta^* B_\beta$ can be empty and that F can be the trivial group. However, if $\prod_\beta^* B_\beta$ is not empty, then F must have infinite index in G; for otherwise some power of an element of $\prod_\beta^* B_\beta$ (other than the identity) would belong to F.

***Theorem 1.7.2* (Schreier).** *Let H be a subgroup of finite index μ in a free group G of finite rank R. Then the rank r of H is also finite and*

$$r = 1 + \mu(R - 1).$$

We note that the rank of a free group is the number of free generators of the group. For proofs of both these theorems see Kurosh (1960).

***Theorem 1.7.3* (Nielsen).** *Let the group G have identity element e and be the free product of cyclic groups G_i of orders m_i $(1 \leq i \leq n)$. Then the commutator group G' is a free group of index $m = m_1m_2 \ldots m_n$ in G and the rank of G' is*

$$1 + m\left\{-1 + \sum_{i=1}^{n}\left(1 - \frac{1}{m_i}\right)\right\}.$$

G' is generated by the set of all commutators $[xax^{-1}, xbx^{-1}]$, where $a \in G_i$, $b \in G_j$, $a \neq e \neq b$, $1 \leq i \leq j \leq n$ and

$$x = a_1 \ldots a_{i-1}a_{i+1} \ldots a_{j-1},$$

for any a_k in G_k $(1 \leq k \leq n)$. The factor group G/G' is isomorphic to the direct product of the cyclic groups $G_1, G_2, \ldots, G_n$.

See Nielsen (1948) and also Lyndon (1973); for the commutator notation see (1.3.12). Theorem 1.7.3 provides an alternative method of proving theorems 1.3.1 and 1.3.2.

With the help of theorem 1.7.1 we easily derive the following result of Newman (1964).

Theorem 1.7.4. *A subgroup of $\hat{\Gamma}(1)$ is free if and only if it contains no elements of finite order other than the identity. If $\hat{\Gamma}$ is a normal subgroup of $\hat{\Gamma}(1)$ different from $\hat{\Gamma}(1)$, $\hat{\Gamma}^2$ and $\hat{\Gamma}^3$, then $\hat{\Gamma}$ is a free group.*

The following theorem is a simple deduction from Schreier's theorem.

Theorem 1.7.5. *Let $\hat{\Gamma}$ be a free subgroup of $\hat{\Gamma}(1)$ of finite index μ. Then $\mu \equiv 0 \pmod 6$ and $\hat{\Gamma}$ has rank $1+\frac{1}{6}\mu$. In particular, the index of any normal subgroup of $\hat{\Gamma}(1)$ other than $\hat{\Gamma}(1)$, $\hat{\Gamma}^2$ and $\hat{\Gamma}^3$ is divisible by 6.*

Proof. We shall prove the last sentence by analytic methods in chapter 2. The fact that $1+\frac{1}{6}\mu$ is the rank of $\hat{\Gamma}$ is due to Mason (1969).

Write $\hat{\Delta} = \hat{\Gamma} \cap \hat{\Gamma}'(1)$ and put

$$\lambda = [\hat{\Gamma}:\hat{\Delta}], \quad \nu = [\hat{\Gamma}'(1):\hat{\Delta}],$$

so that

$$\lambda\mu = [\hat{\Gamma}(1):\hat{\Delta}] = 6\nu.$$

By theorem 1.7.2 applied to $\hat{\Delta}$ as a subgroup of $\hat{\Gamma}$ and $\hat{\Gamma}'(1)$ we have

$$1+(r-1)\lambda = 1+\nu,$$

where r is the rank of $\hat{\Gamma}$. Thus, $\mu = 6(r-1)$, from which the theorem follows.

Fig. 3 displays all the normal subgroups between $\hat{\Gamma}(1)$ and $\hat{\Gamma}(6)$. The two groups $\hat{\Gamma}^{2\prime}$ and $\hat{\Gamma}^{3\prime}$ are the commutator groups of $\hat{\Gamma}^2$ and $\hat{\Gamma}^3$. By applying theorem 1.7.3 their indices in the groups immediately above them can be verified.

A complete study of all subgroups between $\hat{\Gamma}(1)$ and $\hat{\Gamma}(4)$ has been made by Petersson (1963). His list contains thirty groups, including $\hat{\Gamma}(4)$ and $\hat{\Gamma}(1)$; among these thirty groups there are contained, of course, the six groups in fig. 1. Petersson (1953) has also studied subgroups of finite index μ in $\hat{\Gamma}(1)$ for which

$$\hat{\Gamma}(1) = \hat{\Gamma} \cdot \bigcup_{k=0}^{\mu-1} U^k;$$

such a subgroup $\hat{\Gamma}$ is called a *cycloidal* subgroup of $\hat{\Gamma}(1)$.

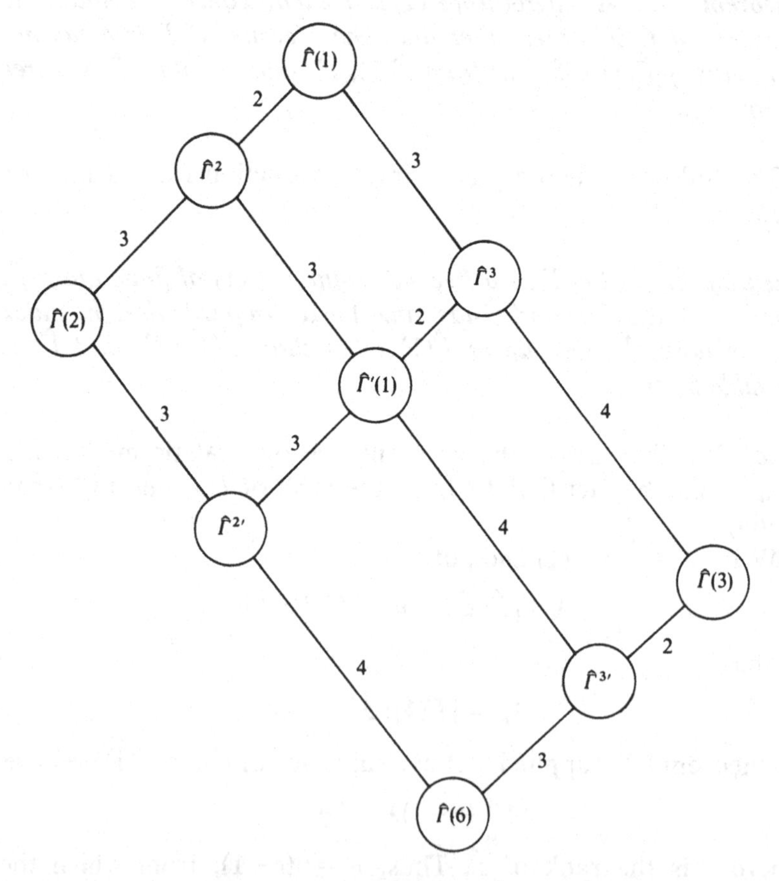

Fig. 3. Normal subgroups between $\hat{\Gamma}(1)$ and $\hat{\Gamma}(6)$.

We have already remarked that $\hat{G}(2)$ is isomorphic to the dihedral group of order 6 and that $\hat{G}(3)$ is isomorphic to the alternating group on four symbols, which is the tetrahedral group. It can be shown that $\hat{G}(4)$ and $\hat{G}(5)$ are isomorphic to the octahedral and icosahedral groups, respectively. In fact, for $n = 2$, 3, 4, 5, $\hat{G}(n)$ is generated by

$$u = U\hat{\Gamma}(n), \quad v = V\hat{\Gamma}(n),$$

where

$$u^n = v^2 = (uv)^3 = e,$$

e being the group identity.

A great deal of information about the modulary groups $\hat{G}(n)$ and $\hat{G}(p^k)$ can be found in Klein and Fricke (1890, 1892) and in Vivanti (1906, 1910); see also Frasch (1933). A complete classification of all the normal subgroups of $\hat{G}(n)$ has been given by McQuillan (1965).

The definition of level given in §1.4 is due to Wohlfahrt (1964). Previously the concept of level was only defined for congruence groups. Thus a congruence subgroup Γ of $\Gamma(1)$ was defined to be of level n if $\Gamma(n) \subseteq \Gamma$ and if n is the least positive integer for which this inclusion is valid. That the two definitions agree for congruence groups is a consequence of the relations

$$\Delta(n)\Gamma(qn) = \Gamma(n), \quad \hat{\Delta}(n)\hat{\Gamma}(qn) = \hat{\Gamma}(n), \tag{1.7.10}$$

which were essentially proved by Wohlfahrt; see also Rankin (1969) and Leutbecher (1970). Here q is any positive integer.

If we now assume that $\operatorname{lev} \Gamma = n$, so that $\Delta(n) \subseteq \Gamma \subseteq \Gamma(1)$, and that Γ is a congruence group, then, for some $q \in \mathbb{Z}^+$, $\Gamma(q) \subseteq \Gamma$. From (1.7.10) we deduce that $\Gamma(n) \subseteq \Gamma$ and it is clear that n is the least positive integer for which this holds. Thus the two definitions agree.

It has been proved by various authors that $\Delta(n) = \Gamma(n)$ for $1 \leq n \leq 5$ and that $[\Gamma(n) : \Delta(n)] = \infty$ for $n \geq 6$. In fact $\Delta(6) = \Gamma''(1)$ and $[\Gamma(n) : \Delta(n)\Gamma(n)] = \infty$ for $n \geq 6$. Similar results hold in the inhomogeneous case. It follows that every subgroup of level 2, 3, 4 or 5 must be a congruence subgroup. Also the groups described in §§1.5, 6 are the only subgroups of levels 2 and 3.

There exist subgroups of finite index in $\hat{\Gamma}(1)$ that are not congruence subgroups. That this is the case was first stated by

Klein in 1879; see Klein and Fricke (1890) (pp. 308, 418, 659–63), where a number of examples of groups and their associated functions are given. Proofs were given by Fricke (1887) and Pick (1887), who produced an example of such a subgroup of index 54. All the subgroups of $\hat{\Gamma}(1)$ of index $\mu \leq 7$ are listed in Rankin (1969). For $\mu \leq 6$ these are all congruence subgroups, but 28 of the 42 subgroups of index 7 are not congruence subgroups and fall into four classes each containing seven conjugate subgroups. In recent years various authors have discovered large families of 'non-congruence' subgroups. See, for example, Reiner (1958), Newman (1965, 1968), Rankin (1967*b*) and Mason (1969).

It is possible to give a rather complicated formula for the number of subgroups of $\hat{\Gamma}(1)$ of given index μ and to give methods, in certain cases, for specifying the number of normal subgroups; see Newman (1967).

2: Mapping properties

2.1. Conformal mappings. We begin by recalling some properties of meromorphic functions on the extended complex plane.

A *domain* $\mathbb{D}$ is a subset of $\bar{\mathbb{C}}$ that is open and connected in the topology on $\bar{\mathbb{C}}$; in particular, $\mathbb{H}$ is a domain.

Let f be a function defined on a domain $\mathbb{D}$. In order to discuss the behaviour of f at ∞, when $\infty \in \mathbb{D}$, we make use of the homeomorphism

$$z^* = 1/z$$

between neighbourhoods of $z = \infty$ and $z^* = 0$, and write

$$f^*(z^*) = f(z),$$

so that f^* is defined on a neighbourhood of 0 when $\infty \in \mathbb{D}$.

The statement that f is *holomorphic* on $\mathbb{D}$ means that (i) f maps $\mathbb{D}$ into $\mathbb{C}$, (ii) f is differentiable (with a finite derivative) on $\mathbb{D} - \{\infty\}$, (iii) if $\infty \in \mathbb{D}$, f^* is differentiable on some neighbourhood of 0. Note that, when $\infty \in \mathbb{D}$, the derivatives of f and f' are related by the equation

$$z^* f^{*\prime}(z^*) = -z f'(z)$$

on punctured neighbourhoods of $z^* = 0$ and $z = \infty$.

We say that f is holomorphic at a point $p \in \bar{\mathbb{C}}$ if f is holomorphic on some domain $\mathbb{D}$ containing p. In particular, f is holomorphic at ∞ if f^* is holomorphic on some domain containing 0; such a domain can always be taken in the form $|z^*| < r$ for some $r > 0$.

We say that f is meromorphic at a point $p \in \mathbb{C}$ if there exists an integer $q \geq 0$ such that h is holomorphic at p, where

$$h(z) := (z - p)^q f(z);$$

this implies in particular that h is continuous at p and that

$$h(p) = \lim_{z \to p} (z - p)^q f(z).$$

We say that f is meromorphic at ∞ if f^* is meromorphic at 0. A point p at which f is meromorphic but not holomorphic (so that

$q>0$) is called a *pole* of f. The smallest value of q that satisfies the conditions is called the *order* of the pole. If we can take $q=1$, the pole is simple. At a pole p of f we have $f(p)=\infty$.

More generally, if f is meromorphic at $p \in \mathbb{C}$, we define $\operatorname{ord}(f, p)$ to be the greatest integer q such that $(z-p)^{-q}f(z)$ is holomorphic at p. If $q>0$, q is the order of the zero of f at p; if $q<0$, $-q$ is the order of the pole of f at p.

Suppose that f maps a domain $\mathbb{D}$ onto a subset $\mathbb{D}_1$ of $\bar{\mathbb{C}}$. we say that f maps $\mathbb{D}$ *conformally* onto $\mathbb{D}_1$, or that f is conformal on $\mathbb{D}$, if and only if f is meromorphic on $\mathbb{D}$ and f is a bijective map of $\mathbb{D}$ onto $\mathbb{D}_1$. If f is conformal on the domain $\mathbb{D}$, it follows that its derivative f' exists and is non-zero at all points of $\mathbb{D}$ that are not poles of f. Also, if p is a pole of f in $\mathbb{D}$, then p is a simple pole. Further, the inverse function f^{-1} of a conformal mapping f is again conformal. The proofs of these facts involve the use of Rouché's theorem.

Since a holomorphic function is continuous, it follows that every conformal mapping f is a homeomorphism of its domain $\mathbb{D}$ onto $f(\mathbb{D})$, which is also a domain.

Conformal mappings have the property that angles between intersecting curves in the domain $\mathbb{D}$ are preserved in $f(\mathbb{D})$ both as regards magnitude and direction.

We now consider the particular case when

$$f(z):=T(z)=\frac{az+b}{cz+d},$$

where $T \in \hat{\Theta}$. As we saw in §1.1, T is a bijection of $\bar{\mathbb{C}}$ onto itself.

If $c=0$, then

$$f(z)=\frac{a}{d}z+\frac{b}{d}.$$

The mapping T is holomorphic on $\mathbb{C}$ and there is a simple pole at ∞. If $c \neq 0$, the mapping T is holomorphic on $\bar{\mathbb{C}}-\{-d/c\}$ and there is a simple pole at $-d/c$. That is, in every case T is holomorphic on $\bar{\mathbb{C}}-\{-d/c\}$ with a simple pole at $-d/c$. Hence T maps $\bar{\mathbb{C}}$ conformally onto $\bar{\mathbb{C}}$. Note also that, for $z \neq -d/c$ and $z \neq \infty$,

$$T'(z)=\frac{1}{(cz+d)^2}. \tag{2.1.1}$$

The converse result that, if f maps $\bar{\mathbb{C}}$ conformally onto $\bar{\mathbb{C}}$, then $f \in \hat{\Theta}$, also holds.

It can be shown similarly that f maps $\mathbb{C}$ conformally onto $\mathbb{C}$ if and only if $f \in \hat{\Theta}$ and $f(\infty)=\infty$. Further, f maps $\mathbb{H}$ conformally onto $\mathbb{H}$ if and only if $f \in \hat{\Omega}$. In these results the 'if' parts are straightforward, while the 'only if' parts require the use of Schwarz's lemma.

Since we shall not require the 'only if' parts, we content ourselves by showing that, if $T \in \hat{\Omega}$, then $T\mathbb{H}=\mathbb{H}$. For this purpose we write

$$w =: u+iv = T(z) = \frac{az+b}{cz+d}, \quad z =: x+iy,$$

where u, v, x and y are real. Now

$$v = \text{Im } T(z) = \frac{y}{|cz+d|^2}, \tag{2.1.2}$$

and conversely

$$y = \frac{v}{|cw-a|^2}. \tag{2.1.3}$$

It follows that $w \in \mathbb{H}$ if and only if $z \in \mathbb{H}$.

We note also that, if $T \in \hat{\Omega}$, then

$$T\bar{\mathbb{R}} = \bar{\mathbb{R}}, \quad T\bar{\mathbb{H}} = \bar{\mathbb{H}}$$

and, if $T \in \hat{\Gamma}(1)$,

$$T\mathbb{P} = \mathbb{P}, \quad T\mathbb{H}' = \mathbb{H}'.$$

Further, circles and straight lines are mapped by T onto circles or straight lines.

It is sometimes convenient to write the transformation $T \in \hat{\Theta}$ in a different form. Let z_1 and z_2 be two different finite complex numbers having finite images $w_j = Tz_j$ $(j=1, 2)$. Then the transformation

$$w = \frac{az+b}{cz+d}$$

can be written in the form

$$\frac{w-w_1}{w-w_2} = K\frac{z-z_1}{z-z_2}, \tag{2.1.4}$$

where K is a constant. Further, if z_3 is another finite point having a known finite image w_3, then K is uniquely determined, namely

$$K = \frac{w_3-w_1}{w_3-w_2}\bigg/\frac{z_3-z_1}{z_3-z_2}.$$

Alternatively, since $\infty = T(-d/c)$ we have

$$K = \frac{cz_2+d}{cz_1+d}, \tag{2.1.5}$$

and this holds also if $c=0$.

2.2. Fixed points. For any $T \in \Theta$ and $z \in \mathbb{C}$ we define

$$T:z = cz + d. \tag{2.2.1}$$

Observe that, although Tz and $(-T)z$ are the same,

$$(-T):z = -(T:z).$$

In particular, $I:z = 1$, $(-I):z = -1$, and, more generally,

$$\pm U^n : z = \pm 1 \quad (n \in \mathbb{Z}).$$

By (1.1.4), for any $S, T \in \hat{\Theta}$ and $z \neq \infty, T^{-1}\infty$,

$$\begin{aligned} ST:z &= (\gamma a + \delta c)z + \gamma b + \delta d = (\gamma Tz + \delta)(cz + d) \\ &= (S:Tz)(T:z). \end{aligned}$$

We thus have the important identity

$$ST:z = (S:Tz)(T:z). \tag{2.2.2}$$

A particular case of this is

$$1 = (T^{-1}:Tz)(T:z). \tag{2.2.3}$$

The equation (2.2.2) holds, in particular, for any $S, T \in \Omega$ and all $z \in \mathbb{H}$. If $T \in \Omega$ and $z \in \mathbb{H}$, $T:z$ is finite and non-zero.

A point $z \in \bar{\mathbb{C}}$ is called a *fixed point* of a mapping $T \in \hat{\Theta}$ if and only if $Tz = z$. We suppose in the first place that $c \neq 0$, so that $z \neq \infty$. It then follows by induction from (2.2.2, 3) that

$$T^n : z = (T:z)^n \tag{2.2.4}$$

for any $n \in \mathbb{Z}$, where T is the associated matrix.

The equation $Tz = z$ is equivalent to

$$cz^2 + (d-a)z - b = 0,$$

which has two, not necessarily distinct, roots, namely

$$z_1, z_2 := \frac{(a-d) \pm [(a+d)^2 - 4]^{\frac{1}{2}}}{2c}. \tag{2.2.5}$$

We note also that

$$z_1z_2=-b/c, \quad z_1+z_2=(a-d)/c,$$

so that

$$(T:z_1)(T:z_2)=(cz_1+d)(cz_2+d)=1. \tag{2.2.6}$$

It is clear that, when $T\in\hat{\Omega}$, the nature of the roots z_1, z_2 depends upon the sign of the real number $(a+d)^2-4$. If (i) $|\text{tr } T|<2$, then $(a+d)^2-4<0$ and z_1 and z_2 are conjugate complex numbers, one of which, say z_1, lies in $\mathbb{H}$. T is then called an *elliptic transformation* and z_1 and z_2 are called elliptic fixed points. If (ii) $|\text{tr } T|=2$, then $z_1=z_2$ and we have one real fixed point. T is called a *parabolic transformation* and z_1 is called a parabolic fixed point. Finally (iii), if $|\text{tr } T|>2$, then z_1 and z_2 are distinct real numbers and T is called a *hyperbolic transformation* and z_1 and z_2 are called hyperbolic fixed points.

We now examine these possibilities in greater detail, for the case when $T\in\Gamma(1)$, still making the assumption that $c\neq 0$. Our object is to express the mapping T in the form (2.1.4), i.e.

$$\frac{w-w_1}{w-w_2}=K\frac{z-z_1}{z-z_2}=\frac{T:z_2}{T:z_1}\cdot\frac{z-z_1}{z-z_2}, \tag{2.2.7}$$

where this is possible; see (2.1.5). Here $w=Tz$ and $w_j=Tz_j$ $(j=1,2)$. Note that, although $T:z_2$ and $T:z_1$ change sign when we replace the matrix T by $-T$, their ratio remains unchanged.

(i) *Elliptic transformations*. Here $|\text{tr } T|<2$, and, by theorem 1.2.3, there are two possibilities:

(*a*) $\text{tr } T=0$, $T=\pm L^{-1}VL$ for some $L\in\Gamma(1)$,

or

(*b*) $\text{tr } T=\pm 1$, $T=\pm L^{-1}P^sL$ for $s=1, 2$ and some $L\in\Gamma(1)$.

In case (*a*) $Tz=z$ is equivalent to $V(Lz)=Lz$; i.e. Lz is a fixed point for V and so $Lz_1=i$, $Lz_2=-i$. Hence

$$z_1=L^{-1}i, \quad z_2=L^{-1}(-i)=\bar{z}_1. \tag{2.2.8}$$

Here the bar denotes the complex conjugate. Since $T^2=V^2=-I$, we have, by (2.2.4), $(T:z_j)^2=-1$ $(j=1,2)$ and therefore, by

(2.2.6), $K=(T:z_2)/(T:z_1)=-1$. Hence

$$\frac{w-z_1}{w-z_2}=-\frac{z-z_1}{z-z_2}. \qquad (2.2.9)$$

The transformation T is of order 2.

In case (b), $Tz=z$ is equivalent to $P^s(Lz)=Lz$, where $s=1$ or 2. An elementary calculation shows that both P and P^2 have fixed points

$$\rho=e^{2\pi i/3} \qquad (2.2.10)$$

and $\bar{\rho}$, so that $Lz_1=\rho$, $Lz_2=\bar{\rho}$. Hence

$$z_1=L^{-1}\rho, \quad z_2=L^{-1}\bar{\rho}. \qquad (2.2.11)$$

Since $T^3=\pm P^{3s}=\pm I$, (2.2.4) gives $T:z_1=\pm\rho$ or $\pm\rho^2$ and so $T:z_2=\pm\rho^2$ or $\pm\rho$, by (2.2.6). Hence $K=\rho$ or ρ^2, and the transformation takes the form

$$\frac{w-z_1}{w-z_2}=K\frac{z-z_1}{z-z_2} \quad (K=\rho \text{ or } \rho^2). \qquad (2.2.12)$$

It is a transformation of order 3.

(ii) *Parabolic transformations.* Here $\operatorname{tr} T=\pm 2$ and, by theorem 1.2.3, $T=\pm L^{-1}U^qL$ for some $q\in\mathbb{Z}$ $(q\neq 0)$ and $L\in\Gamma(1)$. Thus $Tz=z$ is equivalent to $U^q(Lz)=Lz$, i.e. $(Lz)+q=Lz$. Hence $Lz=\infty$ and the single fixed point is

$$z_1=L^{-1}\infty. \qquad (2.2.13)$$

Since $c\neq 0$, z_1 is a finite rational number.

Put $a+d=2\varepsilon$, where $\varepsilon=\pm 1$. Then, by (2.2.5), $z_1=(a-d)/(2c)$ so that $T:z_1=\varepsilon=\frac{1}{2}\operatorname{tr} T$. If $w=Tz$,

$$\frac{1}{w-z_1}=\frac{cz+d}{(a-cz_1)(z-z_1)}=\frac{c(z-z_1)+\varepsilon}{\varepsilon(z-z_1)}.$$

Hence the transformation T can be expressed in the form

$$\frac{1}{w-z_1}=\frac{1}{z-z_1}+c\varepsilon. \qquad (2.2.14)$$

So far we have assumed that $c\neq 0$. When $c=0$, $T=\pm U^q$ for some $q\in\mathbb{Z}$ and $\operatorname{tr} T=\pm 2$. If $q=0$ we obtain the identical transformation under which every point is fixed. When $q\neq 0$, $T=$

$\pm L^{-1}U^qL$ (with $L=I$) and we include T among the set of parabolic transformations. There is just one fixed point, namely $z_1=\infty$. In place of (2.2.14), the transformation T takes the canonical form

$$w=z+q. \tag{2.2.15}$$

(iii) *Hyperbolic transformations.* Here $|\text{tr } T|>2$ and, since z_2 is real, (2.1.5) and (2.2.6) give $K=(T:z_2)^2>0$. The transformation has the canonical form (2.2.7) with $K>0$; clearly $K\neq 1$, since $z_1\neq z_2$.

From the above analysis we see that the transformations $T\in\hat{\Gamma}(1)$ can be divided into five classes:

1. *The identity transformation*, whose matrix $T=\pm I$.
2. *Elliptic transformations of order 2.* The matrix T of such a mapping is conjugate to $\pm V$ and satisfies $T^2=-I$. We denote by

$$\mathbb{E}_2=\{z\in\mathbb{C}: z=L^{-1}i, L\in\hat{\Gamma}(1)\} \tag{2.2.16}$$

the set of all elliptic fixed points of order 2 in $\mathbb{H}$.

3. *Elliptic transformations of order 3.* The matrix T of such a mapping is conjugate to $\pm P$ or $\pm P^2$ and satisfies $T^3=\pm I$. We denote by

$$\mathbb{E}_3=\{z\in\mathbb{C}: z=L^{-1}\rho, L\in\hat{\Gamma}(1)\} \tag{2.2.17}$$

the set of all elliptic fixed points of order 3 in $\mathbb{H}$. Here ρ is defined by (2.2.10).

4. *Parabolic transformations.* The matrix T of such a mapping is conjugate to $\pm U^q$ $(q\in\mathbb{Z}, q\neq 0)$. The set of all parabolic fixed points is $\mathbb{P}$, since $\mathbb{P}=\{z\in\bar{\mathbb{C}}: z=L^{-1}\infty, L\in\hat{\Gamma}(1)\}$. Parabolic fixed points are also called *cusps* for a reason that will be clear later on.
5. *Hyperbolic transformations.* The fixed points of such transformations are less important in the theory. It is easy to see that they are all irrational numbers. Hyperbolic transformations are of infinite order.

Note that $T\mathbb{E}_2=\mathbb{E}_2$, $T\mathbb{E}_3=\mathbb{E}_3$ for all $T\in\hat{\Gamma}(1)$.
We also write

$$\mathbb{E}=\mathbb{E}_2\cup\mathbb{E}_3. \tag{2.2.18}$$

We now suppose that Γ is a subgroup of $\Gamma(1)$. The mappings $T\in\hat{\Gamma}$ can be divided into five classes in a similar way, but some of these classes may be empty. Thus $\hat{\Gamma}$ will contain elliptic transformations only if it contains a mapping conjugate to V or P. For $m=2$, 3 we denote by $\mathbb{E}_m(\Gamma)$ the set of all fixed points in $\mathbb{H}$ of elliptic transformations of order m belonging to $\hat{\Gamma}$. It follows that

$$\mathbb{E}_2(\Gamma)=\{z\in\mathbb{C}: z=L^{-1}i, L\in\hat{\Gamma}(1), L^{-1}VL\in\hat{\Gamma}\}, \quad (2.2.19)$$

$$\mathbb{E}_3(\Gamma)=\{z\in\mathbb{C}: z=L^{-1}\rho, L\in\hat{\Gamma}(1), L^{-1}PL\in\hat{\Gamma}\}. \quad (2.2.20)$$

Note that we can omit $L^{-1}P^2L$ since $L^{-1}P^2L\in\hat{\Gamma}$ if and only if $L^{-1}PL\in\hat{\Gamma}$ and has the same fixed points. We write

$$\mathbb{E}(\Gamma)=\mathbb{E}_2(\Gamma)\cup\mathbb{E}_3(\Gamma). \quad (2.2.21)$$

If $\hat{\Gamma}$ is normal in $\hat{\Gamma}(1)$ it is clear that $\mathbb{E}_m(\Gamma)$ is either $\mathbb{E}_m$ or the null set $\varnothing$.

We now suppose that z is any point of $\mathbb{H}'$. The *stabilizer* of $z \pmod{\Gamma}$ is defined to be the subset Γ_z of Γ consisting of all $T\in\Gamma$ for which $Tz=z$. Clearly Γ_z is a subgroup of Γ. The stabilizer of $z \pmod{\Gamma(1)}$ is denoted by $\Gamma_z(1)$; the corresponding inhomogeneous groups are denoted by $\hat{\Gamma}_z$ and $\hat{\Gamma}_z(1)$. Evidently $\hat{\Gamma}_z$ is a subgroup of $\hat{\Gamma}_z(1)$. Further, if $L\in\hat{\Gamma}(1)$ then

$$L^{-1}\hat{\Gamma}_{Lz}L=(L^{-1}\hat{\Gamma}L)_z. \quad (2.2.22)$$

The preceding discussion of fixed points shows that

$$\hat{\Gamma}_\infty(1)=\hat{\Gamma}_U$$

in the notation of §1.2. Also $\hat{\Gamma}_i(1)$ and $\hat{\Gamma}_\rho(1)$ are the cyclic groups of orders 2 and 3 generated by the mappings V and P, respectively. It follows immediately from this and (2.2.22) that

$$\hat{\Gamma}_z(1)=\begin{cases} L^{-1}\hat{\Gamma}_U L, & \text{when } z=L^{-1}\infty,\\ L^{-1}\hat{\Gamma}_i(1)L, & \text{when } z=L^{-1}i,\\ L^{-1}\hat{\Gamma}_\rho(1)L, & \text{when } z=L^{-1}\rho,\\ \hat{\Lambda}=\{I\} & \text{otherwise.}\end{cases} \quad (2.2.23)$$

Here L is any member of $\hat{\Gamma}(1)$, and $z\in\mathbb{H}'$. Note that, in every case, $\hat{\Gamma}_z(1)$ is a cyclic group.

We now define the *order* of $z \pmod{\Gamma}$ to be

$$n(z,\Gamma):=[\hat{\Gamma}_z(1):\hat{\Gamma}_z]. \quad (2.2.24)$$

The index on the right is possibly infinite. However, if $\hat{\Gamma}$ has finite index in $\hat{\Gamma}(1)$ then $n(z, \Gamma)$ is finite. For if the cyclic group $\hat{\Gamma}_z(1)$ is generated by S it follows that $S^n \in \hat{\Gamma}$ for some finite positive integer n, and $n(z, \Gamma)$ is in fact the least such integer n. We note also that, if we take $\Gamma_1 = \hat{\Gamma}(1)$, $\Gamma_2 = \hat{\Gamma}$ in theorem 1.1.2, with S as above then, for $1 \leq i \leq m$,

$$\sigma_i = n(z, L_i^{-1}\Gamma L_i) = n(L_i z, \Gamma). \tag{2.2.25}$$

The transformation S is parabolic, elliptic or the identity according as $z \in \mathbb{P}$, $\mathbb{E}$ or $\mathbb{H}' - \mathbb{P} \cup \mathbb{E}$, respectively; in fact, we can take $S = L^{-1}UL, L^{-1}VL, L^{-1}PL$ and I in the four cases listed in (2.2.23). In the latter case $n(z, r) = 1$. When $z \in \mathbb{P}$, we also call $n(z, \Gamma)$ the *width of the cusp* z (mod Γ); if $z = L^{-1}\infty$, $\hat{\Gamma}_z$ is generated by $L^{-1}U^nL$, where $n = n(z, r)$. When $z \in \mathbb{E}_m$ $(m = 2, 3)$, $n(z, \Gamma) = 1$ or m according as z does or does not belong to $\mathbb{E}_m(\Gamma)$.

It can be shown similarly that, when $\hat{\Gamma}$ has finite index in $\hat{\Gamma}(1)$, some positive power of every hyperbolic transformation belongs to $\hat{\Gamma}$, and is, of course, hyperbolic. Hence $\hat{\Gamma}$ always contains hyperbolic transformations.

We note, in conclusion, that if $\Gamma = \Gamma'(1)$ or $\Gamma(N)$ for $N > 1$, then $\mathbb{E}(\Gamma) = \varnothing$ and so $n(z, \Gamma) = m$ if $z \in \mathbb{E}_m$ $(m = 2, 3)$. Further, if $z \in \mathbb{P}$,

$$n(z, \Gamma'(1)) = 6 \quad \text{and} \quad n(z, \Gamma(N)) = N.$$

2.3. Fundamental regions. Let Γ be a subgroup of $\Gamma(1)$, so that the mappings T in $\hat{\Gamma}$ map $\mathbb{H}'$ onto itself. In what follows we shall only be interested in subsets of $\mathbb{H}'$, so that we are not, for example, interested in hyperbolic fixed points.

Two points z_1, z_2 in $\mathbb{H}'$ are said to be *congruent*, or *equivalent*, (mod Γ), if there exists a $T \in \hat{\Gamma}$ such that

$$z_2 = Tz_1.$$

It is easily verified that this is an equivalence relation, and we write

$$z_2 \equiv z_1 \pmod{\Gamma}.$$

The equivalence class containing a point $z \in \mathbb{H}'$ is called the *orbit* of z (mod Γ) and is denoted by $\hat{\Gamma}z$; instead of $\hat{\Gamma}(1)z$ we may write $[z]$. Thus $\mathbb{E}_2 = [i]$, $\mathbb{E}_3 = [\rho]$ and, if $\hat{\Gamma}$ is of finite index in $\hat{\Gamma}(1)$, $\mathbb{P} = \hat{\Gamma}\infty = [\infty]$. Clearly $n(z_1, \Gamma) = n(z_2, \Gamma)$ when $z_1 \equiv z_2 \pmod{\Gamma}$.

A subset $\mathbb{F}$ of $\mathbb{H}'$ is called a *proper fundamental region* for $\hat{\Gamma}$ if $\mathbb{F}$ contains exactly one point from each orbit $\hat{\Gamma}z$. By giving $\mathbb{F}$ the quotient topology induced by the topology on $\mathbb{H}$ (compactified suitably at points of $\mathbb{P}$) and the equivalence relation, $\mathbb{F}$ can be made into a connected Hausdorff space which is, in fact, the Riemann surface associated with the group $\hat{\Gamma}$. We shall not, however, use any Riemann surface theory, although we may mention this theory at various points. In practice it is usually convenient to impose further conditions on $\mathbb{F}$, such as that it is a simply connected subset of $\mathbb{H}'$ and is bounded by curves of a prescribed form.

Theorem 2.3.1. *Let $\mathbb{F}$ be a proper fundamental region for a subgroup $\hat{\Gamma}$ of $\hat{\Gamma}(1)$ and suppose that $\mathbb{F} = \bigcup_{n=1}^{\infty} \mathbb{F}_n$, where the sets $\mathbb{F}_n$ are disjoint. Then, if $T_n \in \hat{\Gamma}$ for each $n \in \mathbb{Z}^+$, the set $\bigcup_{n=1}^{\infty} T_n \mathbb{F}_n$ is also a proper fundamental region for $\hat{\Gamma}$.*

This is obvious, as all we have done is to choose Tz as a representative of $\hat{\Gamma}z$ rather than z for some $T \in \hat{\Gamma}$. The theorem is useful since it enables us to piece together fundamental regions in alternative ways that may be convenient for special purposes. Usually the number of non-null regions $\mathbb{F}_n$ is finite.

Theorem 2.3.2. *Let $\mathbb{F}$ be a proper fundamental region for a subgroup $\hat{\Gamma}$ of $\hat{\Gamma}(1)$ and suppose that $T \in \hat{\Gamma}(1)$. Then* (i) *$T\mathbb{F}$ is a proper fundamental region for the conjugate group $T\hat{\Gamma}T^{-1}$.* (ii) *In particular, if $T \in \hat{\Gamma}$ and $T \neq \pm I$, then $T\mathbb{F}$ is a proper fundamental region for $\hat{\Gamma}$ and $\mathbb{F} \cap T\mathbb{F}$ is either empty or consists of a single point ζ, which is a fixed point for T.* (iii) *A fixed point ζ for a mapping $T \in \hat{\Gamma}$ cannot be an interior point of $\mathbb{F}$.* (iv) *The regions $T\mathbb{F}$ for $T \in \hat{\Gamma}$ cover $\mathbb{H}'$ without overlapping; if T_1 and T_2 are different transformations in $\hat{\Gamma}$, then $T_1\mathbb{F}$ and $T_2\mathbb{F}$ have at most one point in common.*

Proof. (i) If $z \in \mathbb{H}'$ and $T \in \hat{\Gamma}(1)$, then $T^{-1}z \in \mathbb{H}'$ and so there exists an $S \in \hat{\Gamma}$ such that $ST^{-1}z \in \mathbb{F}$. Then $TST^{-1}z \in T\mathbb{F}$. Further, if $TS_1T^{-1}z$ and $TS_2T^{-1}z$ are two points of $T\mathbb{F}$, where S_1 and S_2 are in $\hat{\Gamma}$, then $S_1T^{-1}z$ and $S_2T^{-1}z$ are points in the same orbit $\hat{\Gamma}T^{-1}z$ lying in $\mathbb{F}$ and so are identical. The original points $TS_1T^{-1}z$ and $TS_2T^{-1}z$ are therefore also identical.

(ii) Let $T \in \hat{\Gamma}$, $T \neq \pm I$ and suppose that $\zeta \in \mathbb{F} \cap T\mathbb{F}$. Then $\zeta \in \mathbb{F}$ and $T^{-1}\zeta \in \mathbb{F}$ and, since these points are congruent (mod Γ), $\zeta = T^{-1}\zeta$; i.e. $T\zeta = \zeta$. Hence ζ is a fixed point for T and there is only one such point in $\mathbb{H}'$.

(iii) If ζ is an interior point of $\mathbb{F}$, then there exists a neighbourhood $\mathbb{N}$ of ζ with $\mathbb{N} \subseteq \mathbb{F}$. But $T\mathbb{N}$ is a neighbourhood of $T\zeta = \zeta$ and so therefore is

$$\mathbb{N}' = \mathbb{N} \cap T\mathbb{N}.$$

But $\mathbb{N}' \subseteq \mathbb{N} \subseteq \mathbb{F}$ and $\mathbb{N}' \subseteq T\mathbb{N} \subseteq T\mathbb{F}$, so that $\mathbb{N}' \subseteq \mathbb{F} \cap T\mathbb{F}$, which is false since $\mathbb{F} \cap T\mathbb{F} = \{\xi\}$.

(iv) That $\mathbb{H}' \subseteq \bigcup_{T \in \Gamma} T\mathbb{F}$ is obvious. Further $\zeta \in T_1\mathbb{F} \cap T_2\mathbb{F}$ if and only if $T_1^{-1}\zeta \in \mathbb{F} \cap T_1^{-1}T_2\mathbb{F}$ and the last part follows from this and (ii).

When $\hat{\Gamma} = \hat{\Gamma}_{U^k}$ it is easy to find a fundamental region.

Theorem 2.3.3. *Let* $n \in \mathbb{Z}^+$, $\delta \geq 0$ *and put*

$$\mathbb{S}_k(\delta) := \{z \in \mathbb{H}' : -\tfrac{1}{2}k \leq \operatorname{Re} z < \tfrac{1}{2}k, \operatorname{Im} z \geq \delta\}. \tag{2.3.1}$$

Also put

$$\mathbb{S}_k := \mathbb{S}_k(0). \tag{2.3.2}$$

Then $\mathbb{S}_k$ *is a proper fundamental region for* $\hat{\Gamma}_{U^k}$. *Further, for each* $\zeta \in \mathbb{H}'$, $[\zeta] \cap \mathbb{S}_k(\delta)$ *is a finite set when* $\delta > 0$.

Proof. Since $\hat{\Gamma}_{U^k}$ consists of the transformations

$$w = z + kn \quad (n \in \mathbb{Z}),$$

it is obvious that $\mathbb{S}_k$ is a fundamental region for $\hat{\Gamma}_{U^k}$.

Take $\zeta \in \mathbb{H}'$ and $\delta > 0$, and put $\zeta = \xi + i\eta$, where ξ, η are real. If $\eta = 0$, then $[\zeta] \cap \mathbb{S}_k(\delta) = \varnothing$, so that we may assume that $\eta > 0$. If $T\zeta \in \mathbb{S}_k(\delta)$ for any $T \in \hat{\Gamma}(1)$, then, by (2.1.2),

$$c^2(\xi^2 + \eta^2) + 2cd\xi + d^2 \leq \eta/\delta.$$

For given ξ, η, δ the number of pairs of integers c, d that satisfy this inequality is finite; for the inequality states that the point (c, d) lies in a certain ellipse. For each coprime pair of integers c, d satisfying the inequality we can find a matrix $T' \in \Gamma(1)$ with second row $[c, d]$; any other such matrix T is given by $T = U^nT'$ for some $n \in \mathbb{E}$. If $T\zeta \in \mathbb{S}_k(\delta)$, then $T'\zeta + n \in \mathbb{S}_k(\delta)$ and this can hold for at most k different values of n. This completes the proof.

Corollary 2.3.3. *The set* $\mathbb{E}$ *is a countable subset of isolated points of* $\mathbb{H}$.

It follows immediately from the theorem that each point of $\mathbb{E}$ is isolated. Further,

$$\mathbb{E}=\bigcup_{n=1}^{\infty}\{([i]\cup[\rho])\cap\mathbb{S}_n(1/n)\}$$

from which countability follows, as each subset $([i]\cup[\rho])\cap\mathbb{S}_n(1/n)$ is finite.

Let $\hat{\Gamma}$ be a subgroup of $\hat{\Gamma}(1)$. A subset $\mathbb{F}$ of $\mathbb{H}'$ is called a *fundamental region* for $\hat{\Gamma}$ if $\mathbb{F}$ contains at least one point of every orbit $\hat{\Gamma}z$ $(z\in\mathbb{H}')$ and exactly one point whenever $z\notin\mathbb{E}\cup\mathbb{F}$. A proper fundamental region is therefore a fundamental region, and a fundamental region only differs from a proper fundamental region in the possible inclusion of a countable number of fixed points of $\hat{\Gamma}(1)$.

Theorem 2.3.4. *Theorem 2.3.1 holds with the word 'proper' omitted. So does theorem 2.3.2, except that in* (*ii*) *and* (*iv*) *the two different fundamental regions may intersect in more than one point of* $\mathbb{E}\cup\mathbb{P}$.

This follows immediately. We note that in the proof of part (iii) of theorem 2.3.2, $\mathbb{F}\cap T\mathbb{F}$ is a set of isolated points and so cannot contain $\mathbb{N}'$.

Theorem 2.3.5. *Let* $\hat{\Gamma}_1$ *and* $\hat{\Gamma}_2$ *be subgroups of* $\hat{\Gamma}(1)$ *and suppose that* $\hat{\Gamma}\subseteq\hat{\Gamma}_1=\hat{\Gamma}_2\cdot\mathcal{R}$. *Then, if* $\mathbb{F}_1$ *is a fundamental region for* $\hat{\Gamma}_1$,

$$\mathbb{F}_2=\bigcup_{T\in\mathcal{R}}T\mathbb{F}_1$$

is a fundamental region for $\hat{\Gamma}_2$.

Proof. Let $z\in\mathbb{H}'$; then there exists an $S_1\in\hat{\Gamma}_1$ such that $S_1z\in\mathbb{F}_1$. Write $S_1^{-1}=S_2^{-1}T$, where $S_2\in\hat{\Gamma}_2$ and $T\in\mathcal{R}$. Then $S_2z=TS_1z\in T\mathbb{F}_1\subseteq\mathbb{F}_2$.

Conversely, suppose that z and $z'\in\mathbb{F}_2$, where $z'=S_2z$ for $S_2\in\hat{\Gamma}_2$, and that neither z nor z' is a fixed point of $\hat{\Gamma}(1)$. Then $z\in T\mathbb{F}_1$, $z'\in T'\mathbb{F}_1$, where $T,T'\in\mathcal{R}$. Hence $T^{-1}z$ and $T'^{-1}z'$ are congruent $(\text{mod}\,\Gamma_1)$, lie in $\mathbb{F}_1$ and are not fixed points of $\hat{\Gamma}(1)$. They are therefore identical; i.e. $z=S_2^{-1}z'=S_2^{-1}T'T^{-1}z$. Since z is not a fixed point, we must have $S_2^{-1}T'T^{-1}=\pm I$; i.e. $T'\in\hat{\Gamma}_2T$. This implies that $T'=T$ and $S_2=\pm I$; i.e. $z'=z$.

***Corollary* 2.3.5.** *Under similar assumptions, if $\hat{\Gamma}_1 = \mathscr{L} \cdot \hat{\Gamma}_2$, $\bigcup_{T \in \mathscr{L}} T^{-1}\mathbb{F}_1$ is a fundamental region for $\hat{\Gamma}_2$.*

This is proved similarly.

2.4. Construction of fundamental regions for $\hat{\Gamma}(1)$ and its subgroups. We denote by $\mathbb{F}_I$ the set

$$\mathbb{F}_I = \mathbb{F}^{(1)} \cup \mathbb{F}^{(2)}, \tag{2.4.1}$$

where

$$\mathbb{F}^{(1)} := \{z \in \bar{\mathbb{C}} : -\tfrac{1}{2} \le \operatorname{Re} z \le 0, |z| \ge 1\} \tag{2.4.2}$$

and

$$\mathbb{F}^{(2)} := \{z \in \mathbb{C} : 0 < \operatorname{Re} z < \tfrac{1}{2}, |z| > 1\}. \tag{2.4.3}$$

We include ∞ in $\mathbb{F}^{(1)}$, but not in $\mathbb{F}^{(2)}$. The closure of any transform $T\mathbb{F}^{(1)}$ or $T\mathbb{F}^{(2)}$ $(T \in \hat{\Gamma}(1))$ we call a *triangle.*

***Theorem* 2.4.1.** *$\mathbb{F}_I$ is a proper fundamental region for $\hat{\Gamma}(1)$.*

Proof. Let ζ be any point of $\mathbb{H}'$. We show first that some member of the orbit $[\zeta]$ lies in $\mathbb{F}_I$. For this purpose we may assume that ζ is not congruent to any point on the frontier $\partial\mathbb{F}_I$ of $\mathbb{F}_I$, since every such point z either lies in $\mathbb{F}_I$, or else one of the congruent points $z-1$, $-1/z$ does. We call $\eta = \operatorname{Im} \zeta$ the *height* of ζ.

We observe first that, for some $n \in \mathbb{Z}$, $\zeta_1 = U^n\zeta \in \mathbb{S}_1$ (see (2.3.2)) and has the same height as ζ. If $\zeta_1 \notin \mathbb{F}_I$, then $|\zeta| < 1$ and so $V\zeta_1 \in [\zeta]$ and has height greater than ζ; for

$$\operatorname{Im}(-1/\zeta_1) = (\operatorname{Im} \zeta_1)/|\zeta_1|^2.$$

Further, for some $m \in \mathbb{Z}$, $\zeta_2 = U^m V\zeta_1 = U^m V U^n \zeta \in \mathbb{S}_1$ and has height greater than η. Either $\zeta_2 \in \mathbb{F}_I$ or else we can continue the process and find a congruent point $\zeta_3 \in \mathbb{S}_1$ of greater height than ζ_2. Since $\mathbb{S}_1(\eta) \cap [\zeta]$ is finite, the process ultimately terminates after a finite number k of stages in the finding of a point $\zeta_k \in [\zeta] \cap \mathbb{F}_I$.

It remains to prove that each orbit contains only one point of $\mathbb{F}_I$. For suppose that z_1 and z_2 are different congruent points of $\mathbb{F}_I$. Then both are finite and we may assume that

$$y_2 = \operatorname{Im} z_2 \ge y_1 = \operatorname{Im} z_1.$$

Let $z_2 = Tz_1$, where $T \in \Gamma(1)$, so that

$$y_2 = \frac{y_1}{|cz_1 + d|^2}.$$

Hence $|cz_1 + d| \leq 1$.

We cannot have $|c| \geq 2$, since no circle of radius $r \leq \frac{1}{2}$ and orthogonal to $\mathbb{R}$ meets $\mathbb{F}_I$; also, if $c = 0$, then $T = U^k$, which is clearly impossible. We may therefore assume that $c = 1$. The only circles of unit radius centred at points of $\mathbb{Z}$ that meet $\mathbb{F}_I$ are the circles

$$|z| = 1 \quad \text{and} \quad |z + 1| = 1.$$

There are thus two cases:

$$\text{(i) } c = 1, d = 0, |z_1| = 1, \quad \text{(ii) } c = d = 1, z_1 = \rho.$$

In case (i) we must have $T = U^k V$ and either $k = 0$, $z_1 = z_2 = i$, or $k = -1$ and $z_1 = z_2 = \rho$. In case (ii), $T = U^k P$ and we must have $k = 0$, $z_1 = z_2 = \rho$. Hence, in both cases $z_1 = z_2$ and this completes the proof of the theorem.

It follows from theorem 2.3.1 that

$$\hat{\mathbb{F}}_I = \mathbb{F}^{(1)} \cup \{U^{-1}\mathbb{F}^{(2)}\} \tag{2.4.4}$$

is also a proper fundamental region for $\hat{\Gamma}(1)$. For each $T \in \hat{\Gamma}(1)$ we write

$$\mathbb{F}_T = T\mathbb{F}_I, \quad \hat{\mathbb{F}}_T = T\hat{\mathbb{F}}_I. \tag{2.4.5}$$

It follows from theorem 2.3.2 that

$$\mathbb{H}' = \bigcup_{T \in \hat{\Gamma}(1)} \mathbb{F}_T = \bigcup_{T \in \hat{\Gamma}(1)} \hat{\mathbb{F}}_T.$$

We note that the boundary (frontier) of $\mathbb{F}_I$ consists of four 'sides' l_U, Ul_U, l_V and Vl_V, where

$$l_U = \{z = x + yi : x = -\tfrac{1}{2}, \tfrac{1}{2}\sqrt{3} \leq y\} \tag{2.4.6}$$

and

$$l_V = \{z = x + yi : -\tfrac{1}{2} \leq x \leq 0, |z| = 1, y > 0\}. \tag{2.4.7}$$

l_U and l_V are contained in $\mathbb{F}_I$ but $Ul_U - \{\infty\}$ and $Vl_V - \{i\}$ are not.

Fig. 4 shows how the regions $\mathbb{F}_T$ fit together. The angles between sides contained in the regions in question are marked. That the

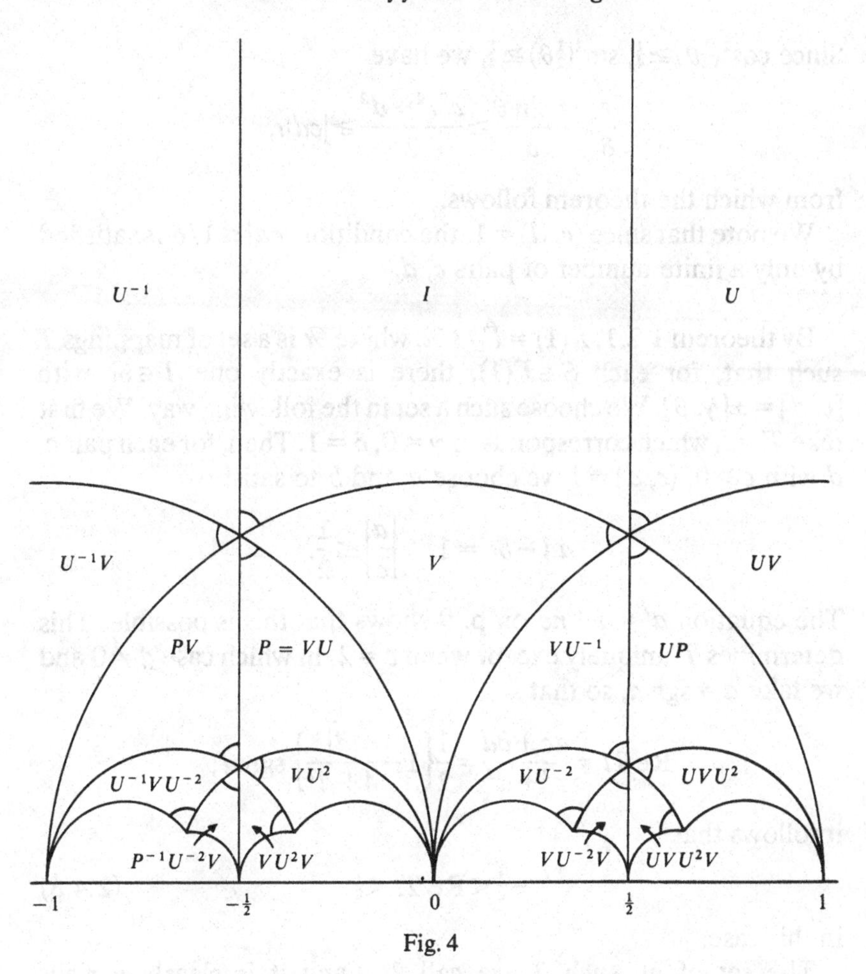

Fig. 4

regions cluster closer and closer to the real axis is shown by the following theorem.

Theorem 2.4.2. *If $\mathbb{F}_T$ contains a point ζ with $\eta = \operatorname{Im} \zeta \geq \delta$, then $|cd| \leq 1/\delta$.*

Proof. Suppose that $\zeta = Tz$ is such a point, so that $z \in \mathbb{F}_I$. Write $r = |z|$, $z = r e^{i\theta}$ where $\frac{1}{3}\pi \leq \theta \leq \frac{2}{3}\pi$. If $\eta \geq \delta$ we have, by (2.1.2),

$$|cz + d|^2 = (cx + d)^2 + c^2 y^2 \leq y/\delta;$$

i.e.

$$(cr + d)^2 \cos^2(\tfrac{1}{2}\theta) + (cr - d)^2 \sin^2(\tfrac{1}{2}\theta) \leq (r/\delta) \sin \theta.$$

Since $\cos^2(\frac{1}{2}\theta) \geq \frac{1}{4}$, $\sin^2(\frac{1}{2}\theta) \geq \frac{1}{4}$, we have

$$\frac{r}{\delta} \geq \frac{r \sin \theta}{\delta} \geq \frac{c^2 r^2 + d^2}{2} \geq |cd| r,$$

from which the theorem follows.

We note that since $(c, d) = 1$, the condition $|cd| \leq 1/\delta$ is satisfied by only a finite number of pairs c, d.

By theorem 1.2.1, $\hat{\Gamma}(1) = \hat{\Gamma}_U \cdot \mathcal{R}$, where $\mathcal{R}$ is a set of mappings T such that, for each $S \in \Gamma(1)$, there is exactly one $T \in \mathcal{R}$ with $[c, d] = \pm[\gamma, \delta]$. We choose such a set in the following way. We first take $T = I$, which corresponds to $\gamma = 0$, $\delta = 1$. Then, for each pair c, d with $c > 0$, $(c, d) = 1$ we choose a and b to satisfy

$$ad - bc = 1, \quad \left|\frac{a}{c}\right| \leq \frac{1}{2}.$$

The equation $a' = a + nc$ on p. 9 shows that this is possible. This determines T uniquely except when $c = 2$, in which case $d \neq 0$ and we take $a = \operatorname{sgn} d$, so that

$$\operatorname{Re} Ti = \frac{ac + bd}{c^2 + d^2} = \frac{1}{2}\left\{1 - \frac{|d|}{4 + d^2}\right\} \operatorname{sgn} d;$$

it follows that

$$-\tfrac{1}{2} < \operatorname{Re} Ti < \tfrac{1}{2} \tag{2.4.8}$$

in this case.

The set of all such T we call $\mathcal{R}_0$, and it is clearly a right transversal of $\hat{\Gamma}_U$ in $\hat{\Gamma}(1)$.

We now apply theorem 2.3.5 with this choice of $\mathcal{R}_0$ and $\Gamma_1 = \hat{\Gamma}(1)$, $\Gamma_2 = \hat{\Gamma}_U$, taking $\mathbb{F}_1 = \mathbb{F}_I$. We obtain a region

$$\mathbb{S}_U = \bigcup_{T \in \mathcal{R}_0} T\mathbb{F}_I = \bigcup_{T \in \mathcal{R}_0} \mathbb{F}_T, \tag{2.4.9}$$

which is a fundamental region for Γ_U. We show that $\mathbb{S}_U$ differs from the proper fundamental region $\mathbb{S}_1$ of theorem 2.3.3 only in the respect that the straight line segment

$$\lambda := \{z = x + iy : x = -\tfrac{1}{2}, 0 \leq y < \tfrac{1}{2}\}$$

in $\mathbb{S}_1$ is replaced by the congruent segment $U\lambda$, and the fixed point $Wi = \frac{1}{2}(1 + i)$ is added.

To prove this we need only observe from fig. 4 that the line $l=\{z:x=-\frac{1}{2}\}$ is made up of sides of the regions $\mathbb{F}_I$, $\mathbb{F}_P$ and $\mathbb{F}_{VU^2V}$ and that no points of λ belong to any of these regions, while $U\lambda$ and $\frac{1}{2}(1+i)$ are contained in $\mathbb{F}_{VU^{-1}}\cup\mathbb{F}_{VU^{-2}V}$. Each of the five regions mentioned is a subset of $\mathbb{S}_U$. In particular, the lines l and Ul contain no interior points of any region $\mathbb{F}_T$. Hence every region $\mathbb{F}_T$ is either wholly contained in $\mathbb{S}_U$ or has no points in common with it except possibly the nine fixed points that lie on l and Ul; these are ∞, ρ, $-\rho^2$, $\frac{1}{2}(1-i)$, $\frac{1}{2}(1+i)$, $\frac{1}{6}(\rho-2)$, $\frac{1}{6}(\rho+4)$, $-\frac{1}{2}$ and $\frac{1}{2}$. Hence, if $|a/c|=|T\infty|<\frac{1}{2}$, then $\mathbb{F}_T\subseteq\mathbb{S}_U$, while, if $|T\infty|=\frac{1}{2}$, this is also true by our choice of T in (2.4.8) to make $|\text{Re } Ti|<\frac{1}{2}$.

We now construct a fundamental region for a subgroup $\hat{\Gamma}$ of $\hat{\Gamma}(1)$ of finite index μ by applying theorem 2.3.5 with

$$\hat{\Gamma}_1=\hat{\Gamma}(1),\quad \hat{\Gamma}_2=\hat{\Gamma}.$$

We take $\mathbb{F}_1$ to be $\mathbb{F}_I$. The regions that we shall progressively construct will have boundaries consisting of a finite number of arcs of circles and segments of straight lines, these lines and circles being orthogonal to the real axis. Each arc or segment is called a *side* and has two endpoints called *vertices* which are points of $\mathbb{E}\cup\mathbb{P}$. Two such regions are said to be adjacent at a side l if their interiors are disjoint and if l is a side of each region.

The *vertex angle* at a vertex of a region is the interior angle between the sides meeting at the vertex in question. If ζ is a vertex of a fundamental region $\mathbb{F}$ for Γ, the *vertex set* $\mathbb{V}(\zeta,\mathbb{F})$ is defined to be the set of all vertices of $\mathbb{F}$ that are congruent to $\zeta \pmod \Gamma$. If $z_1\equiv z_2 \pmod \Gamma$, where $z_1\in\mathbb{E}\cup\mathbb{P}$, then z_1 and z_2 have the same order $(\text{mod } \Gamma)$. Hence points in the same vertex set have the same order $(\text{mod } \Gamma)$.

For every choice of $\mathcal{R}$ in theorem 2.3.5 we get a fundamental region $\mathbb{F}_2$ for $\hat{\Gamma}$, but it may happen that, if $\mathcal{R}$ is not chosen suitably, $\mathbb{F}_2$ will consist of several disjoint components. It is possible, however, to choose $\mathcal{R}$ in such a way that $\mathbb{F}_2$ is a connected subset of $\bar{\mathbb{C}}$. To do this we proceed as follows.

Let ζ_1 be any point of $\mathbb{P}$ and take any $L_1\in\hat{\Gamma}(1)$ such that $\zeta_1=L_1\infty$. Let $n_1=n(\zeta_1,\Gamma)$, the order of $\zeta_1 \pmod \Gamma$ (see (2.2.24)), so that the mappings $L_1U^{k_1}$ $(0\leq k_1\leq n_1)$ belongs to different right cosets of $\hat{\Gamma}$ in $\hat{\Gamma}(1)$. Write

$$\mathbb{D}_1:=\bigcup_{k_1=0}^{n_1-1} L_1U^{k_1}\mathbb{F}_I.$$

$\mathbb{D}_1$ is formed from n_1 adjacent fundamental regions for $\hat{\Gamma}(1)$ and contains ζ_1 as a vertex. If $n_1 < \mu$, we choose, if possible, an $L_2 \in \Gamma(1)$ such that $L_2\, \mathbb{F}_I$ is adjacent to $\mathbb{D}_1$ and such that L_2 belongs to none of the n_1 cosets mentioned above. Let n_2 be the order of $\zeta_2 = L_2\infty \pmod{\Gamma}$. Then the mappings $L_2U^{k_2}$ $(0 \le k_2 < n_2)$, together with $L_1U^{k_1}$ $(0 \le k_1 < n_1)$ all belong to different right cosets, since $L_2U^sL_1^{-1} \notin \hat{\Gamma}$ for all $s \in \mathbb{Z}$. Write

$$\mathbb{D}_2 := \bigcup_{k_2=0}^{n_2-1} L_2U^{k_2}\mathbb{F}_I.$$

Then $\mathbb{D}_2$ is formed from n_2 adjacent fundamental regions for $\hat{\Gamma}(1)$ and is adjacent to $\mathbb{D}_1$. We proceed in this way. We ultimately obtain a connected set

$$\mathbb{F} := \bigcup_{i=1}^{\lambda} \mathbb{D}_1,$$

where

$$\mathbb{D}_1 := \bigcup_{k_i=0}^{n_i-1} L_iU^{k_i}\mathbb{F}_I,$$

$n_i = n(\zeta_i, \Gamma)$ and $\zeta_i = L_i\infty$ $(1 \le i \le \lambda)$. The mappings $L_iU^{k_i}$ $(0 \le k_i < n_i, 1 \le i \le \lambda)$ all belong to different right cosets of $\hat{\Gamma}$ in $\hat{\Gamma}(1)$ and

$$\mu' := \sum_{i=1}^{\lambda} n_i \le \mu.$$

Further, if $\mathbb{F}_T$ is adjacent to $\mathbb{F}$, for any $T \in \hat{\Gamma}(1)$, then $T \in \Sigma$, where Σ is the union of the μ' right cosets mentioned above.

Now $\mathbb{F}$ is a subset of a fundamental region for $\hat{\Gamma}$ and so, by theorem 2.3.4, the different regions $S\mathbb{F}$ for $S \in \hat{\Gamma}$ can only overlap at points of $\mathbb{E} \cup \mathbb{P}$. Further, each region $S\mathbb{F}$ has a finite number of sides and is adjacent at each side to some other region $S'\mathbb{F}$ $(S' \in \hat{\Gamma})$. It follows that, if $\zeta \in \mathbb{E}$, the finite number (2 or 6) of regions $\mathbb{F}_T$ that meet at ζ either are all contained in

$$\mathbb{H}_1 := \bigcup_{T \in \Sigma} \mathbb{F}_T,$$

or all contained in

$$\mathbb{H}_2 := \bigcup_{T \in \hat{\Gamma}(1) - \Sigma} \mathbb{F}_T.$$

Thus every point $z \in \mathbb{H}$, whether it is an interior or a boundary point of a region $\mathbb{F}_T$, is an interior point of $\mathbb{H}_1$ or $\mathbb{H}_2$. Since $\mathbb{H}$ is a connected open subset of $\bar{\mathbb{C}}$ and $\mathbb{H} \cap \mathbb{H}_1 \neq \varnothing$, it follows that $\mathbb{H} \cap \mathbb{H}_2 = \varnothing$. Hence $\Sigma = \hat{\Gamma}(1)$ and it follows that

$$\sum_{i=1}^{\lambda} n_i = \mu \tag{2.4.10}$$

and that $\mathbb{F}$ is a fundamental region for $\hat{\Gamma}$.

We note also that the cusps ζ_1 and ζ_2 are incongruent (mod Γ). For otherwise, for some $S \in \hat{\Gamma}$, $L_1\infty = SL_2\infty$, which implies that $L_1 U^k L_2^{-1} \in \hat{\Gamma}$ for some $k \in \mathbb{Z}$; this is false. It follows that the λ cusps $\zeta_i = L_i\infty (1 \leq i \leq \lambda)$ of $\mathbb{F}$ are incongruent (mod Γ). Hence the orbit $[\infty]$ (mod $\Gamma(1)$) splits up into λ different orbits $\hat{\Gamma}L_i\infty$ (mod Γ) $(1 \leq i \leq \lambda)$. The number $\lambda = \lambda(\Gamma)$ depends only on $\hat{\Gamma}$ and is called the number of cusps of the group $\hat{\Gamma}$. Note that each of the λ orbits $\hat{\Gamma}L_i\infty$ is represented by a single vertex of $\mathbb{F}$ and so, for $1 \leq i \leq \lambda$, $\mathbb{V}(\zeta_i, \mathbb{F})$ consists of the single point ζ_i. In the general case, for arbitrary $\mathcal{R}$, λ is the number of different parabolic vertex sets.

It is clear that the same process could have been carried out by using $\hat{\mathbb{F}}_I$ in place of $\mathbb{F}_I$. Further, instead of grouping fundamental regions $\mathbb{F}_T$ at parabolic fixed points, we could have grouped them together at points of $\mathbb{E}_2$ (using $\mathbb{F}_I$) or $\mathbb{E}_3$ (using $\hat{\mathbb{F}}_I$); for the above analysis goes through similarly with V or P in place of U.

For $m = 2, 3$ let $\mathbb{E}_m$ split into $\varepsilon_m = \varepsilon_m(\Gamma)$ orbits $\hat{\Gamma}\zeta_i^{(m)}$ (mod Γ). Then we obtain, analogously to (2.4.10),

$$\mu = \sum_{i=1}^{\varepsilon_2} n_i^{(2)} = \sum_{i=1}^{\varepsilon_3} n_i^{(3)}, \tag{2.4.11}$$

where $n_i^{(m)} = n(\zeta_i^{(m)}, \Gamma)$; i.e. $n_i^{(m)}$ is 1 or m according as the point $\zeta_i^{(m)}$ is or is not a fixed point for $\hat{\Gamma}$. The number of incongruent sets of elliptic fixed points of order m for $\hat{\Gamma}$ is denoted by $e_m = e_m(\Gamma)$ $(m = 2, 3)$. Clearly

$$e_m\left(1-\frac{1}{m}\right) = \sum_{i=1}^{\varepsilon_m}\left(1-\frac{n_i^{(m)}}{m}\right) = \varepsilon_m - \frac{\mu}{m}. \tag{2.4.12}$$

Note that the equations (2.4.10, 11) are particular cases of (1.1.15) in the cases $S = U$, V and P.

We have therefore proved the following theorem.

Theorem 2.4.3. *Let $\hat{\Gamma}$ be a subgroup of $\hat{\Gamma}(1)$ of finite index μ and suppose that $\hat{\Gamma}(1)=\hat{\Gamma}\cdot\mathcal{R}$. Then* (i) *the regions*

$$\mathbb{F}:=\bigcup_{T\in\mathcal{R}}\mathbb{F}_T \quad \text{and} \quad \hat{\mathbb{F}}:=\bigcup_{T\in\mathcal{R}}\hat{\mathbb{F}}_T$$

are fundamental regions for $\hat{\Gamma}$. Further the order of any vertex of $\mathbb{F}$ (mod Γ) *is equal to half the number of triangles in $\bar{\mathbb{F}}$ containing points of its vertex set.*

(ii) *$\mathcal{R}$ can be chosen so that $\mathbb{F}$ is a connected subset of $\bar{\mathbb{C}}$ and so that one of the following conditions hold:*

(*a*) *Each one of the $\lambda(\Gamma)$ cusps of $\hat{\Gamma}$ is represented by a single vertex ζ_i of $\mathbb{F}$, and $n(\zeta_i,\Gamma)$ fundamental regions for $\hat{\Gamma}(1)$ meet in $\mathbb{F}$ at ζ_i.*

(*b*) *Each of the $e_2(\Gamma)$ orbits* (mod Γ) *of elliptic fixed points in $\mathbb{E}_2(\Gamma)$ is represented by a single vertex of $\mathbb{F}$ of vertex angle π. No points of $\mathbb{E}_2-\mathbb{E}_2(\Gamma)$ lie on the boundary of $\mathbb{F}$.*

(iii) *$\mathcal{R}$ can be chosen so that $\hat{\mathbb{F}}$ is a connected subset of $\bar{\mathbb{C}}$ and so that one of the two following conditions hold:*

(*a*) *Each one of the $\lambda(\Gamma)$ cusps of $\hat{\Gamma}$ is represented by a single vertex ζ_i of $\hat{\mathbb{F}}$ and $n(\zeta_1,\Gamma)$ fundamental regions for $\hat{\Gamma}(1)$ meet in $\hat{\mathbb{F}}$ at ζ_i.*

(*b*) *Each of the $e_3(\Gamma)$ orbits* (mod Γ) *of elliptic fixed points in $\mathbb{E}_3(\Gamma)$ is represented by a single vertex of $\hat{\mathbb{F}}$ of vertex angle $\frac{2}{3}\pi$. No points of $\mathbb{E}_2-\mathbb{E}_3(\Gamma)$ lie on the boundary of $\hat{\mathbb{F}}$.*

Theorem 2.4.4. *Let $\hat{\Gamma}$ be a subgroup of $\hat{\Gamma}(1)$ of finite index μ and suppose that $\hat{\Gamma}(1)=\hat{\Gamma}\cdot\mathcal{R}$ and that*

$$\mathbb{F}=\bigcup_{T\in\mathcal{R}}\mathbb{F}_T.$$

Then (i) *the sides of $\mathbb{F}$ can be grouped into pairs λ_j, λ_j' ($j=1, 2,\ldots,s$) in such a way that $\lambda_j\subseteq\mathbb{F}$, $\lambda_j'\cap\mathbb{F}\subseteq\mathbb{E}\cup\mathbb{P}$ and $\lambda_j'=L_j\lambda_j$, where $L_j\in\hat{\Gamma}$ ($j=1, 2,\ldots,s$).* (ii) *No side in any one pair is congruent* (mod Γ) *to any side in another pair.* (iii) *The regions $L_j^{-1}\mathbb{F}$ and $\mathbb{F}$ are adjacent at λ_j while $\mathbb{F}$ and $L_j\mathbb{F}$ are adjacent at λ_j'.* (iv) *If a point z describes λ_j in such a way that the interior of $\mathbb{F}$ is on the left, then L_jz describes λ_j' with the interior of $\mathbb{F}$ on the right.* (v) *$\hat{\Gamma}$ is generated by the s transformations $L_1, L_2,\ldots,L_s$.*

A similar result holds for the fundamental region $\hat{\mathbb{F}}$ of theorem 2.4.3 (*i*).

Proof. Let $\lambda = Tl$ be a side of $\mathbb{F}$, where $l = l_R$ and R is U or V; see (2.4.6, 7). We suppose that $\lambda \subseteq \mathbb{F}$. Then $\lambda \subseteq \mathbb{F}_T \subseteq \mathbb{F}$ and $T \in \mathscr{R}$. Now $\mathbb{F}_{TR^{-1}}$ and $\mathbb{F}_T$ are adjacent at λ and so, since λ is a side of $\mathbb{F}$, $\mathbb{F}_{TR^{-1}} \nsubseteq \mathbb{F}$. Now, for some $L \in \hat{\Gamma}$ and $S \in \mathscr{R}$

$$TR^{-1} = L^{-1}S$$

and $\mathbb{F}_S \subseteq \mathbb{F}$. Then $\lambda' := L\lambda = SRl$, which is a side of $\mathbb{F}_S$ not contained in $\mathbb{F}_S$. λ' is also a side of $\mathbb{F}$, since the adjacent region $\mathbb{F}_{SR}$ is not contained in $\mathbb{F}$; for SR belongs to the same right coset as T and, if $\mathbb{F}_{SR} \subseteq \mathbb{F}$, then $SR = T$ and so $\mathbb{F}_{TR^{-1}} = \mathbb{F}_S \subseteq \mathbb{F}$, which is false. Since λ' is a side of $\mathbb{F}$, $\lambda = L^{-1}\lambda'$ is a side also of $L^{-1}\mathbb{F}$. Similarly λ' is also a side of $L\mathbb{F}$. Thus with each side λ of $\mathbb{F}$ contained in $\mathbb{F}$ there is associated $L \in \hat{\Gamma}$ such that $\lambda' = L\lambda$ is a side of $\mathbb{F}$ not contained in $\mathbb{F}$.

Each different side λ_j $(j \leq s)$ of $\mathbb{F}$ contained in $\mathbb{F}$ determines in this way a congruent side λ'_j $(j \leq s)$ of $\mathbb{F}$ not contained in $\mathbb{F}$; since $\mathbb{F}$ is a fundamental region for $\hat{\Gamma}$, λ_j and λ'_j are not congruent to λ'_i if $i \neq j$. Further, the method of construction of $\mathbb{F}$ shows that the number of sides λ'_j not contained in $\mathbb{F}$ cannot exceed s. This proves parts (i), (ii) and (iii), and (iv) is obvious by the conformal property of bilinear mappings.

Finally, if $\mathbb{F}, S_1\mathbb{F}, S_2\mathbb{F}, \ldots, S_n\mathbb{F}(S_j \in \hat{\Gamma})$ is any sequence of images of $\mathbb{F}$, each adjacent to its successor, it follows from (iii) that S_n belongs to the group generated by $L_1, L_2, \ldots, L_s$. Also the set of all points z in $\mathbb{H}$ belonging to regions $S_n\mathbb{F}$ that can be reached by such sequences is open, and so also is its complement in $\mathbb{H}$, which must therefore be empty. This completes the proof of the theorem.

The last part of the theorem shows that, if we can construct a fundamental region for a group $\hat{\Gamma}$, we can find mappings in $\hat{\Gamma}$ that generate $\hat{\Gamma}$. For this purpose it is particularly convenient to choose fundamental regions with as few sides as possible, such as are constructed in parts (ii) and (iii) of theorem 2.4.3, since then the number of generators is small. Also we may, if we wish, include λ'_j in $\mathbb{F}$ instead of λ_j, for any value of j. Further, if λ_i and λ_j $(i \neq j)$ are consecutive sides on the same arc or straight line segment, and so are λ'_j and λ'_i, we may count each of $\lambda_i \cup \lambda_j$, $\lambda'_j \cup \lambda'_i$ as a single side. We note also that, if $\mathbb{F}$ is a fundamental region for $\hat{\Gamma}$, we may obtain a proper fundamental region for $\hat{\Gamma}$ by omitting a finite number of points of $\mathbb{E} \cup \mathbb{P}$ from $\mathbb{F}$.

If, in theorem 2.4.4, $\mathcal{R}$ is chosen so that no two of the fundamental regions $\mathbb{F}_T$ are adjacent, then the number of pairs of congruent sides will be 2μ, since each $\mathbb{F}_T$ has four sides. We note, however, that for every choice of $\mathcal{R}$ the arguments used in the proof yield the following:

Corollary 2.4.4. *Under the assumptions of theorem 2.4.4, the 4μ sides of the regions $\mathbb{F}_T$ ($T \in \mathcal{R}$) can be grouped in two families $\mathbb{L}_U$ and $\mathbb{L}_V$ of pairs of sides (λ, λ'), where $\lambda' = S\lambda$ for some $S \in \hat{\Gamma}$. Each family contains μ pairs and, for $R = U$ or V,*

$$\mathbb{L}_R = \{(\lambda, \lambda'): \lambda = Tl_R, \lambda' = T'Rl_R;\ T, T' \in \mathcal{R}, T'RT^{-1} \in \Gamma\}. \tag{2.4.13}$$

Theorems 2.4.3, 4 enable us to construct fundamental regions for groups and to find their generators. We give several examples of this in table 3, but first obtain some general results on normal subgroups of $\hat{\Gamma}(1)$. If $\hat{\Gamma}$ is a normal subgroup of $\hat{\Gamma}(1)$ with finite index μ, then the order $n(z, \Gamma)$ of each point $z \in \mathbb{P}$ (mod Γ) is the same; in fact, for each $z \in \mathbb{P}$, $n(z, \Gamma) = n_\infty$, where n_∞ is the smallest positive integer r for which $U^r \in \hat{\Gamma}$; i.e. $n_\infty = \text{lev}\, \hat{\Gamma}$. In exactly the same way

$$n(z, \Gamma) = n_m \quad \text{for each } z \in \mathbb{E}_m \quad (m = 2, 3),$$

where $n_m = 1$ or m; n_2 and n_3 are respectively the smallest positive integers r for which V^r and $P^r \in \hat{\Gamma}$. We can therefore classify normal subgroups $\hat{\Gamma}$ of $\hat{\Gamma}(1)$ according to their type or *branch schema* $\{n_2, n_3, n_\infty\}$. We have:

Theorem 2.4.5. *If $\hat{\Gamma}$ is a normal subgroup of finite index μ in $\hat{\Gamma}(1)$ and is of branch schema $\{n_2, n_3, n_\infty\}$, then*

$$\mu = n_2\varepsilon_2 = n_3\varepsilon_3 = n_\infty\lambda, \tag{2.4.14}$$

where ε_2, ε_3 and λ are the number of orbits into which $\mathbb{E}_2$, $\mathbb{E}_3$ and $\mathbb{P}$ split (mod Γ), *respectively.*

Further, only the following four branch schemata occur:

(i) $\{1, 1, 1\}$, (ii) $\{2, 1, 2\}$, (iii) $\{1, 3, 3\}$, (iv) $\{2, 3, n\}$.

Only one group exists of each of the first three schemata, namely $\hat{\Gamma}(1)$, $\hat{\Gamma}^2$, and $\hat{\Gamma}^3$. In case (iv) $\mu \equiv 0$ (mod 6). (*Cf. theorem* 1.7.5.)

Proof. The equation (2.4.14) follows from (2.4.10, 11) and shows that, if $\hat{\Gamma}$ has branch schema $\{2, 3, n\}$, then $\mu \equiv 0 \pmod 6$.

We prove that if a group $\hat{\Gamma}$ has schema $\{1, 3, n\}$, then $n = 3$. In the first place, by (2.4.14),

$$\mu = n\lambda = \varepsilon_2 = 3\varepsilon_3,$$

so that $\mu \equiv 0 \pmod 3$. Let the fundamental region $\mathbb{F}$ of $\hat{\Gamma}$, if it exists, be formed by adjoining images of $\hat{\mathbb{F}}_I$. Since I, P, P^2 belong to different right cosets of $\hat{\Gamma}$ in $\hat{\Gamma}(1)$, we may assume that the region

$$\mathbb{F}_3 = \hat{\mathbb{F}}_I \cup \hat{\mathbb{F}}_P \cup \hat{\mathbb{F}}_{P^2} \tag{2.4.15}$$

is a subset of $\mathbb{F}$. Each of its six sides has one endpoint at a point of $\mathbb{E}_2$, and such vertices cannot be interior points of $\mathbb{F}$, since they are fixed points for $\hat{\Gamma}$; further only two triangles in $\mathbb{F}$ can contain such a vertex. It follows that $\mathbb{F} = \mathbb{F}_3$ and that $\mu = 3$.

The transformations L mapping congruent sides of $\mathbb{F}_3$ into each other cannot map any one of these three elliptic fixed points i, $\frac{1}{2}(i-1)$, $i-1$ into another, so that the six sides fall into three pairs $\lambda_\nu, \lambda'_\nu$ $(\nu = 1, 2, 3)$ with $\lambda'_\nu = L_\nu \lambda_\nu$, $L_\nu \in \hat{\Gamma}$ $(\nu = 1, 2, 3)$, where L_1, L_2 and L_3 have i, $\frac{1}{2}(i-1)$ and $i-1$ as fixed points, respectively. We take

$$L_1 := V, \quad L_2 := PVP^{-1} = V_2, \quad L_2 := P^2VP^{-2} = V_1 \tag{2.4.16}$$

and it follows from (1.3.9) that these three mappings do in fact generate the normal subgroup $\hat{\Gamma}^3$ of index 3 in $\hat{\Gamma}(1)$; further I, P, P^2 is a right transversal of $\hat{\Gamma}^3$ in $\hat{\Gamma}(1)$.

It can be shown similarly that, when $n_3 = 1$, the only groups are $\hat{\Gamma}(1)$ and $\hat{\Gamma}^2$. We already know that the latter is a normal subgroup of $\hat{\Gamma}(1)$ of index 2, and arguments of a similar nature to those given for $\hat{\Gamma}^3$ show that it has $\mathbb{F}_I \cup \mathbb{F}_V$ as a fundamental region and is generated by P and $P_1 := V^{-1}PV$. This completes the proof of the theorem.

In fig. 5 the fundamental regions of some of the groups that we have discussed are shown. We note that, since $\{I, P, P^2\}$ is a right transversal for the groups $\hat{\Gamma}_U(2)$, $\hat{\Gamma}_V(2)$, $\hat{\Gamma}_W(2)$ and $\hat{\Gamma}_3$, the region $\mathbb{F}_3$, defined by (2.4.15), is a fundamental region for each of these four groups. The transformations L_j (see theorem 2.4.4) that map a side λ_j into a corresponding side λ'_j are, however, different in each

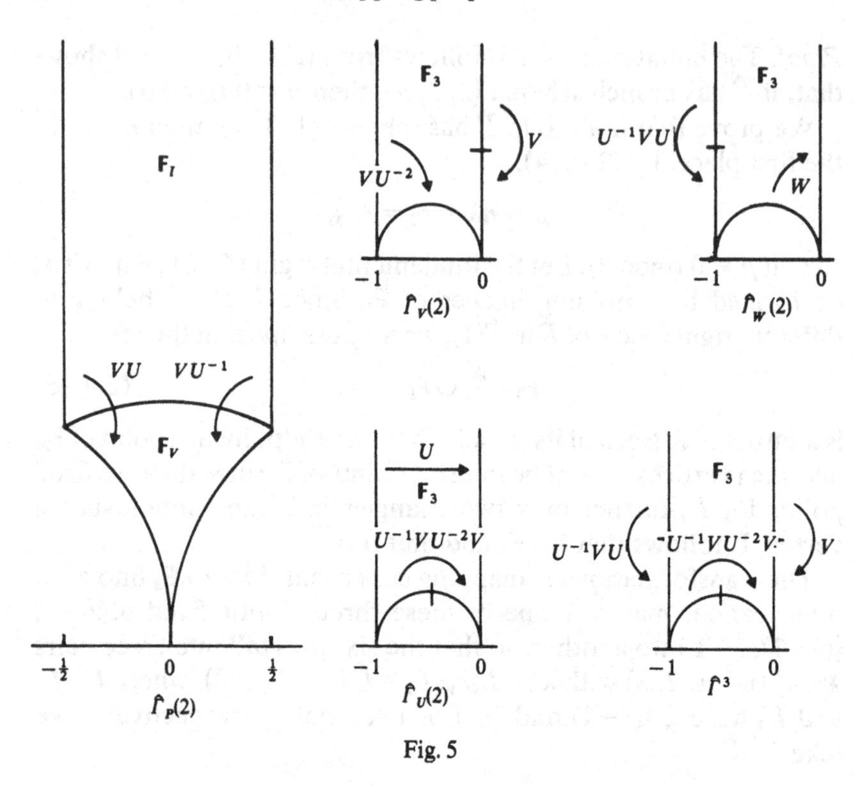

Fig. 5

of the four cases. The generators of the four groups, as given in the penultimate column of table 3, are derived from the mappings L_j illustrated in fig. 5. Theorem 2.4.4 can be applied in a similar way to find the generators of other subgroups of $\hat{\Gamma}(1)$ such as $\hat{\Gamma}(2)$ and $\hat{\Gamma}'(1)$. Note that $\hat{\Gamma}(2)$ has schema {2, 3, 2} and that both $\hat{\Gamma}(6)$ and $\hat{\Gamma}'(1)$ have schema {2, 3, 6}.

Table 4 contains a variety of information about the nine normal subgroups of $\hat{\Gamma}(1)$ lying between $\hat{\Gamma}(6)$ and $\hat{\Gamma}(1)$. The rank of each subgroup is the minimum number of generators. The relations between the groups are illustrated in fig. 3.

We conclude by observing that, for $n > 1$, the principle congruence group $\hat{\Gamma}(n)$ has schema $\{2, 3, n\}$, and that the number $\hat{\lambda}(n)$ of incongruent cusps is

$$\hat{\lambda}(n) = \frac{\hat{\mu}(n)}{n} = \begin{cases} 3 & (n=2), \\ \frac{1}{2}n^2 \prod_{p|n}\left(1-\frac{1}{p^2}\right) & (n>2). \end{cases} \tag{2.4.17}$$

Table 3. Groups of small index

$\hat{\Gamma}$	μ	$\mathbf{F}$	Vertex sets in $\mathbf{E}_2$	Order	Vertex sets in $\mathbf{E}_3$	Order	Vertex sets in $\mathbf{P}$	Order	Generators	Alternative generators
$\hat{\Gamma}(1)$	1	$\mathbf{F}_I$	i	1	$\rho, -\rho^2$	1	∞	1	U, V	V, P
$\hat{\Gamma}^2$ $\hat{\Gamma}_P(2)$	2	$\mathbf{F}_I \cup \mathbf{F}_V$	—	—	ρ $-\rho^2$	1 1	$\infty, 0$	2	VU, VU^{-1}	$P, V^{-1}PV$
$\hat{\Gamma}_U(2)$	3	$\mathbf{F}_3$	$\frac{1}{2}(i-1)$ $i, i-1$	1 2	—	—	∞ $0, -1$	1 2	$U, U^{-1}VU^{-2}V$	U, PVP^{-1}
$\hat{\Gamma}_V(2)$	3	$\mathbf{F}_3$	i $i-1, \frac{1}{2}(i-1)$	1 2	—	—	-1 $0, \infty$	1 2	V, VU^2	V, U^2
$\hat{\Gamma}_W(2)$	3	$\mathbf{F}_3$	$i-1$ $i, \frac{1}{2}(i-1)$	1 2	—	—	0 $-1, \infty$	1 2	$W, U^{-1}VU$	W, U^2
$\hat{\Gamma}^3$	3	$\mathbf{F}_3$	i $i-1$ $\frac{1}{2}(i-1)$	1 1 1	—	—	$0, -1, \infty$	3	$V, U^{-1}VU,$ $U^{-1}VU^{-2}V$	$V, PVP^{-1}, P^{-1}VP$
$\hat{\Gamma}(2)$	6	$\mathbf{F}_3 \cup U\mathbf{F}_3$	—	—	—	—	0 ∞	2 2	U^2, VU^2V	
$\hat{\Gamma}(1)$	6	$\mathbf{F}_3 \cup U\mathbf{F}_3$	—	—	—	—	$1, -1$ $0, \infty, 1, -1$	2 6	UW, WU	

Table 4. Normal subgroups between $\hat{\Gamma}(1)$ and $\hat{\Gamma}(6)$

$\hat{\Gamma}$	μ	Rank	n_∞	λ	Generators	Congruence relations for $T \in \Gamma$: $\pm T \equiv$
$\hat{\Gamma}(1)$	1	2	1	1	$V=\begin{bmatrix}0 & -1\\1 & 0\end{bmatrix}$, $P=\begin{bmatrix}0 & -1\\1 & 1\end{bmatrix}$	
$\hat{\Gamma}^2$	2	2	2	1	$P=\begin{bmatrix}0 & -1\\1 & 1\end{bmatrix}$, $V^{-1}PV=\begin{bmatrix}1 & -1\\1 & 0\end{bmatrix}=P_1$	$\begin{bmatrix}1 & 0\\0 & 1\end{bmatrix}, \begin{bmatrix}0 & -1\\1 & 1\end{bmatrix}, \begin{bmatrix}-1 & -1\\1 & 0\end{bmatrix}=P^2$ (mod 2). See also (1.7.1)
$\hat{\Gamma}^3$	3	3	3	1	$V=\begin{bmatrix}0 & -1\\1 & 0\end{bmatrix}$, $P^{-1}VP=\begin{bmatrix}-1 & -2\\1 & 1\end{bmatrix}$, $P^{-2}VP=\begin{bmatrix}-1 & -1\\2 & 1\end{bmatrix}$	$\begin{bmatrix}1 & 0\\0 & 1\end{bmatrix}, \begin{bmatrix}0 & -1\\1 & 0\end{bmatrix}, \begin{bmatrix}-1 & -1\\2 & 1\end{bmatrix}, \begin{bmatrix}1 & 2\\-1 & -1\end{bmatrix}$ (mod 3). See also (1.7.2)
$\hat{\Gamma}'(1)$	6	2	6	1	$UW=[V,P]=\begin{bmatrix}2 & 1\\1 & 1\end{bmatrix}$, $WU=[V,P^2]=\begin{bmatrix}1 & 1\\1 & 2\end{bmatrix}$	See (1.7.3)
$\hat{\Gamma}(2)$	6	2	2	3	$U^2=\begin{bmatrix}1 & 2\\0 & 1\end{bmatrix}$, $V^{-1}U^{-2}V=\begin{bmatrix}1 & 0\\2 & 1\end{bmatrix}$	$\begin{bmatrix}1 & 0\\2 & 1\end{bmatrix}$ (mod 2)
$\hat{\Gamma}(3)$	12	3	3	4	$U^3=\begin{bmatrix}1 & 3\\0 & 1\end{bmatrix}$, $P^{-1}U^3P=\begin{bmatrix}4 & 3\\-3 & -2\end{bmatrix}$, $P^{-2}U^3P^2=\begin{bmatrix}1 & 0\\-3 & 1\end{bmatrix}$	$\begin{bmatrix}1 & 0\\0 & 1\end{bmatrix}$ (mod 3)

$\hat{\Gamma}^{2\prime}$	18	4	6	3	$[P, VPV] = \begin{bmatrix} 1 & -2 \\ -2 & 5 \end{bmatrix}$, $[P, VP^2V] = \begin{bmatrix} 3 & -2 \\ -4 & 3 \end{bmatrix}$, $[P^2, VPV] = \begin{bmatrix} 3 & -4 \\ -2 & 3 \end{bmatrix}$, $[P^2, VP^2V] = \begin{bmatrix} 5 & -2 \\ -2 & 1 \end{bmatrix}$	$\begin{bmatrix} 1 & 0 \\ 0 & 1 \end{bmatrix}, \begin{bmatrix} 1 & -2 \\ -2 & 5 \end{bmatrix}, \begin{bmatrix} 5 & -2 \\ -2 & 1 \end{bmatrix}, \begin{bmatrix} 3 & -2 \\ -4 & 3 \end{bmatrix} \pmod 6$
$\hat{\Gamma}^{3\prime}$	24	5	6	4	$(VP^2VP)^2 = \begin{bmatrix} 2 & 3 \\ 3 & 5 \end{bmatrix}$, $PVP^2(VP^2VP)^2PVP^2 = \begin{bmatrix} -11 & -3 \\ 15 & 4 \end{bmatrix}$, $(VPVP^2)^2 = \begin{bmatrix} 5 & 3 \\ 3 & 2 \end{bmatrix}$, $V(P^2VP^2VP^2)^2V = \begin{bmatrix} 1 & -3 \\ 3 & -8 \end{bmatrix}$, $(P^2VP^2VP^2)^2 = \begin{bmatrix} 8 & 3 \\ -3 & -1 \end{bmatrix}$,	$\begin{bmatrix} 1 & 0 \\ 0 & 1 \end{bmatrix}, \begin{bmatrix} 2 & 3 \\ 3 & 5 \end{bmatrix}, \begin{bmatrix} 5 & 3 \\ 3 & 2 \end{bmatrix} \pmod 6$
$\hat{\Gamma}(6)$	72	13	6	12		$\begin{bmatrix} 1 & 0 \\ 0 & 1 \end{bmatrix} \pmod 6$

2.5. Further results. The method that we have used for obtaining fundamental regions for $\hat{\Gamma}(1)$ and its subgroups is convenient for the purposes we have in mind. However, if we had been considering the more general case of an arbitrary discontinuous subgroup $\hat{\Gamma}$ of $\hat{\Omega}$, we should probably have found it advantageous to introduce hyperbolic geometry and to adopt a slightly different definition of a fundamental region. A simple account of this more general theory can be found, for example, in the book by Lehner (1966), or the more advanced treatises by Fricke and Klein (1926) and Lehner (1964) may be consulted. We content ourselves here by sketching briefly how this general theory could be applied to a subgroup $\hat{\Gamma}$ of finite index μ in $\hat{\Gamma}(1)$; in this sketch we shall introduce a number of concepts that we do not define.

The *hyperbolic length ds* of an element of arc in $\mathbb{H}$ is defined by

$$(ds)^2 := y^{-2}\{(dx)^2 + (dy)^2\}, \qquad (2.5.1)$$

so that the hyperbolic length of a piecewise differentiable curve C in $\mathbb{H}$ is

$$L(C) := \int_C ds = \int_C \left\{\left(\frac{dx}{dt}\right)^2 + \left(\frac{dy}{dt}\right)^2\right\}^{\frac{1}{2}} \frac{dt}{y}. \qquad (2.5.2)$$

In a similar way, the *hyperbolic area* $A(E)$ of a measurable subset E of $\mathbb{H}$ is defined to be

$$A(E) := \iint_E y^{-2}\, dx\, dy; \qquad (2.5.3)$$

this can be infinite even when the ordinary Euclidean area of E is finite.

A curve in $\mathbb{H}$ is called a *hyperbolic straight line* if it is either a semicircle in $\mathbb{H}$ centred at a point of $\mathbb{R}$, or is an ordinary straight line in $\mathbb{H}$ that is orthogonal to $\mathbb{R}$. If z_1 and z_2 are distinct points of $\mathbb{H}$, there is a unique hyperbolic straight line passing through z_1 and z_2 and the arc of this hyperbolic straight line joining z_1 and z_2 is the curve C of smallest hyperbolic length joining z_1 to z_2. The *hyperbolic distance* $d(z_1, z_2)$ of z_1 from z_2 is defined to be $L(C)$; if $z_1 = z_2$, we put $d(z_1, z_2) = 0$.

It can be shown that $d(z_1, z_2)$ defines a metric on $\mathbb{H}$, and the corresponding metric topology on $\mathbb{H}$ is identical with the natural topology on $\mathbb{H}$. It is convenient to extend this topology from $\mathbb{H}$ to $\mathbb{H}'$

by giving each point of $\mathbb{P}$ a suitable base of neighbourhoods; see Lehner (1964), chapter 4.

The hyperbolic metric has the useful property that $L(TC) = L(C)$ and $A(TE) = A(E)$ for every $T \in \hat{\Omega}$. Thus every fundamental region for $\hat{\Gamma}$ has the same hyperbolic area. Further, this hyperbolic area is easy to evaluate by using the Gauss–Bonnet formula, which states that the area of a triangle bounded by hyperbolic line segments is $\pi - (\alpha + \beta + \gamma)$, where α, β and γ are the interior angles between the sides. This gives, for example, for a fundamental region $\mathbb{F}$ for $\hat{\Gamma}$

$$A(\mathbb{F}) = \mu A(\mathbb{F}_I) = \tfrac{1}{3}\mu\pi, \tag{2.5.4}$$

since $\mathbb{F}_I$ has interior angles 0, $\frac{1}{3}\pi$ and $\frac{1}{3}\pi$.

Different definitions of a fundamental region are given by different authors. What we have called a proper fundamental region is called a fundamental set by Lehner (1966) and Schoeneberg (1974). In his two books Lehner defines a subset $\mathbb{D}$ of $\mathbb{H}$ to be a fundamental region for $\hat{\Gamma}$ if (i) $\mathbb{D}$ is open in $\mathbb{H}$, (ii) no two distinct points of $\mathbb{D}$ are congruent modulo Γ and (iii) every point of $\mathbb{H}$ is congruent to a point of the closure $\bar{\mathbb{D}}$ of $\mathbb{D}$ in $\mathbb{H}$. On the other hand, Macbeath (1961) takes $\mathbb{D}$ to be closed and modifies (ii) accordingly. The interior of either of the fundamental regions $\mathbb{F}$ or $\hat{\mathbb{F}}$ of theorem 2.4.3 is a fundamental region according to Lehner.

Let $z_0 \in \mathbb{H} - \mathbb{E}(\Gamma)$ and define $\mathbb{D} = \mathbb{D}(z_0)$ to be the set of all points $z \in \mathbb{H}$ such that

$$d(z, z_0) < d(z, Tz_0)$$

for all $T \in \hat{\Gamma}$ except $T = \pm I$. It can be shown that $\mathbb{D}$ is a fundamental region in the sense of Lehner and that it is bounded by a finite number of segments of hyperbolic straight lines. This fundamental region is called a *normal polygon* or a *Dirichlet region*.

For theoretical purposes the normal polygon has many advantages as a fundamental region, but in individual cases it is not so easy to construct as the fundamental regions set up in theorem 2.4.3. An alternative method due to Ford (1929), which is also easily applicable in particular cases, defines $\mathbb{D}$ to be the region of $\mathbb{H}$ contained in $\mathbb{S}_{n_I}$ and outside all the *isometric circles* $|T:z| = 1$ $(T \in \Gamma, c \neq 0)$; see also Rankin (1954). This again yields a region bounded by segments of hyperbolic straight lines. When applied to $\hat{\Gamma}(1)$ the isometric circle method gives the interior of $\mathbb{F}_I$.

Each subgroup $\hat{\Gamma}$ has associated with it a Riemann surface, which is obtained in the following way. The set of orbits $\hat{\Gamma}z$ $(z \in \mathbb{H}')$ is denoted by $\mathbb{H}'/\Gamma$ and, when given the identification topology induced by congruence modulo Γ, becomes a connected Hausdorff space, which is, in fact, a Riemann surface. Its points are in one-to-one correspondence with those of a proper fundamental region $\mathbb{F}$ for $\hat{\Gamma}$. It is convenient to regard $\mathbb{H}'/\Gamma$ as being constructed from $\mathbb{F}$ by identifying pairs of congruent sides λ_j, λ_j' (see theorem 2.4.4). The Riemann surface $\mathbb{H}'/\Gamma$ has $\mathbb{H}'$ as a branched covering surface. At points of $\mathbb{H}-\mathbb{E}(\Gamma)$ the covering is unbranched; at a point of $\mathbb{E}_k(\Gamma)$ there is a branchpoint of order k; at points of $\mathbb{P}$ there are logarithmic winding points.

The surface $\mathbb{H}'/\Gamma$ is compact and has finite genus $g=g(\hat{\Gamma})$. The images of the triangles $\mathbb{F}^{(1)}$ and $\mathbb{F}^{(2)}$ (see (2.4.2, 3)) provide a natural triangulation of $\mathbb{H}'/\Gamma$ from which it is possible to show that

$$g=1+\tfrac{1}{2}(\mu-\lambda-\varepsilon_2-\varepsilon_3). \tag{2.5.5}$$

Here, as in §2.4, λ is the number of incongruent cusps (mod Γ) and ε_k is the number of incongruent points of $\mathbb{E}_k$ (mod Γ) $(k=2, 3)$; see, for example, Gunning (1962), §4, theorem 5 or Schoeneberg (1974), chapter 4, §7. It follows from (2.4.12) and (2.5.5) that we also have

$$g=1+\frac{1}{2}\left(\frac{\mu}{6}-\lambda-\frac{e_2}{2}-\frac{2e_3}{3}\right). \tag{2.5.6}$$

In particular, when $\hat{\Gamma}$ is a normal subgroup of $\hat{\Gamma}(1)$ with branch schema $\{n_2, n_2, n_\infty\}$, then, by (2.4.14) and (2.5.5),

$$g=1+\frac{1}{2}\mu\left(1-\frac{1}{n_2}-\frac{1}{n_3}-\frac{1}{n_\infty}\right). \tag{2.5.7}$$

A knowledge of the genus of $\hat{\Gamma}$ is useful when applications of the Riemann–Roch theorem are made to find the number of linearly independent modular forms of different kinds for $\hat{\Gamma}$. However, it is possible in many cases to obtain exact results without using deep theorems of this kind, as we shall see. This happens, in particular, when the genus is zero. In this connexion we note that, by (2.5.7), $g=0$ for all the groups in tables 3 and 4 except for $\hat{\Gamma}'(1)$ and $\hat{\Gamma}(6)$, which both have genus 1.

Since, by theorem 2.4.5, $\hat{\Gamma}(n)$ has branch schema $\{2, 3, n\}$ for $n \geq 2$, it follows from (2.5.7) and theorem 1.4.1 that

$$g\{\Gamma(n)\} = 1 + \frac{n^2(n-6)}{24} \prod_{p|n} \left(1 - \frac{1}{p^2}\right) \quad (n \geq 3), \qquad (2.5.8)$$

while $g\{\hat{\Gamma}(2)\} = 0$. It follows that the genus of $\hat{\Gamma}(n)$ is zero for $n \leq 5$.

We conclude by remarking that it can be shown that a *canonical fundamental region* for $\hat{\Gamma}$ can be constructed having $2n + 4g$ sides that are segments of hyperbolic straight lines. Here $n = \lambda + e_2 + e_3$ and the sides follow each other in the order

$$\lambda_1\lambda'_1\lambda_2\lambda'_2 \ldots \lambda_n\lambda'_n\mu_1\nu_1\mu'_1\nu'_1\mu_2\nu_2\mu'_2\nu'_2 \ldots \mu_g\nu_g\mu'_g\nu'_g.$$

Here

$$\lambda'_i = L_i\lambda_i, \quad \mu'_j = M_j\mu_j, \quad \nu'_j = N_j\nu_j \quad (1 \leq i \leq n, 1 \leq j \leq g)$$

where the mappings L_i, M_j, N_j belong to $\hat{\Gamma}$. Further the first e_2 of the L_i are elliptic transformations of order 2, the next e_3 are elliptic transformations of order 3 and the last λ are parabolic transformations; see Lehner (1964), chapter 7. The group $\hat{\Gamma}$ is generated by the $n + 2g$ mappings L_i, M_j, N_j $(1 \leq i \leq n, 1 \leq j \leq g)$. Each elliptic generator L_i satisfies a relation of the form $L_i^k = I$ $(k = 2$ or $3)$; apart from these relations we also have

$$L_1L_2 \ldots L_nM_1N_1M_1^{-1}N_1^{-1}M_2N_2M_2^{-1}N_2^{-1} \ldots M_gN_gM_g^{-1}N_g^{-1} = I, \qquad (2.5.9)$$

and all other relations between the elements of $\hat{\Gamma}$ are consequences of the relations given.

3: Automorphic factors and multiplier systems

3.1. Introduction. In this chapter we study the properties of automorphic factors and multiplier systems. These properties will be needed when we construct modular forms by means of Poincaré series and when we consider special forms associated with the modular group and subgroups of small index.

In §2.2 we defined $T: z$ for any $T \in \Theta$ and $z \in \mathbb{C}$ and obtained the identity (2.2.2):

$$ST: z = (S: Tz)(T: z). \tag{3.1.1}$$

This holds, in particular, for all S, $T \in \Omega$ and $z \in \mathbb{H}$.

Throughout this section we shall suppose that k is a fixed real number, not necessarily an integer. For every non-zero $w \in \mathbb{C}$, we define

$$w^k := \exp(k \log w), \tag{3.1.2}$$

where the principal value of the logarithm is used; i.e.

$$\log w = \log |w| + i \text{ ph } w,$$

where

$$-\pi < \text{ph } w \leq \pi. \tag{3.1.3}$$

Let Γ be a subgroup of Ω. A function ν defined on $\Gamma \times \mathbb{H}$ is called an *automorphic factor* (AF) of weight k on Γ, if the following four conditions are satisfied:

(i) For each $T \in \Gamma$, $\nu(T, z)$ is a holomorphic function of z for $z \in \mathbb{H}$.

(ii) For all $z \in \mathbb{H}$ and all $T \in \Gamma$,

$$|\nu(T, z)| = |T: z|^k. \tag{3.1.4}$$

(iii) For all S, $T \in \Gamma$ and $z \in \mathbb{H}$,

$$\nu(ST, z) = \nu(S, Tz)\nu(T, z). \tag{3.1.5}$$

(iv) If $-I \in \Gamma$, then, for all $T \in \Gamma$ and all $z \in \mathbb{H}$,

$$\nu(-T, z) = \nu(T, z). \tag{3.1.6}$$

The last condition shows that ν can be regarded as a function on $\hat{\Gamma}\times\mathbb{H}$; i.e., T can be regarded as a *mapping*, instead of as a *matrix*, and then condition (iv) can be omitted as being obvious. On the other hand, if ν is an AF on $\Gamma\times\mathbb{H}$ and $-I\notin\Gamma$, condition (iv) can be used to extend the domain of ν to $\Gamma\Lambda\times\mathbb{H}$. Usually we shall assume that $-I\in\Gamma$, and there is no loss of generality in doing so, but there are cases where it is not convenient to make this assumption. The definition that we have given of an automorphic factor is a modification of a more general definition of a factor of automorphy given by Gunning (1955).

If we take $S=T=I$ in (3.1.5), we deduce that

$$\nu(I, z)=1 \quad \text{for all } z\in\mathbb{H}. \tag{3.1.7}$$

The equation (3.1.1) is similar to (3.1.5) and this suggests that $(T:z)^k$ may be an automorphic factor. We therefore put, for all $T\in\Omega$ and all $z\in\mathbb{H}$,

$$\mu(T, z):=(T:z)^k, \tag{3.1.8}$$

where the right-hand side is defined by (3.1.2) to be the principal value. When k is an even integer, $\mu(-T, z)=\mu(T, z)$ and in this case it is easily seen that $\mu(T, z)$ is an AF of weight k on Ω. In other cases $\mu(T, z)$ is not an AF on Ω, although it may be an AF on some subgroup of Ω. We have, by (3.1.8),

$$|\mu(T, z)|=|T:z|^k. \tag{3.1.9}$$

We can use the function $\mu(T, z)$ to investigate the form of an AF $\nu(T, z)$ more closely. Since a holomorphic function of constant modulus on $\mathbb{H}$ must be a constant, it follows from properties (i) and (ii) that

$$\nu(T, z)=v(T)\mu(T, z) \tag{3.1.10}$$

for all $T\in\Gamma$ and all $z\in\mathbb{H}$, where $v(T)$ depends only on the matrix T (and not on z) and

$$|v(T)|=1. \tag{3.1.11}$$

We call $v(T)$ a *multiplier*, and the function v defined by (3.1.10) on Γ is called a *multiplier system* (MS) of weight k.

From (3.1.7, 8, 10) we deduce that

$$v(I)=1. \tag{3.1.12}$$

Further, if $-I \in \Gamma$ if follows from (3.1.6, 7, 8, 10) that

$$v(-I) = e^{-\pi i k}, \tag{3.1.13}$$

since $\text{ph}(-1) = \pi$. The equations (3.1.12, 13) show that, in contrast to the AF ν, the MS v is defined on Γ but not, in general on $\hat{\Gamma}$.

If we substitute from (3.1.10) into (3.1.5) we obtain, for any S, $T \in \Gamma$.

$$v(ST) = \sigma(S, T)v(S)v(T), \tag{3.1.14}$$

where

$$\sigma(S, T) := \frac{\mu(S, Tz)\mu(T, z)}{\mu(ST, z)}. \tag{3.1.15}$$

By (3.1.9),

$$|\sigma(S, T)| = 1, \tag{3.1.16}$$

and $\sigma(S, T)$ is well defined by (3.1.15) for all S and T in Ω; $\sigma(S, T)$ depends on S, T and k, but not on z. When $k \in \mathbb{Z}$, $\sigma(S, T) = 1$ for all $S, T \in \Omega$, by (3.1.1, 8).

Conversely, if we are given a function v defined on Γ and satisfying (3.1.11, 14), and (3.1.13) when $-I \in \Gamma$, we can use (3.1.10) to define an automorphic factor $\nu(T, z)$ of weight k and Γ.

Suppose now that ν is an AF of weight k on a subgroup Γ of Ω and let L be any member of Ω. We show how to define an AF ν^L for the conjugate group $\Gamma^L = L^{-1}\Gamma L$. We note first that, if $L \in \Gamma$, then, by (3.1.5),

$$\nu(L^{-1}TL, z) = \frac{\nu(T, Lz)\nu(L, z)}{\nu(L, L^{-1}TLz)} \quad (T \in \Gamma, z \in \mathbb{H}),$$

and so, by (3.1.10),

$$\nu(L^{-1}TL, z) = \frac{\nu(T, Lz)\mu(L, z)}{\mu(L, L^{-1}TLz)}.$$

The right-hand side of this equation is defined even when $L \notin \Gamma$. Accordingly, we define for any $L \in \Omega$, the conjugate AF ν^L on $\Gamma^L \times \mathbb{H}$ by

$$\nu^L(L^{-1}TL, z) = \frac{\nu(T, Lz)\mu(L, z)}{\mu(L, L^{-1}TLz)} \quad (T \in \Gamma, z \in \mathbb{H}). \tag{3.1.17}$$

With this definition properties (i), (ii) and (iv) are obvious for ν^L. To check (iii), write $S_1 = L^{-1}SL$, $T_1 = L^{-1}TL$, where $S, T \in \Gamma$. Then, for all $z \in \mathbb{H}$, by (3.1.5),

$$\begin{aligned}\nu^L(S_1T_1, z) &= \nu^L(L^{-1}STL, z)\\ &= \frac{\nu(S, TLz)\nu(T, Lz)\mu(L, z)}{\mu(L, L^{-1}STLz)}\\ &= \frac{\nu(S, TLz)\mu(L, L^{-1}TLz)}{\mu(L, L^{-1}STLz)} \cdot \frac{\nu(T, Lz)\mu(L, z)}{\mu(L, L^{-1}TLz)}\\ &= \nu^L(S_1, T_1z)\nu^L(T_1, z),\end{aligned}$$

as required. Thus ν^L is an AF for Γ^L; when $L \in \Gamma$, clearly $\Gamma^L = \Gamma$ and $\nu^L = \nu$. It can happen, however, that $\Gamma^L = \Gamma$ when $L \notin \Gamma$, and it does not then necessarily follow that $\nu^L = \nu$.

We note that it follows easily from (3.1.15, 17) that, for any $L_1, L_2 \in \Omega$,

$$\nu^{L_1L_2} = (\nu^{L_1})^{L_2} \tag{3.1.18}$$

and also that

$$\nu^{(-L)} = \nu^L. \tag{3.1.19}$$

We denote the associated MS to v by v^L. It follows from (3.1.10, 17) that

$$v^L(L^{-1}TL) = v(T)\sigma(T, L)/\sigma(L, L^{-1}TL) \tag{3.1.20}$$

for all $T \in \Gamma$. In particular,

$$v^L(L^{-1}TL) = v(T) \quad \text{for } T \in \Gamma \quad \text{if } k \in \mathbb{Z}. \tag{3.1.21}$$

We note in conclusion that we have not yet proved the existence of automorphic factors or multiplier systems except in the case when k is even, when we can take $\nu(T, z) = \mu(T, z)$ and $v(T) = 1$ for all $T \in \Gamma$ and all $z \in \mathbb{H}$.

3.2. The functions $\sigma(S, T)$ and $w(S, T)$. Because of the formula (3.1.14), the properties of the function $\sigma(S, T)$ are clearly of importance, and we consider this function and the related function $w(S, T)$ in the present section. By (3.1.1, 8, 15) we have, for any $S, T \in \Omega$,

$$\sigma(S, T) = \exp\{2\pi ik\, w(S, T)\}, \tag{3.2.1}$$

where, for any $z \in \mathbb{H}$,

$$w(S, T) := \frac{1}{2\pi}\{\text{ph}(S: Tz) + \text{ph}(T: z) - \text{ph}(ST: z)\}. \quad (3.2.2)$$

The number $w(S, T)$ is an integer and depends on S and T, but not on k or z. Since each phase has modulus less than or equal to π, clearly $|w(S, T)| \le \frac{3}{2}$. Hence

$$w(S, T) = -1, 0 \text{ or } 1. \quad (3.2.3)$$

Now let S_1, S_2 and S_3 be any three members of Ω. Then we have the *associative law*

$$w(S_1, S_2S_3) + w(S_2, S_3) = w(S_1, S_2) + w(S_1S_2, S_3). \quad (3.2.4)$$

This follows immediately from (3.2.2) if we take S_3z in place of z in the expression for $w(S_1, S_2)$.

Further, if we take $z = T^{-1}i$ in (3.2.2), we obtain the explicit formula:

$$w(S, T) = \frac{1}{2\pi}[\text{ph}(\gamma i + \delta) + \text{ph}(ci + a) - \text{ph}\{(\gamma i + \delta)(ci + a)\}] \quad (3.2.5)$$

from which we deduce that

$$w(S, T) = \begin{cases} 1 & \text{if ph}(\gamma i + \delta) + \text{ph}(ci + a) > \pi, \\ 0 & \text{if } -\pi < \text{ph}(\gamma i + \delta) + \text{ph}(ci + a) \le \pi, \\ -1 & \text{if ph}(\gamma i + \delta) + \text{ph}(ci + a) \le -\pi. \end{cases} \quad (3.2.6)$$

This makes it clear that $w(S, T)$ depends only on the second row of S and the first column of T. In particular,

$$w(S, T) = 0 \quad \text{if } a \ge 0, \delta \ge 0 \text{ and } a + \delta > 0. \quad (3.2.7)$$

In order to examine further the properties of $w(S, T)$ we introduce the following two subgroups of Ω, namely

$$\Sigma := \left\{R \in \Omega : R = \begin{bmatrix} 1/\rho & \sigma \\ 0 & \rho \end{bmatrix}, \rho > 0, \sigma \in \mathbb{R}\right\}, \quad (3.2.8)$$

and

$$\Xi := \left\{X \in \Omega : X = \begin{bmatrix} \cos\xi & -\sin\xi \\ \sin\xi & \cos\xi \end{bmatrix}, -\pi < \xi \le \pi\right\}. \quad (3.2.9)$$

The subgroup Ξ is abelian.

Then it follows that

$$\Omega = \Sigma \cdot \Xi = \Xi \cdot \Sigma.$$

In fact, every $T \in \Omega$ is uniquely expressible in each of the forms

$$T = R_1 X_1, \quad T = X_2 R_2, \tag{3.2.10}$$

where $R_1, R_2 \in \Sigma$, $X_1, X_2 \in \Xi$ and, in an obvious notation,

$$\xi_1 = \text{ph}(ci + d), \quad \xi_2 = \text{ph}(ci + a). \tag{3.2.11}$$

Since $ci + d$ and $ci + a$ have the same imaginary part, we deduce that

$$0 < \xi_1 \leq \pi \quad \text{if and only if } 0 < \xi_2 \leq \pi. \tag{3.2.12}$$

It follows that

$$-\pi < \xi_1 \leq 0 \quad \text{if and only if } -\pi < \xi_2 \leq 0$$

and that

$$|\xi_1 - \xi_2| < \pi.$$

For any $X \in \Xi$, as given by (3.2.9), we write

$$\text{ph}\, X := \xi. \tag{3.2.13}$$

It then follows that, if (3.2.10) holds, we have

$$\text{ph}\, X_1 X_2^{-1} = \text{ph}\, X_2^{-1} X_1 = \text{ph}\, X_1 - \text{ph}\, X_2. \tag{3.2.14}$$

Since, for any $S \in \Omega$ and $R \in \Sigma$,

$$\text{ph}(RS : z) = \text{ph}(S : z), \quad \text{ph}(SR : z) = \text{ph}(S : Rz),$$

we deduce from (3.2.2) that, for any $S, T \in \Omega$ and $R \in \Sigma$,

$$w(S, T) = w(RS, T) = w(S, TR) \tag{3.2.15}$$

and

$$w(SR, T) = w(S, RT). \tag{3.2.16}$$

We also note the following particular consequences of (3.2.5) or (3.2.2):

$$w(R, T) = w(T, R) = 0, \tag{3.2.17}$$

$$w(-R, T) = w(T, -R) = \begin{cases} 0 & \text{if } c < 0 \quad \text{or if } c = 0, d > 0 \\ 1 & \text{if } c > 0 \quad \text{or if } c = 0, d < 0 \end{cases} \tag{3.2.18}$$

and

$$w(T, T^{-1}) = w(T^{-1}, T) = \begin{cases} 1 & \text{if } c = 0, d < 0, \\ 0 & \text{otherwise.} \end{cases} \tag{3.2.19}$$

These hold for $R \in \Sigma$ and $T \in \Omega$.

Further, if $S = R_1X_1$, $T = X_2R_2$, where R_1 and R_2 belong to Σ and $X_1, X_2 \in \Xi$, then, by (3.2.5, 15),

$$w(S, T) = w(X_1, X_2) = \frac{1}{2\pi}\{\text{ph } X_1 + \text{ph } X_2 - \text{ph } X_1X_2\}. \tag{3.2.20}$$

We use these results to prove that, for any $R \in \Sigma$ and $T \in \Omega$,

$$w(TRT^{-1}, T) = w(T, T^{-1}RT) = 0. \tag{3.2.21}$$

It is enough to prove that $w(TRT^{-1}, T) = 0$. Take $T = X_2R_2$ as (3.2.10) so that, by (3.2.15),

$$w(TRT^{-1}, T) = w(X_2R_3X_2^{-1}, X_2),$$

where $R_3 = R_2RR_2^{-1} \in \Sigma$. Now $X_2R_3 = R_4X_4$, where $R_4 \in \Sigma$ and $X_4 \in \Xi$. By (3.2.14),

$$\text{ph}(X_4X_2^{-1}) = \text{ph } X_4 - \text{ph } X_2.$$

Hence, by (3.2.15, 20),

$$\begin{aligned} w(TRT^{-1}, T) &= w(X_4X_2^{-1}, X_2) \\ &= \frac{1}{2\pi}\{\text{ph}(X_4X_2^{-1}) + \text{ph } X_2 - \text{ph } X_4\} \\ &= 0. \end{aligned}$$

The proof that $w(T, T^{-1}RT) = 0$ is similar.

We can use (3.2.21) to prove the following generalization:

$$w(T_1RT_1^{-1}, T_1T_2^{-1}) = w(T_1T_2, T_2RT_2^{-1}) \tag{3.2.22}$$

for all $T_1, T_2 \in \Omega$ and $R \in \Sigma$. For, by (3.2.4),

$$\begin{aligned} w(T_1RT_1^{-1}, T_1T_2^{-1}) &= w(T_1R, T_2^{-1}) + w(T_1RT_1^{-1}, T_1) - w(T_1, T_2^{-1}) \\ &= w(T_1, RT_2^{-1}) + w(T_2^{-1}, T_2RT_2^{-1}) \\ &\quad - w(T_1, T_2^{-1}) \\ &= w(T_1T_2^{-1}, T_2RT_2^{-1}). \end{aligned}$$

Finally, if $L, T \in \Omega$ and $R_1 \in \Sigma$, it follows from (3.2.4, 15, 17, 21) that

$$\sigma(L, L^{-1}R_1LT)\sigma(L^{-1}R_1L, T) = \sigma(L, T). \qquad (3.2.23)$$

3.3. Automorphic factors on subgroups of the modular group. In this section we assume that Γ is a subgroup of $\Gamma(1)$ of finite index and that ν is an AF of weight k on Γ. We also assume that $-I \in \Gamma$. We consider the form of $\nu(T, z)$ and the associated multiplier $v(T)$, when T is a parabolic or elliptic matrix.

We note first that, for any $T \in \Omega$ and $q \in \mathbb{Z}$,

$$\mu(TU^q, z) = (cz + d + cq)^k = \mu(T, U^q z). \qquad (3.3.1)$$

Suppose now that $\zeta = L\infty \in \mathbb{P}$, where $L \in \Gamma(1)$. Then $\hat{\Gamma}_\zeta$ is generated by $T := LU^{n_L}L^{-1}$, where $n_L = n(L\infty, \Gamma)$ (see §2.2); for convenience we write n for n_L. Since $U^n \in \Gamma^L = L^{-1}\Gamma L$, we have, by (3.1.10),

$$\nu^L(U^n, z) = v^L(U^n)(U^n : z)^k = v^L(U^n) =: e^{2\pi i \kappa_L}, \qquad (3.3.2)$$

say, where $0 \leq \kappa_L < 1$. Hence, by (3.1.20),

$$\begin{aligned} e^{2\pi i \kappa_L} &= v(T)\sigma(LU^nL^{-1}, L)/\sigma(L, U^n) \\ &= v(T) = v(LU^{n_L}L^{-1}), \end{aligned} \qquad (3.3.3)$$

by (3.2.1, 17, 21).

For a given AF ν, the number κ_L depends only on the orbit $\hat{\Gamma}\zeta$ of the cusp $\zeta = L\infty$. To show this, suppose that $M\infty$ also belongs to $\hat{\Gamma}\zeta$, where $M \in \Gamma(1)$, so that $M = SLU^q$, where $q \in \mathbb{Z}$ and $S \in \Gamma$. Then $n_M = n_L = n$, and we have

$$MU^nM^{-1} = STS^{-1}.$$

Hence

$$\begin{aligned} e^{2\pi i \kappa_M} &= v(MU^nM^{-1}) = v(STS^{-1}) \\ &= v(ST)v(S^{-1})\sigma(ST, S^{-1}) \\ &= v(T)v(S^{-1})v(S)\sigma(ST, S^{-1})\sigma(S, T) \\ &= v(T)\frac{\sigma(ST, S^{-1})\sigma(S, T)}{\sigma(S^{-1}, S)}. \end{aligned}$$

Now, by (3.2.4), with $S_1 = ST$, $S_2 = S^{-1}$ and $S_3 = S$,

$$w(ST, S^{-1}) + w(S, T) - w(S^{-1}, S) = w(S, T) - w(STS^{-1}, S)$$
$$= 0,$$

by (3.2.22) with $R = U^n$, $T_1 = SL$ and $T_2 = L$. It follows that

$$e^{2\pi i\kappa_M} = v(T) = e^{2\pi i\kappa_L},$$

so that $\kappa_M = \kappa_L$. We call κ_L the (*cusp*) *parameter* associated with the cusp $L\infty$ and the AF ν, and write

$$\kappa_L =: \kappa(L\infty, \Gamma, v).$$

By induction from (3.1.5) and (3.3.2, 3) we have, for any $m \in \mathbb{Z}$,

$$e^{2\pi im\kappa_L} = v^L(U^{mn_L}, z) = v^L(U^{mn_L}) = v(LU^{mn_L}L^{-1}). \quad (3.3.4)$$

We suppose that Γ has an elliptic fixed point $\zeta = L\rho$ (see (2.2.10)), so that $T := LPL^{-1} \in \Gamma$ and $T\zeta = \zeta$. By (3.1.1),

$$(L:P\rho)(P:\rho) = LP:\rho = (LPL^{-1}:L\rho)(L:\rho)$$

so that, since $P\rho = \rho$,

$$T:\zeta = LPL^{-1}:L\rho = P:\rho = 1 + \rho = e^{\pi i/3}. \quad (3.3.5)$$

By induction from (3.1.5),

$$\nu(T^r, \zeta) = \{\nu(T, \zeta)\}^r \quad (r \in \mathbb{Z}). \quad (3.3.6)$$

Now $T^3 = -I$ and $\nu(-I, z) = 1$, so that (3.3.5) shows that for some unique integer m_L $(0 \leq m_L < 3)$,

$$\rho^{m_L} = \nu(T, \zeta) = (T:\zeta)^k v(T) = e^{\pi ik/3} v(T). \quad (3.3.7)$$

Hence

$$v(T) = v(LPL^{-1}) = \rho^{m_L} e^{-\pi i\kappa/3}. \quad (3.3.8)$$

For any $r \in \mathbb{Z}$,

$$T^r:\zeta = (T:\zeta)^r = e^{\pi ir/3}.$$

Therefore by (3.3.6),

$$v(LP^rL^{-1}) = v(T^r) = \nu(T^r, \zeta)/(T^r:\zeta)^k$$
$$= \rho^{rm_L} e^{-\pi ikr/3}, \quad (3.3.9)$$

provided that $-2 \leq r \leq 3$; for then $\text{ph}(T^r:\zeta) = -\pi r/3$.

It can be shown similarly that, if Li is a fixed point for Γ, then $LVL^{-1}\in\Gamma$ and

$$v(LVL^{-1})=(-1)^{rs_L}e^{-\pi ikr/2}\quad(-1\leq r\leq 2)\qquad(3.3.10)$$

for some integer s_L $(0\leq s_L<2)$.

We call m_L and s_L the parameters associated with the automorphic factors (or multiplier system) and the elliptic fixed points $L\rho$ and Li, respectively. It is easy to show that they depend only on the orbits $\Gamma L\rho$ and ΓLi, respectively. For example, if $M=SLP^q$, where $q\in\mathbb{Z}$ and $S\in\Gamma$, then, by (3.3.7) and (3.1.17) (with $z=L\rho$, $L=S$, and T replaced by STS^{-1}),

$$\rho^{m_M}=\nu(STS^{-1},SL\rho)=\nu(T,L\rho)=\rho^{m_L},$$

since $\nu^S=\nu$; then $m_M=m_L$.

We now consider how many different multiplier systems of weight k can be defined on the group Γ. Suppose that v_1 and v_2 are two multiplier systems of the same dimension on Γ, and write $\chi(T)=v_2(T)/v_1(T)$ for all $T\in\Gamma$. Then, by (3.1.14),

$$\chi(ST)=\chi(S)\chi(T)$$

for all $S,T\in\Gamma$ and, by (3.1.13), $\chi(-I)=1$.It follows that χ is a (linear) character of the group Γ with the property that $\chi(-T)=\chi(T)$ for all $T\in\Gamma$; i.e. χ is a character on the inhomogeneous group $\hat{\Gamma}$.

Conversely, if χ is any character on the inhomogeneous group $\hat{\Gamma}$ and if v_1 is any MS on Γ of weight k, then so is v_2, where $v_2(T)=\chi(T)v_1(T)$ $(T\in\Gamma)$. The number of different multiplier systems of weight k on Γ is therefore either zero, or else is equal to h, the number of characters on $\hat{\Gamma}$. Clearly h is equal to the order of the factor group $\hat{\Gamma}/\hat{\Gamma}'$, where $\hat{\Gamma}'$ is the commutator subgroup of $\hat{\Gamma}$; h may be infinite.

In particular, if k is an even integer, then μ is an AF for every group Γ, so that, in this case, the number of different multiplier systems is h. The MS v associated with the AF μ is called a *constant multiplier system*, since

$$v(T)=1\quad\text{for all } T\in\Gamma.\qquad(3.3.11)$$

For this MS, $\kappa_L=0$ for all $L\in\Gamma(1)$, by (3.3.3). Further, by (3.3.9, 10),

$$m_L\equiv\tfrac{1}{2}k\ (\mathrm{mod}\ 3)\quad\text{if } L\rho\in\mathbb{E}_3(\Gamma)\quad(k\text{ even})\qquad(3.3.12)$$

and

$$s_L \equiv \tfrac{1}{2}k \pmod 2 \quad \text{if } Li \in \mathbb{E}_2(\Gamma) \quad (k \text{ even}). \tag{3.3.13}$$

We denote the constant MS by the figure 1.

In certain cases we can draw similar conclusions when k is an odd integer, and we make this assumption for the remainder of the present section.

If $-I \notin \Gamma$, then μ is an AF on Γ and the associated MS v is constant on Γ since it satisfies (3.3.11).

If $-I \in \Gamma$, however, things are not quite so simple. This often arises when we are given the inhomogeneous group $\hat{\Gamma}$ and Γ is taken to be the set of all matrices associated with mappings in $\hat{\Gamma}$. If we can find a subgroup Γ^* of index 2 in Γ, which does not contain $-I$, then $\hat{\Gamma} \cong \Gamma^*$ and we can define an AF ν and a corresponding MS v on Γ as follows:

$$\begin{gathered}\nu(T,z)=\mu(T,z)\ (T\in\Gamma^*),\quad \nu(T,z)=-\mu(T,z)\ (T\in\Gamma-\Gamma^*),\\ v(T)=1\ (T\in\Gamma^*),\quad v(T)=-1\ (T\in\Gamma-\Gamma^*).\end{gathered} \tag{3.3.14}$$

It is clear that these equations do in fact define an AF and MS on Γ; further v is a constant MS on Γ^*.

Since $-I$ is the square of every elliptic matrix $\pm L^{-1}VL$, it follows that a necessary condition for such a subgroup Γ^* to exist is that

$$\mathbb{E}_2(\Gamma) = \varnothing. \tag{3.3.15}$$

Petersson (1938*a*) (III, Satz 18, p. 566) has proved a general theorem from which it follows that (3.3.15) is also a sufficient condition. This can also be deduced, in the case in which we are interested, from theorem 1.7.1 with $G = \hat{\Gamma}(1)$ and $H = \hat{\Gamma}$. For the groups B_β (if they exist) have order 3 and are generated by mappings $L_\beta^{-1}P^2L_\beta$, where $L_\beta \in \hat{\Gamma}(1)$,while the free group F (if it exists) is generated by mappings $T_1, T_2, \ldots, T_r$, say. Accordingly, we can define Γ^* to be the group generated by the matrices $L_\beta^{-1}P^2L_\beta$ and the matrices $\pm T_1, \pm T_2, \ldots, \pm T_r$ for any fixed choice of the signs $\pm$. It is clear that $-I \notin \Gamma^*$ and that there are 2^r different choices of Γ^*.

For example, when

$$\hat{\Gamma} = \hat{\Gamma}^2,\ \hat{\Gamma}(2),\ \hat{\Gamma}(n)\ (n>2),$$

we may take

$$\Gamma^* = \Gamma^4,\ \Gamma^*(2),\ \Gamma(n)\ (n>2),$$

respectively. For $\hat{\Gamma}(n)$ $(n>2)$ there are $2^{1+\mu/6}$ different choices for Γ^* of which $\Gamma(n)$ is the simplest; here $\mu = \hat{\mu}(n)$ and is given by theorem 1.4.1. In the case of $\hat{\Gamma}(2)$, we note that

$$\Gamma^*(2) = \langle U^2\rangle * \langle W^2\rangle \qquad (3.3.16)$$

(see (1.5.4)), and that there are four possible choices for Γ^*, namely

$$\Gamma^*(2),\ P^{-1}\Gamma^*(2)P \text{ and } P^{-2}\Gamma^*(2)P^2, \qquad (3.3.17)$$

and

$$\tilde{\Gamma}(2) = \langle -U^2\rangle * \langle -W^2\rangle, \qquad (3.3.18)$$

which is normal in $\Gamma(1)$. The *core* of $\Gamma^*(2)$, i.e. the intersection of the conjugate groups (3.3.17), is $\Gamma(4)$. An alternative definition of $\tilde{\Gamma}(2)$ is

$$\tilde{\Gamma}(2) = \{T \in \Gamma(1): a \equiv d \equiv \varepsilon \pmod 4,\ b - c \equiv \varepsilon - 1 \pmod 4,\ \varepsilon = \pm 1\}, \qquad (3.3.19)$$

so that the associated MS $\tilde{v}$ is given by

$$\tilde{v}(T) = (-1)^{\frac{1}{2}(a-b+c-1)} \quad \text{for } T \in \Gamma(2). \qquad (3.3.20)$$

This multiplier system $\tilde{v}$ is identical with each conjugate $\tilde{v}^L$, by (3.1.21). On the other hand, the three multiplier systems v, v^P, v^{P^2} defined on $\Gamma(2)$ by (3.3.14) with $\Gamma^* = \Gamma^*(2)$ are different and take the common value 1 on the core $\Gamma(4)$. Further properties of these groups and their multiplier systems are given in §5.6.

Fig. 6 displays the relationship between some of these groups of level 2 and 4. All the groups illustrated are normal subgroups of $\Gamma(1)$, with the exception of $\Gamma^*(2)$, $\Gamma_V(2)$ and $\Gamma_V(4)$. In order to avoid complicating the figure, these three groups have been selected from their families of conjugates, since they are of particular importance in the study of the theta function $\vartheta_3(z)$. It may be noted that $\Gamma_V(2)$ is the normalizer in $\Gamma(1)$ of both $\Gamma_V(4)$ and $\Gamma^*(2)$. That $\tilde{\Gamma}(2) \subseteq \Gamma^4$ follows because

$$-U^2 = P_1^{-2}P^{-2}, \quad -W^2 = P_1^2P_2^2.$$

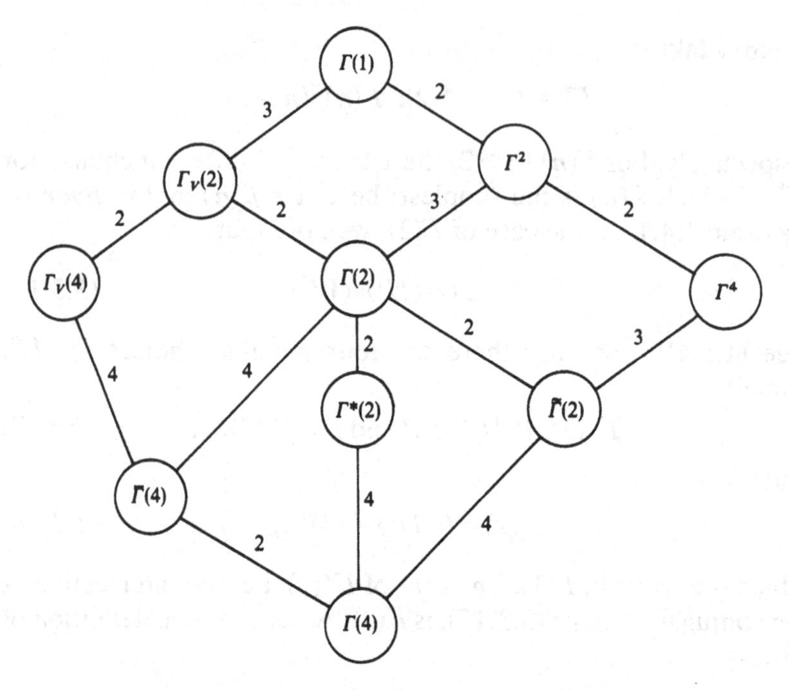

Fig. 6. Homogeneous groups between $\Gamma(1)$ and $\Gamma(4)$.

3.4. Multiplier systems on $\Gamma(1)$, Γ^2 and Γ^3. We now take $\Gamma=\Gamma(1)$ and assume that ν is an AF of weight k for $\Gamma(1)$, where $k\in\mathbb{R}$. The associated MS we denote by v, and we put

$$\nu(U,z)=v(U)=:\omega. \tag{3.4.1}$$

For all $L\in\Gamma(1)$, $\nu^L=\nu$, $v^L=v$ and $n_L=1$. Further, $\kappa_L=\kappa$, $m_L=m$ and $s_L=s$, say, for all $L\in\Gamma(1)$, and

$$e^{2\pi i\kappa}=v(U)=v(LUL^{-1})=\omega. \tag{3.4.2}$$

Since $P=VU$, it follows from (3.1.5) that

$$\nu(P,z)=\omega\nu(V,Uz),$$

so that, by (3.1.10),

$$(z+1)^k v(P)=\omega(z+1)^k v(V). \tag{3.4.3}$$

By (3.3.9, 10),

$$v(P)=\rho^m e^{-\pi ik/3}, \quad v(V)=(-1)^s e^{-\pi ik/2} \tag{3.4.4}$$

and therefore, by (3.4.3),

$$\omega e^{-\pi ik/6} = (-1)^s \rho^m = :\chi(U), \tag{3.4.5}$$

where $\chi(U)$ is a sixth root of unity; the reason for this notation will be explained later. Hence, by (3.1.13) and (3.3.4, 9, 10),

$$v(-I) = \omega^{-6}, \quad v(U^m) = \omega^m \ (m \in \mathbb{Z}), \tag{3.4.6}$$

$$v(P^r) = \omega^{-2r} \ (-2 \le r \le 3), \quad v(V^r) = \omega^{-3r} \ (-1 \le r \le 2). \tag{3.4.7}$$

Since $W = UP$ and, by (3.2.17), $w(U, P) = 0$, we deduce that $v(W) = \omega^{-1}$ and it is easily shown by induction, with the help of (3.2.7), that

$$v(W^m) = \omega^{-m} \quad (m \in \mathbb{Z}). \tag{3.4.8}$$

Since $\sigma(S, T)$ is a power of $e^{2\pi ik} = \omega^{12}$, it follows that, for every $T \in \Gamma(1)$, there exists an integer $r(T)$ such that

$$v(T) = \omega^{r(T)}. \tag{3.4.9}$$

Equation (3.4.5) defines $\chi(U)$ as the sixth root of unity $(-1)^s \rho^m$. Now, by theorem 1.3.2, the factor group $\hat{\Gamma}(1)/\hat{\Gamma}'(1)$ is a cyclic group generated by the coset containing U. We use this to extend the domain of the function χ to the whole group $\Gamma(1)$; we define

$$\chi(T) = \{\chi(U)\}^m \quad \text{when} \quad T \in \pm U^m \Gamma'(1). \tag{3.4.10}$$

It follows that χ is a character on $\Gamma(1)$ with the property that $\chi(-I) = 1$. Hence, if one MS of weight k exists, then exactly six such multiplier systems exist, one for each choice of character, i.e. one for each choice of $\chi(U)$. For any such choice, $v(T)$ is given by (3.4.9), where ω is defined in terms of $\chi(U)$ by (3.4.5). We shall show in §6.4 that a MS of weight k on $\Gamma(1)$ does exist, and it follows that there are exactly six such multiplier systems for each k.

Similar arguments can be applied to subgroups of $\Gamma(1)$. For example, if $\Gamma = \Gamma^2 = \Gamma_P(2)$, we obtain, from (3.3.9),

$$v(P) = e^{-\pi ik/3} \chi(P), \quad v(P_1) = e^{-\pi ik/3} \chi(P_1) \tag{3.4.11}$$

where $P_1 = V^{-1}PV$. Here $\chi(P)$ and $\chi(P_1)$ are cube roots of unity. Since any MS on $\Gamma(1)$ is also a MS on any subgroup, it follows that there are exactly nine different multiplier systems on Γ^2. We note that the commutator factor group $\hat{\Gamma}^2/\hat{\Gamma}^{2\prime}$ is isomorphic to the direct product of two cyclic groups of order 3.

It can be shown similarly from (3.3.10) that there are eight multiplier systems on Γ^3, for each $k \in \mathbb{R}$, and that

$$v(V)=e^{-\pi ik/2}\chi(V), \quad v(V_j)=e^{-\pi ik/2}\chi(V_j) \qquad (3.4.12)$$

for $j=1,2$ (see (1.3.10)), where $\chi^2(V)=\chi^2(V_1)=\chi^2(V_2)=1$. In fact, $\hat{\Gamma}^3/\hat{\Gamma}^{3\prime}$ is isomorphic to the direct product of three cyclic groups of order 2.

For the remaining groups in tables 3 and 4 the commutator subgroups have infinite index, so that there are infinitely many multiplier systems of weight k for each of these subgroups.

We conclude this section by determining explicitly the integer $r(T)$ appearing in the formula (3.4.9) for $v(T)$, where

$$\omega=v(U)=e^{\pi ik/6}(-1)^s\rho^m=e^{\pi ik/6}\chi(U). \qquad (3.4.13)$$

We note that, since $VP^2=-W$ and $VP^4=U$, it follows from theorem 1.2.5 that every $T\in\Gamma(1)$ can be expressed uniquely in the form

$$T=P^{2p}\,S\,V^q \quad (0\le p\le 2, 0\le q\le 3), \qquad (3.4.14)$$

where S is of the form

$$S=U^{a_1}W^{b_1}U^{a_2}W^{b_2}\dots U^{a_m}W^{b_m} \quad (a_j\ge 0, b_j\ge 0; 1\le j\le m). \qquad (3.4.15)$$

The matrices S form a free semigroup with generators U and W; we denote this semigroup by Γ^+. The entries of each matrix $S\in\Gamma^+$ are non-negative integers. Since it is easily verified that, if $S\in\Gamma^+$, then the matrix $P^{2p}\,S\,V^q$ in (3.4.14) has non-negative entries only when $p=q=0$, it follows that

$$\Gamma^+=\{S\in\Gamma(1)\colon \alpha\ge 0, \beta\ge 0, \gamma\ge 0, \delta\ge 0\}. \qquad (3.4.16)$$

We therefore have

$$\Gamma(1)=\langle P^2\rangle\cdot\Gamma^+\cdot\langle V\rangle. \qquad (3.4.17)$$

Instead of (3.4.14) we shall find it convenient to use the equivalent representation

$$T=(-1)^rP^s\,S\,V^{-t} \quad (0\le r\le 1, 0\le s\le 2, 0\le t\le 1), \qquad (3.4.18)$$

where S is given by (3.4.15), and we shall put

$$a_0=a_0(S):=\sum_{j=1}^{m}a_j, \quad b_0=b_0(S):=\sum_{j=1}^{m}b_j. \qquad (3.4.19)$$

We note that, if

$$S=R_1X_1=X_2R_2,\quad P=R_3X_3,\quad P^2=R_4X_4,\quad R_2V^{-1}=X_5R_5,$$

where $R_j\in\Sigma$, $X_j\in\Xi$ $(1\le j\le 5)$, then

$$\text{ph}\,X_1\in[0,\tfrac{1}{2}\pi],\quad \text{ph}\,X_2\in[0,\tfrac{1}{2}\pi],\quad \text{ph}\,X_3=\tfrac{1}{4}\pi,$$
$$\text{ph}\,X_4=\tfrac{1}{2}\pi,\quad \text{ph}\,V^{-1}=-\tfrac{1}{2}\pi,\quad \text{ph}\,X_s\in[-\pi,0],$$

so that, by (3.2.15, 20), $w(S_1, S_2)=0$ for all $S_1, S_2\in\Gamma^+$ and

$$w(S, V^{-1})=w(P,S)=w(P^2,S)=w(P,SV^{-1})=w(P^2,SV^{-1})=0. \tag{3.4.20}$$

It follows by induction from (3.1.14), (3.2.7) and (3.4.6, 8, 19) that $r(S)=a_0-b_0$, if S is given by (3.4.15) and, by (3.4.7, 20),

$$r(P^s\,S\,V^{-t})=-2s+a_0-b_0+3t.$$

It remains to evaluate $r(-T)$, where $T=P^s\,S\,V^{-t}$. We have

$$v(-T)=v(-I)v(T)\sigma(-I,T)=\omega^{-6}\sigma(-I,T)v(T).$$

By (3.2.18), $\sigma(-I,T)=1$ except when $c>0$ or when $c=0$, $d<0$, in which cases $\sigma(-I,T)=\omega^{12}$. In table 5 we list the values of c, d and $\sigma(-I,T)$, in terms of the entries of S for the matrix $T=P^s\,S\,V^{-t}$. We note, for example, that, for $s=2$, $t=1$ we cannot have $\alpha=0$ since then $\beta\gamma=-1$, which is impossible for $S\in\Gamma^+$.

Table 5

s	0	1	2	0	1	2
t	0	0	0	1	1	1
c	γ	$\alpha+\gamma$	α	$-\delta$	$-(\beta+\delta)$	$-\beta$
d	δ	$\beta+\delta$	β	γ	$\alpha+\gamma$	α
$\sigma(-I,T)$		ω^{12}	ω^{12}	1	1	1

When $s=t=0$, we have $\sigma(-I,T)=\omega^{12}$ except for $T=S=U^\beta$ $(\beta\ge 0)$, when $\sigma(-I,T)=1$. We therefore have, for T expressed in the form (3.4.18),

$$r(T)=\begin{cases} 6-2s+a_0-b_0 & \text{if } r=1, 0\le s\le 2, t=0, s+b_0>0,\\ -6r-2s+a_0-b_0+3t & \text{otherwise.}\end{cases} \tag{3.4.21}$$

3.5. Further results. The most complete treatment of multiplier systems is to be found in Petersson (1938*a*); see, in particular, I §2 and III, §§6, 7. Petersson's theory includes the case when the dimension is complex and applies to any horocyclic group (*Grenzkreisgruppe*). An earlier less complete discussion for real dimension is given in Petersson (1930). It is possible to tabulate the values of $w(S, T)$ for all signs of the parameters γ, δ, a, c and $\gamma a+\delta c$; see Satz 4 on pp. 44–5 of the first paper.

Other authors, such as Rademacher (1955), van Lint (1957, 1958) and Lehner (1968), have studied multiplier systems associated with particular groups and particular modular forms.

The fact that (see (1.3.10))

$$w(V_1, V_2)=0, \quad w(V_2, V_1)=1 \tag{3.5.1}$$

shows that it is not necessarily the case that

$$w(S, T)=w(T, S) \tag{3.5.2}$$

holds for all S, T in $\Gamma(1)$. However, (3.5.2) holds whenever S and T commute. To show this we may assume, in view of (3.2.17, 18), that neither S nor T is $\pm I$. Let ζ be a fixed point of S lying in $\bar{\mathbb{H}}$. Since

$$ST\zeta=TS\zeta=T\zeta,$$

$T\zeta$ is also a fixed point of S in $\bar{\mathbb{H}}$; so is $T^2\zeta$, and clearly $T^2\zeta=\zeta$, since S has at most two fixed points. We show that

$$T\zeta=\zeta. \tag{3.5.3}$$

This is obvious if S is elliptic or parabolic. Hence we may assume that S is hyperbolic, and therefore T^2 is hyperbolic since it has ζ as a fixed point. It follows that T is hyperbolic and that ζ is one of its fixed points; i.e. (3.5.3) holds.

Now, by (3.2.2),

$$w(S, T)=\frac{1}{2\pi i}\{\log(S: Tz)+\log(T: z)-\log(ST: z)\},$$

so that

$$w(S, T)-w(T, S)=\frac{1}{2\pi i}\{\log(S: Tz)+\log(T: z)-\log(T: Sz) -\log(S: z)\}.$$

If we let z tend to ζ in $\mathbb{H}$ we deduce (3.5.2).

The relation (3.2.4) can be expressed by stating that w is a 2-cocycle of the trivial Γ-module $\mathbb{Z}$, where Γ is a subgroup of $\Gamma(1)$; i.e. $w \in Z^2(\Gamma, \mathbb{Z})$. The cohomology class of w consists of all members $w^* \in Z^2(\Gamma, \mathbb{Z})$ such that

$$w^* - w \in B^2(\Gamma, \mathbb{Z}), \tag{3.5.4}$$

the group of 2-coboundaries. Here $f \in B^2(\Gamma, \mathbb{Z})$ when

$$f(ST) = \varphi(S) + \varphi(T) - \varphi(ST)$$

for some $\varphi: \Gamma \to \mathbb{Z}$. We call w^* *symmetric* if

$$w^*(S, T) = w^*(T, S)$$

for all S and T in Γ, and *semisymmetric* if this holds whenever S and T commute. We recall that we have just shown that w is semisymmetric.

Leutbecher (1970) has shown, in a rather more general setting, that the cohomology class of an element of $Z^2(\Gamma, \mathbb{Z})$ contains a symmetric cocycle if and only if the element is semisymmetric; see also Wohlfahrt (1972). It follows that we can find a symmetric w^* such that (3.5.4) holds, where w is the function studied in §§3.2–4. The multiplier system v^* associated with w^* has the convenient property that $v^*(ST) = v^*(TS)$ for all S and T, but this is offset by the disadvantage that, in the definitions of v^* and w^*, it is necessary to choose $\mathrm{ph}(T: z)$ to depend upon the first row of T as well as on the second.

4: General properties of modular forms

4.1. Definitions and general theorems. The object of this chapter is to define what is meant by modular forms and functions and to discuss the distribution of their zeros and poles in a fundamental region. The dimensions of various vector spaces of modular forms are investigated and simple estimates for the magnitude of their Fourier coefficients are given. Algebraic, as well as linear, relations between modular functions are also considered.

Throughout this chapter we assume that Γ is a subgroup $\Gamma(1)$ of finite index μ and let ν be an automorphic factor of weight k on Γ, where $k \in \mathbb{R}$. We write v for the associated multiplier system and assume that $-I \in \Gamma$. Note that $[\Gamma(1):\Gamma] = \mu = [\hat{\Gamma}(1):\hat{\Gamma}]$.

We denote by $M'(\Gamma, k, v)$ the set of all functions f that possess the following two properties:

(I) f is a meromorphic function on $\mathbb{H}$.

(II) For all $T \in \Gamma$ and all $z \in \mathbb{H}$ (including poles of f)

$$f(Tz) = \nu(T, z) f(z). \tag{4.1.1}$$

We recall that

$$\nu(T, z) = (T:z)^k v(T), \tag{4.1.2}$$

where the multiplier $v(T)$ is of unit modulus and depends on T and k but not on z, and $\nu(T, z)$ is holomorphic on $\mathbb{H}$.

The *zero function* 0, which takes the value zero at all points of $\mathbb{H}$, clearly belongs to $M'(\Gamma, k, v)$ for all $k \in \mathbb{R}$ and all multiplier systems v. A function f belonging to $M'(\Gamma, k, v)$ is called an *unrestricted modular form* of weight† k for the group Γ; the adjective 'unrestricted' is used since we shall find it convenient to impose further conditions on f later on. We do not yet know whether any

† We follow Ogg (1973) in using this terminology. The classical usage is to call f a form of *dimension* $-k$ (rather than weight k), but this conflicts with the use of the word dimension in vector-space theory. On the other hand, the use of the word weight does not avoid all ambiguity, since some authors call f a form of weight $\frac{1}{2}k$.

unrestricted modular forms, other than the zero form, exist and it will only be in the next chapter that we shall construct such forms.

If $f \in M'(\Gamma, k, v)$ and $L \in \Gamma(1)$, the L-transform f_L of f is defined by

$$f_L(z) = f(z)|L = \{\mu(L, z)\}^{-1} f(Lz). \tag{4.1.3}$$

Here, as in (3.1.8), $\mu(L, z) = (L:z)^k$. Observe that the *stroke operator* $|$ depends on k.

Theorem 4.1.1. *Suppose that $f \in M'(\Gamma, k, v)$ and that $L, L_1, L_2 \in \Gamma(1)$. Then we have:*

(i) *$f_L \in M'(L^{-1}\Gamma L, k, v^L)$.*

(ii) *$f|(L_1L_2) = \sigma(L_1, L_2)(f|L_1)|L_2$, where $\sigma(L_1, L_2)$ is defined by (3.1.15).*

(iii) *If $T \in \Gamma$, $f_{TL} = \sigma(T, L)v(T)f_L$; in particular, $f_T = v(T)f$, and $f_{-L} = e^{\pm \pi i k} f_L$.*

(iv) *If $\zeta = L\infty$, then*

$$f_L(z + n_L) = e^{2\pi i \kappa_L} f_L(z) \tag{4.1.4}$$

for all $z \in \mathbb{H}$, where $n_L = n(L\infty, \Gamma)$, the width of the cusp ζ (mod Γ), and κ_L is its parameter.

Proof. (i) F_L is clearly meromorphic on $\mathbb{H}$. For $T \in \Gamma$ and $z \in \mathbb{H}$,

$$\begin{aligned} f_L(L^{-1}TLz) &= \{\mu(L, L^{-1}TLz)\}^{-1} f(TLz) \\ &= \{\mu(L, L^{-1}TLz)\}^{-1} \nu(T, Lz) f(Lz) \\ &= \nu^L(L^{-1}TL, z) f_L(z), \end{aligned}$$

by (3.1.17).

(ii) By (4.1.1, 3) and (3.1.15),

$$\begin{aligned} (f(z)|L_1)|L_2 = f_{L_1}(z)|L_2 &= \{\mu(L_2, z)\}^{-1} f_{L_1}(L_2 z) \\ &= \{\mu(L_2, z)\mu(L_1, L_2 z)\}^{-1} f(L_1L_2 z) \\ &= \mu(L_1L_2, z)\{\mu(L_2, z)\mu(L_1, L_2 z)\}^{-1} f_{L_1L_2}(z) \\ &= f_{L_1L_2}(z)/\sigma(L_1, L_2). \end{aligned}$$

(iii) This follows from (ii) since

$$\begin{aligned} f_T(z) = \{\mu(T, z)\}^{-1} f(Tz) &= \{\nu(T, z)/\mu(T, z)\} f(z) \\ &= v(T) f(z). \end{aligned}$$

The result for f_{-L} uses (3.1.13) and (3.2.18).

(iv) Since $U^{n_L} \in L^{-1}\Gamma L$, we have, by (i) and (3.3.2),

$$f_L(z+n_L)=f_L(U^{n_L}z)=v^L(U^{n_L}, z)f_L(z)=e^{2\pi i\kappa_L}f_L(z).$$

Part (iii) of the theorem shows that, if constant factors of unit modulus are neglected, there is one transform of f for each right coset of $\hat{\Gamma}$ in $\hat{\Gamma}(1)$. We saw in the proof of theorem 2.4.3 that these μ right cosets can be grouped into $\lambda=\lambda(\Gamma)$ families $\hat{\Gamma}LU^q$ $(0 \leq q < n_L)$, each family corresponding to one of the orbits $\hat{\Gamma}L\infty$ into which the cusps are partitioned modulo Γ. Now, for any $q \in \mathbb{Z}$, by (4.1.3) and (3.3.1),

$$f_{LU^q}(z)=f_L(z+q). \tag{4.1.5}$$

Thus f_{LU^q} is merely a 'translation' of f_L. Hence there are really only λ essentially different transforms of f (some of which may, in particular cases, turn out to be the same). Every other transform of f is obtained from one of these λ functions by multiplication by a constant of unit modulus and by replacing z by $z+q$ for some integer q.

If we write

$$f_L^*(z)=e^{-2\pi i\kappa_L z/n_L}f_L(z), \tag{4.1.6}$$

it follows from (4.1.4) that

$$f_L^*(z+n_L)=f_L^*(z). \tag{4.1.7}$$

Write

$$t=t_L=e^{2\pi iz/n_L}. \tag{4.1.8}$$

We call t a *local uniformizing variable* (LUV) at the cusp $L\infty$. If we put

$$f_L^*(z)=F_L(t), \tag{4.1.9}$$

$F_L(t)$ is uniquely defined for $0<|t|<1$ and is a meromorphic function of t on this punctured neighbourhood of the origin. In particular, if f_L is holomorphic on $\{z: z \in \mathbb{H}, \operatorname{Im} z > \eta\}$ for some $\eta \geq 0$, then F_L is holomorphic on $\{t: 0<|t|<e^{-2\pi\eta/n_L}\}$ and has a convergent Laurent expansion on this punctured neighbourhood of $t=0$. This expansion may contain an infinite number of negative powers of t.

We are now in a position to state the further restrictions that will be placed on modular forms. Let $f \in M'(\Gamma, k, v)$. Then f is called a modular form of weight k for the group Γ, with AF ν (or, equivalently, with MS v), if f satisfies the following additional condition:

(III) For each $L \in \Gamma(1)$, the function F_L defined by (4.1.6, 9) is a meromorphic function at $t = 0$. I.e., for each $L \in \Gamma(1)$, there exists a $\delta_L > 0$ and an integer N_L such that

$$F_L(t) = \sum_{m=N_L}^{\infty} a_m(L) t^m \tag{4.1.10}$$

for $0 < |t| < \delta_L$; hence

$$f_L(z) = e^{2\pi i \kappa_L z/n_L} \sum_{m=N_L}^{\infty} a_m(L)\, e^{2\pi i m z/n_L} \tag{4.1.11}$$

for $\operatorname{Im} z > \{n_L \log(1/\delta_L)\}/(2\pi) =: \eta_L$.

The class of all functions f satisfying (I), (II) and (III) we denote by $M(\Gamma, k, v)$. Our earlier remarks show that essentially only λ functions f_L are affected by the condition (III). The condition affects the behaviour of f_L near the point ∞; i.e., condition (III) is a restriction on the behaviour of the unrestricted modular form f near each cusp.

Since the behaviour of f_{L_1} at $L_2\infty$ is determined by the behaviour of $f_{L_1L_2}$ at ∞, we note that, if $f \in M(\Gamma, k, v)$, then $f_L \in M(L^{-1}\Gamma L, k, v^L)$ for each $L \in \Gamma(1)$.

We note that, in particular, $0 \in M(\Gamma, k, v)$ where 0 denotes the zero form. If $f \in M(\Gamma, k, v)$ and $f \neq 0$, we can take N_L in (4.1.10) to be such that $a_{N_L}(L) \neq 0$. The power series for $f_L(z)$ then begins with a non-zero constant multiple of $t^{\kappa_L + N_L}$ and we call $\kappa_L + N_L$ the *order of f at the cusp $L\infty$* (mod Γ) and write

$$\kappa_L + N_L =: \operatorname{ord}(f, L\infty, \Gamma). \tag{4.1.12}$$

If $\zeta \in \mathbb{H}$, we let

$$\operatorname{ord}(f, \zeta, \Gamma) := \frac{\operatorname{ord}(f, \zeta)}{|\hat{\Gamma}_\zeta|} \tag{4.1.13}$$

where $\hat{\Gamma}_\zeta$ is the stabilizer of ζ (mod Γ). It follows that

$$\operatorname{ord}(f, \zeta, \Gamma) = \begin{cases} \dfrac{1}{m} \operatorname{ord}(f, \zeta) & \text{if } \zeta \in \mathbb{E}_m(\Gamma), \\ \operatorname{ord}(f, \zeta) & \text{if } \zeta \notin \mathbb{E}_m(\Gamma),\ \zeta \in \mathbb{H}. \end{cases} \tag{4.1.14}$$

Theorem 4.1.2. *Suppose that $f \in M(\Gamma, k, v)$ and that $f \neq 0$. Then we have*:

(i) *If* $\zeta_1 \equiv \zeta_2 \pmod{\Gamma}$, $\mathrm{ord}(f, \zeta_1, \Gamma) = \mathrm{ord}(f, \zeta_2, \Gamma)$.

(ii) *If* $\zeta = Li \in \mathbb{E}_2(\Gamma)$, *where* $L \in \Gamma(1)$, *then*

$$s_L + 2\,\mathrm{ord}(f, \zeta, \Gamma) \equiv 0 \pmod{2},$$

where s_L is defined in (3.3.10). If $v(LVL^{-1}) = 1$, then $\frac{1}{2}k \in \mathbb{Z}$ and

$$\tfrac{1}{2}k + 2\,\mathrm{ord}(f, \zeta, \Gamma) \equiv 0 \pmod{2}.$$

(iii) *If* $\zeta = L\rho \in \mathbb{E}_3(\Gamma)$, *where* $L \in \Gamma(1)$, *then*

$$m_L + 3\,\mathrm{ord}(f, \zeta, \Gamma) \equiv 0 \pmod{3},$$

where m_L is defined in (3.3.9). If $v\,(LPL^{-1}) = 1$, then $\frac{1}{2}k \in \mathbb{Z}$ and

$$\tfrac{1}{2}k + 3\,\mathrm{ord}(f, \zeta, \Gamma) \equiv 0 \pmod{3}.$$

Proof. (i) If $\zeta_1 \in \mathbb{P}$ the result follows, since N_L and κ_L depend on the orbit $\Gamma\zeta_1$. Suppose therefore that $\zeta_2 = T\zeta_1 \in \mathbb{H}$, where $T \in \Gamma$, and put $w = Tz$. Then for z in a neighbourhood of ζ, by (2.1.4, 5),

$$\left(\frac{w-\zeta_2}{w-\bar{\zeta}_2}\right)^{-n} f(w) = \left(\frac{z-\zeta_1}{z-\bar{\zeta}_1}\right)^{-n} \left(\frac{T:\zeta_1}{T:\bar{\zeta}_1}\right)^{n} f(Tz)$$

$$= v\,(T, z)\left(\frac{T:\zeta_1}{T:\bar{\zeta}_1}\right)^{n}\left(\frac{z-\zeta_1}{z-\bar{\zeta}_1}\right)^{-n} f(z) \tag{4.1.15}$$

for any $n \in \mathbb{Z}$. It follows from this that $\mathrm{ord}(f, \zeta_1) = \mathrm{ord}(f, \zeta)$ and so $\mathrm{ord}(f, \zeta_1, \Gamma) = \mathrm{ord}(f, \zeta_2, \Gamma)$.

We now assume that $\zeta = L\rho \in \mathbb{E}_3(\Gamma)$, where $L \in \Gamma(1)$, so that $T := LPL^{-1} \in \Gamma$ and $T\zeta = \zeta$. Write

$$\tau = \frac{z-\zeta}{z-\bar{\zeta}},$$

so that $|\tau| < 1$ for $z \in \mathbb{H}$. By (2.1.4, 5) and (3.3.5),

$$\frac{w-T\zeta}{w-T\bar{\zeta}} = \frac{T:\bar{\zeta}}{T:\zeta}\cdot\frac{z-\zeta}{z-\bar{\zeta}} = \rho^2\tau. \tag{4.1.16}$$

Put

$$\phi(\tau) = \left(\frac{z-\bar{\zeta}}{\zeta-\bar{\zeta}}\right)^{k}\left(\frac{z-\zeta}{z-\bar{\zeta}}\right)^{-n} f(z), \tag{4.1.17}$$

where $n=\mathrm{ord}(f,\zeta)$, so that ϕ is holomorphic and non-zero on a neighbourhood $\mathbb{N}$ of $\tau=0$. We may choose $\mathbb{N}$ sufficiently small so that, for $\tau\in\mathbb{N}$,

$$\left|\mathrm{ph}\frac{z-\bar{\zeta}}{\zeta-\bar{\zeta}}\right|<\frac{\pi}{2},\quad -\frac{\pi}{2}<\frac{\pi}{3}-\mathrm{ph}(T:z)=\mathrm{ph}\frac{e^{\pi i/3}}{T:z}<\frac{\pi}{2}. \tag{4.1.18}$$

Now

$$\frac{w-\bar{\zeta}}{\zeta-\bar{\zeta}}=\frac{z-\bar{\zeta}}{\zeta-\bar{\zeta}}\cdot\frac{1}{(T:z)(T:\bar{\zeta})}=\frac{z-\bar{\zeta}}{\zeta-\bar{\zeta}}\cdot\frac{e^{\pi i/3}}{T:z},$$

so that, by (4.1.18),

$$\left(\frac{w-\bar{\zeta}}{\zeta-\bar{\zeta}}\right)^k=\left(\frac{z-\bar{\zeta}}{\zeta-\bar{\zeta}}\right)^k\left(\frac{e^{\pi i/3}}{T:z}\right)^k=\left(\frac{z-\bar{\zeta}}{\zeta-\bar{\zeta}}\right)^k\frac{e^{\pi ik/3}}{(T:z)^k}.$$

It follows from (4.1.15–17) and (3.3.9) that

$$\phi(\rho^2\tau)=\frac{e^{\pi ik/3}}{(T:z)^k}\rho^n v(T,z)\phi(\tau)=\rho^{n+m_L}\phi(\tau).$$

If we put $\tau=0$, we deduce that

$$\rho^{n+m_L}=1,$$

from which part (iii) of the theorem follows; that $\frac{1}{2}k\in\mathbb{Z}$ and $m_L\equiv\frac{1}{2}k \pmod 3$, when $v(LPL^{-1})=1$, follows from (3.3.9) or (3.3.12). Further, for $\tau\in\mathbb{N}, \phi(\rho^2\tau)=\phi(\tau)$ and so ϕ must be expressible as a power series in

$$t=\tau^3=\left(\frac{z-\zeta}{z-\bar{\zeta}}\right)^3.$$

We call t a local uniformizing variable at ζ. We have

$$\left(\frac{z-\bar{\zeta}}{\zeta-\bar{\zeta}}\right)^k f(z)=t^{n/3}\sum_{r=0}^{\infty}c_r t^r\quad(c_0\neq 0). \tag{4.1.19}$$

The LUV t is the natural variable in terms of which to measure the order of f at ζ, which is why we have defined $\mathrm{ord}(f,\zeta,\Gamma)$ to be $n/3$.

Part (ii) of the theorem is proved similarly. In (4.1.16) ρ^2 is replaced by -1 and (3.3.10) gives $(-1)^{n+s_L}=1$. Also (4.1.19) holds with $n/3$ replaced by $n/2$, where the LUV t is defined by $t=\tau^2$.

Parts (ii) and (iii) of theorem 4.1.2 place restrictions on $\text{ord}(f, \zeta)$ at an elliptic fixed point. We can put these into a more convenient form as follows. If ν is an AF of weight k for Γ we define a function ε on $\mathbb{E}(\Gamma)$ as follows:

If $\zeta = Li \in \mathbb{E}_2(\Gamma)$ we put $\varepsilon(\zeta) = \varepsilon(\zeta, \nu) = 0$ or $\frac{1}{2}$ according as s_L is 0 or 1. If $\zeta = L\rho \in \mathbb{E}_3(\Gamma)$ we put $\varepsilon(\zeta) = \varepsilon(\zeta, \nu) = 0, \frac{1}{3}$ or $\frac{2}{3}$ according as $m_L = 0$, 2 or 1. Then $\varepsilon(\zeta)$ depends only on the orbit $\hat{\Gamma}\zeta$ and is non-negative for all $\zeta \in \mathbb{E}(\Gamma)$. We have

Corollary 4.1.2. *If $\zeta \in \mathbb{E}(\Gamma)$ then* $\text{ord}(f, \zeta, \Gamma) - \varepsilon(\zeta)$ *is an integer, so that $\varepsilon(\zeta)$ is the fractional part of* $\text{ord}(f, \zeta, \Gamma)$. *In particular, if f is holomorphic at ζ then* $\text{ord}(f, \zeta, \Gamma) \geq \varepsilon(\zeta)$.

The first part is a direct consequence of the congruences for s_L and m_L in the theorem. If f is holomorphic at ζ, $\text{ord}(f, \zeta, \Gamma) \geq 0$ and so

$$\text{ord}(f, \zeta, \Gamma) - \varepsilon(\zeta) \geq -\varepsilon(\zeta) > -1.$$

When k is an even integer and the MS is constant, m_L and s_L can be replaced by $k/2$ in the congruences and it follows that

$$\varepsilon(\zeta) = \left\{\frac{k}{2}\left(1 - \frac{1}{m}\right)\right\} \quad (\zeta \in \mathbb{E}_m(\Gamma)), \tag{4.1.20}$$

where the braces denote the fractional part.†

We denote by $H(\Gamma, k, v)$ the subset of $M(\Gamma, k, v)$ consisting of all forms f that are holomorphic on $\mathbb{H}$. If $f \in H(\Gamma, k, v)$ and $L \in \Gamma(1)$, the expansion (4.1.11) is valid for all $z \in \mathbb{H}$. If $f \in M(\Gamma, k, v)$ and $\text{ord}(f, \zeta, \Gamma) \geq 0$ for all $\zeta \in \mathbb{H}'$ we call f an *entire modular form* and denote by $\{\Gamma, k, v\}$ the class of all such forms, including the zero form in this class. If, in addition, $\text{ord}(f, \zeta, \Gamma) > 0$ for all $\zeta \in \mathbb{P}$, or if $f = 0$, we call f a *cusp form*; the class of all cusp forms is denoted by $\{\Gamma, k, v\}_0$. If $\zeta = L\infty \in \mathbb{P}$ and $\text{ord}(f, \zeta, \Gamma) = \kappa_L + N_L = n$, we say that f has a zero at ζ if $n > 0$ and a pole at ζ if $n < 0$; the values taken by f at ζ in the two cases are 0 and ∞, respectively. If $n = 0$ we can say that f does not vanish and is not infinite at ζ. It may not, however, be meaningful to say that f takes some particular value a_0 at ζ when $n = 0$. For we have defined the behaviour of f at ζ in terms of the behaviour of f_L at ∞. Different choices of L for which $\zeta = L\infty$

† If $x \in \mathbb{R}$, then $\{x\} := x - [x]$, where $[x]$ is the integral part of x; i.e. $[x]$ is the greatest integer that does not exceed x.

give, in general, different constant terms $a_0(L)$ in the power series (4.1.11) differing by factors of unit modulus. Note that

$$\{\Gamma, k, v\}_0 \subseteq \{\Gamma, k, v\} \subseteq H(\Gamma, k, v) \subseteq M(\Gamma, k, v).$$

If $f \in M(\Gamma, k, v)$ and $\mathbb{F}$ is a fundamental region for $\hat{\Gamma}$ given by

$$\mathbb{F} = \bigcup_{L \in \mathcal{R}} \mathbb{F}_L \text{ or } \bigcup_{L \in \mathcal{R}} \hat{\mathbb{F}}_L,$$

where $\hat{\Gamma}(1) = \hat{\Gamma} \cdot \mathcal{R}$, then f has only a finite number of zeros and poles in $\mathbb{F}$. It is enough to prove that, for each $L \in \mathcal{R}$, f has only a finite number of zeros and poles in $\mathbb{F}_L$, i.e. that f_L has only a finite number of zeros and poles in $\mathbb{F}_I$. Now the expansions (4.1.10, 11) show that $f_L(z)$ has only a finite number of zeros and poles in the part of $\mathbb{F}_I$ for which $\operatorname{Im} z \geq 2\eta_L$, while the part of $\bar{\mathbb{F}}_I$ for which $\operatorname{Im} z \leq 2\eta_L$ is compact and so can contain only a finite number of zeros and poles. Hence $\operatorname{ord}(f, \zeta, \Gamma)$ is non-zero at only a finite number of points ζ belonging to a proper fundamental region $\mathbb{F}^*$ for $\hat{\Gamma}$. We define the *total order* $\operatorname{ord}(f, \Gamma)$ of f (mod Γ) to be the sum of $\operatorname{ord}(f, \zeta, \Gamma)$ for all points $\zeta \in \mathbb{F}^*$. It is clear that $\operatorname{ord}(f, \Gamma)$ is independent of the choice of $\mathbb{F}^*$. We shall show that it depends only on the index μ of $\hat{\Gamma}$ in $\hat{\Gamma}(1)$ and on the weight k.

We have

$$\operatorname{ord}(f, \Gamma) = \sum_{z \in \mathbb{F}_I} \sum_{\zeta \in \mathbb{F}^* \cap [z]} \operatorname{ord}(f, \zeta, \Gamma). \tag{4.1.21}$$

If $z \in \mathbb{F}_I$ and S generates $\hat{\Gamma}_z(1)$, then, by theorem 1.1.2, with $\Gamma_1 = \hat{\Gamma}(1)$, $\Gamma_2 = \hat{\Gamma}$,

$$[z] = \hat{\Gamma}(1)z = \hat{\Gamma} \bigcup_{i=1}^{m} \mathcal{S}_i z = \hat{\Gamma} \bigcup_{i=1}^{m} L_i z,$$

where the m points $L_i z$ $(1 \leq i \leq m)$ are incongruent (mod Γ). Put $n_i = n(L_i z, \Gamma)$. Then

$$\begin{aligned}
\sum_{\zeta \in \mathcal{R} \cap [z]} \operatorname{ord}(f, \zeta, \Gamma) &= \sum_{i=1}^{m} \operatorname{ord}(f, L_i z, \Gamma) \\
&= \sum_{i=1}^{m} \sum_{j=0}^{n_i - 1} \frac{1}{n_i} \operatorname{ord}(f, L_i S^j z, \Gamma) \\
&= \sum_{L \in \mathcal{R}} \frac{\operatorname{ord}(f, Lz, \Gamma)}{n(Lz, \Gamma)}
\end{aligned} \tag{4.1.22}$$

where $\hat{\Gamma}(1)=\hat{\Gamma}\cdot\mathcal{R}$. This follows, in the first place, for $\mathcal{R}=\bigcup_{i=1}^{m}\mathcal{S}_i$, but is true for any transversal $\mathcal{R}$.

In particular, if $z=\infty$, so that $[z]=\mathbb{P}$, we have by (4.1.12, 22)

$$\sum_{\zeta\in F^*\cap\mathbb{P}}\operatorname{ord}(f,\zeta,\Gamma)=\sum_{L\in\mathcal{R}}(\kappa_L+N_L)/n_L, \tag{4.1.23}$$

where, as usual, $n_L=n(L\infty,\Gamma)$. When $z\in\mathbb{H}$,

$$\sum_{\zeta\in F^*\cap[z]}\operatorname{ord}(f,\zeta,\Gamma)=\frac{1}{|\hat{\Gamma}_z(1)|}\sum_{L\in\mathcal{R}}\operatorname{ord}(f,Lz), \tag{4.1.24}$$

since

$$|\hat{\Gamma}_{Lz}|n(Lz,\Gamma)=|\hat{\Gamma}_{Lz}(1)|=|\hat{\Gamma}_z(1)|.$$

Theorem 4.1.3. *Suppose that $f\in M(\Gamma,k,v)$, $f\neq 0$ and that $\Gamma(1)=\Gamma\cdot\mathcal{R}$. Let*

$$g(z)=g(z;f,\mathcal{R}):=\prod_{L\in\mathcal{R}}f_L(z). \tag{4.1.25}$$

Then $g\in M(\Gamma(1),\mu k,v^*)$, where μ is the index of $\hat{\Gamma}$ in $\hat{\Gamma}(1)$ and the MS v^* is associated with the AF ν^* defined by

$$\nu^*(T,z)=\prod_{L\in\mathcal{R}}\frac{\nu(S,L_1z)\mu(L_1,z)}{\mu(L,Tz)} \tag{4.1.26}$$

where $LT=SL_1$ and $S\in\Gamma$, $L_1\in\mathcal{R}$.

Proof. For any $T\in\Gamma(1)$ we have

$$g(Tz)=\prod_{L\in\mathcal{R}}f_L(Tz)=\prod_{L\in\mathcal{R}}\{\mu(L,Tz)\}^{-1}f(LTz).$$

Now $LT=SL_1$, where $S\in\Gamma$ and $L_1\in\mathcal{R}$ are uniquely determined by this equation. Hence

$$g(Tz)=\prod_{L\in\mathcal{R}}\nu(S,L_1z)\{\mu(L,Tz)\}^{-1}f(L_1z)$$

$$=\prod_{L\in\mathcal{R}}\nu(S,L_1z)\{\mu(L_1,z)/\mu(L,Tz)\}f_{L_1}(z).$$

Now L_1 runs through $\mathcal{R}$ as L does, and so

$$g(Tz)=\nu^*(T,z)g(z), \tag{4.1.27}$$

where

$$\nu^*(T, z) = \prod_{L \in \mathcal{R}} \frac{\nu(S, L_1 z)\mu(L_1, z)}{\mu(L, Tz)}.$$

We note that $\nu^*(T, z)$ is uniquely determined by T, for fixed $\mathcal{R}$, and that by (3.1.10, 15), $\nu^*(T, z) = v^*(T)\mu(T, z)$, where

$$v^*(T) = \prod_{L \in \mathcal{R}} \frac{v(S)\sigma(S, L_1)}{\sigma(L, T)} \quad (T \in \Gamma(1)).$$

Thus $|v^*(T)| = 1$; to verify that ν^* is an AF for $\Gamma(1)$ it suffices to prove that $\nu^*(ST, z) = \nu^*(S, Tz)\nu^*(T, z)$ for all $S, T \in \Gamma(1)$. By (4.1.27),

$$\nu^*(ST, z)g(z) = g(STz) = \nu^*(s, Tz)g(Tz)$$
$$= \nu^*(S, Tz)\nu^*(T, z)g(z),$$

and the required result follows since $g \neq 0$.

Since g is clearly meromorphic on $\mathbb{H}$, it remains to prove that g satisfies condition (III). By (4.1.11, 25),

$$g(z) = e^{2\pi i q z} \sum_{m=0}^{\infty} b_m e^{2\pi i m z/n} \quad (b_0 \neq 0),$$

where q is real, n is the least common multiple of the λ numbers $n_L (L \in \mathcal{R})$, and $\operatorname{Im} z > \eta > 0$, say. Write $t = e^{2\pi i z/n}$ and put

$$G(t) = \sum_{m=0}^{\infty} b_m t^m = e^{-2\pi i q z} g(z),$$

the series being convergent for sufficiently small t.

Since $g \in M'(\Gamma(1), k\mu, v^*)$ it follows from theorem 4.1.1(iv) that there exists a κ with $0 \leq \kappa < 1$ such that

$$g(z+1) = e^{2\pi i \kappa} g(z) \quad (z \in \mathbb{H}).$$

Hence

$$e^{2\pi i(\kappa + qz)} G(t) = g(z+1) = e^{2\pi i q(z+1)} G(t e^{2\pi i/n}).$$

This shows that $q = \kappa + N$, where $N \in \mathbb{Z}$, and that $b_m = 0$ if $m \not\equiv 0 \pmod{n}$. Hence $G(t)$ is, in fact, a power series in $t^n = e^{2\pi i z}$, and we have

$$g(z) = e^{2\pi i(\kappa + N)z} \sum_{m=0}^{\infty} B_m e^{2\pi i m z} \quad (B_0 \neq 0)$$

for $\mathrm{Im}\, z > \eta$. It follows that $g \in M(\Gamma(1), k\mu, v^*)$ and this completes the proof of theorem 4.1.3.

Further, from (4.1.11, 12, 23) we have

$$\mathrm{ord}(g, \infty, \Gamma(1)) = \kappa + N = q = \sum_{L \in \mathfrak{R}} (\kappa_L + N_L)/n_L$$

$$= \sum_{\zeta \in \mathbb{F}^* \cap \mathbb{P}} \mathrm{ord}(f, \zeta, \Gamma)$$

where $\mathbb{F}^*$ is a proper fundamental region for $\hat{\Gamma}$. Also, when $z \in \mathbb{H}$,

$$\mathrm{ord}(g, z, \Gamma(1)) = \frac{\mathrm{ord}(g, z)}{|\hat{\Gamma}_z(1)|}$$

$$= \sum_{L \in \mathfrak{R}} \frac{\mathrm{ord}(f, Lz)}{|\hat{\Gamma}_z(1)|}$$

$$= \sum_{\zeta \in \mathbb{F}^* \cap [z]} \mathrm{ord}(f, \zeta, \Gamma)$$

by (4.1.13, 24, 25). It follows from (4.1.21) that

$$\mathrm{ord}(g, \Gamma(1)) = \mathrm{ord}(f, \Gamma). \qquad (4.1.28)$$

We are now in a position to prove

Theorem 4.1.4. *If* $f \in M(\Gamma, k, v)$ *and* $f \neq 0$, *then* $\mathrm{ord}(f, \Gamma) = \mu k/12$, *where* μ *is the index of* $\hat{\Gamma}$ *in* $\hat{\Gamma}(1)$.

Note that the value of $\mathrm{ord}(f, \Gamma)$ is independent of the AF v although it depends on k.

Proof. It follows from (4.1.28) that it is only necessary to prove that $\mathrm{ord}(f, \Gamma(1)) = k/12$ for a function $f \in M(\Gamma(1), k, v)$.

We prove the theorem by integrating $f'(z)/f(z)$ round the boundary of $\mathbb{F}_I$, which we first modify as follows. We replace $\mathbb{F}_I$ by the compact subset $\mathbb{F}_\varepsilon$ defined for sufficiently small $\varepsilon > 0$ by

$$\mathbb{F}_\varepsilon = \{z \colon z \in \bar{\mathbb{F}}_I, y \leq 1/\varepsilon, |z - \rho| \geq \varepsilon, |z + \rho^2| \geq \varepsilon, |z - i| \geq \varepsilon\},$$

so that all fixed points have been excluded. The boundary of $\mathbb{F}_\varepsilon$ then consists (see (2.4.6, 7)) of parts of l_U, l_V, Ul_U, Vl_V and arcs $\lambda_1, \lambda_2, \lambda_3, \lambda_4$ on which

$$y = 1/\varepsilon, \quad |z - \rho| = \varepsilon, \quad |z + \rho^2| = \varepsilon, \quad |z - i| = \varepsilon,$$

respectively. We can take ε sufficiently small so that $\text{ord}(f, \zeta) = 0$ for all ζ on $\lambda_1, \lambda_2, \lambda_3, \lambda_4$ and in $\mathbb{F}_I - \mathbb{F}_\varepsilon - \{\infty, \rho, -\rho^2, i\}$. However, $\text{ord}(f, \zeta)$ may be non-zero at a finite number of points on the parts of l_U, l_V, Ul_U, Vl_V that are sides of $\mathbb{F}_\varepsilon$. Each such ζ on l_U or l_V has a congruent point on Ul_U or Vl_V, and conversely. We include the former and exclude the latter by indenting the boundary along semicircular arcs of radius ε, and ε can be chosen small enough so that none of these semicircles overlap each other or have points in common with more than one side of $\mathbb{F}_\varepsilon$; also we can arrange that no zeros or poles lie on these semicircles. The resulting region we denote by $\mathbb{F}'_\varepsilon$ (see fig. 7). Its boundary is composed of arcs $\lambda_1, l'_U, \lambda_2,$

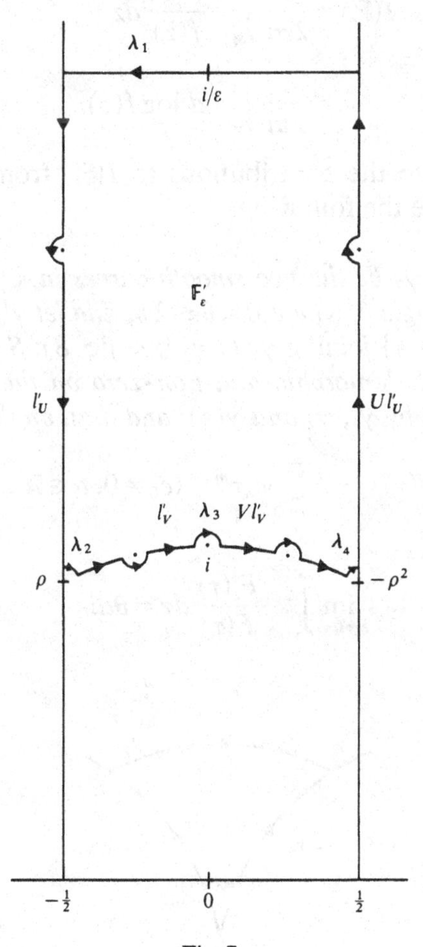

Fig. 7

l'_V, λ_4, Vl'_V, λ_3, Ul'_U, which we suppose described in that order, keeping the interior of $\mathbb{F}'_\varepsilon$ on the left. Then, clearly

$$\begin{aligned}\operatorname{ord}(f, \Gamma(1)) &= \operatorname{ord}(f, \infty, \Gamma(1)) + \operatorname{ord}(f, i, \Gamma(1)) + \operatorname{ord}(f, \rho, \Gamma(1)) \\ &\quad + I(\mathbb{F}'_\varepsilon) \\ &= \kappa + N + \tfrac{1}{2}n_2 + \tfrac{1}{3}n_3 + I(\mathbb{F}'_\varepsilon),\end{aligned} \tag{4.1.29}$$

where

$$\kappa + N = \operatorname{ord}(f, \infty, \Gamma(1)) \quad (0 \le \kappa < 1),$$
$$n_2 = \operatorname{ord}(f, i), \quad n_3 = \operatorname{ord}(f, \rho)$$

and

$$\begin{aligned}I(\mathbb{F}'_\varepsilon) &= \frac{1}{2\pi i}\int_{\partial\mathbb{F}'_\varepsilon} \frac{f'(z)}{f(z)}\,dz \\ &= \frac{1}{2\pi i}\int_{\partial\mathbb{F}'_\varepsilon} d\log f(z).\end{aligned}$$

We now compute the contributions to $I(\mathbb{F}'_\varepsilon)$ from the different sides. We require the following:

Lemma. *Let γ_1, γ_2 be the two smooth curves in $\mathbb{C}$ meeting at the origin O at an angle θ, where $0 < \theta \le 2\pi$, and let $\gamma(\varepsilon)$ be the arc of the circle $\{\tau : |\tau| = \varepsilon\}$ joining γ_1 to γ_2 (see fig. 8). Suppose that, for $0 < \varepsilon \le \varepsilon_0$, F is holomorphic and non-zero on the interior of the domain enclosed by γ_1, γ_2 and $\gamma(\varepsilon)$, and that, on this domain,*

$$F(\tau) = \tau^\alpha \sum_{m=0}^{\infty} c_n \tau^m \quad (c_0 \ne 0, \alpha \in \mathbb{R}).$$

Then

$$\lim_{\varepsilon \to 0+} \int_{\gamma(\varepsilon)} \frac{F'(\tau)}{F(\tau)}\,d\tau = \theta\alpha i.$$

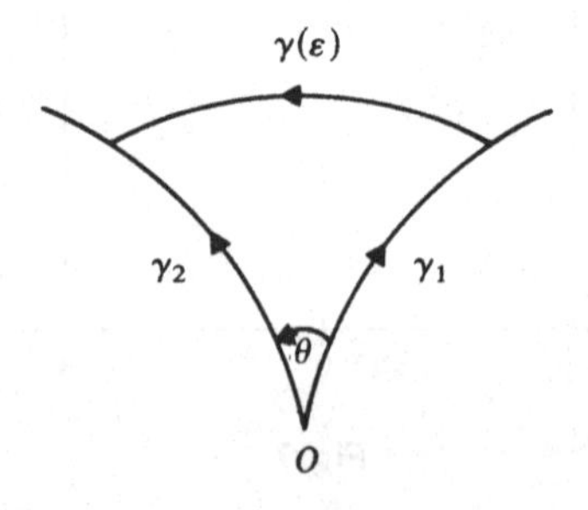

Fig. 8

Proof. This is immediate since, for small ε,

$$\frac{F'(\tau)}{F(\tau)}=\frac{\alpha}{\tau}+O(\varepsilon).$$

We apply this lemma to $f(z)$ on $\lambda_1, \lambda_2, \lambda_3, \lambda_4$ as follows:

$$\lambda_1:\quad \tau=e^{2\pi iz}, \alpha=\kappa+N, \theta=2\pi$$

$$\lambda_2:\quad \tau=z-\rho, \alpha=n_3, \theta=\tfrac{1}{3}\pi.$$

$$\lambda_4:\quad \tau=z-i, \alpha=n_2, \theta=\pi.$$

$$\lambda_5:\quad \tau=z+\rho^2, \alpha=n_3, \theta=\tfrac{1}{3}\pi.$$

The direction of integration is the reverse of that given in the lemma, so that

$$\frac{1}{2\pi i}\sum_{m=1}^{4}\int_{\lambda_m}\frac{f'(z)}{f(z)}\,dz=-(\kappa+N+\tfrac{1}{2}n_2+\tfrac{1}{3}n_3)+o(1) \qquad (4.1.30)$$

as $\varepsilon\to 0+$.

Further, since $f(Uz)=v(U)f(z)$ and $f(Vz)=v(V)z^k f(z)$, we have

$$\frac{f'(z+1)}{f(z+1)}=\frac{f'(z)}{f(z)},\quad \frac{f'(Vz)}{f(Vz)}=z^2\left\{\frac{f'(z)}{f(z)}+\frac{k}{z}\right\}.$$

Hence

$$\frac{1}{2\pi i}\int_{l'_U}\frac{f'(z)}{f(z)}\,dz+\frac{1}{2\pi i}\int_{Ul'_U}\frac{f'(z)}{f(z)}\,dz=0,$$

$$\frac{1}{2\pi i}\int_{l'_V}\frac{f'(z)}{f(z)}\,dz+\frac{1}{2\pi i}\int_{Vl'_V}\frac{f'(z)}{f(z)}\,dz=-\frac{k}{2\pi i}\int_{l'_V}\frac{dz}{z}$$

$$=-\frac{k}{2\pi}\{\text{ph}\, i-\text{ph}\,\rho\}+o(1)$$

$$=\frac{k}{12}+o(1).$$

From these results and (4.1.29, 30) we obtain

$$\text{ord}(f,\Gamma(1))=\frac{k}{12}+o(1),$$

from which the theorem follows, since the left-hand side does not depend on ε.

The theorem could have been proved directly for $\hat{\Gamma}$, without the use of theorem 4.1.3, by integrating round the boundary of a fundamental region $\mathbb{F}$ for $\hat{\Gamma}$ of the type constructed in §2.4.

4.2. Dimensions of spaces of modular forms. It is clear that the families $M(\Gamma, k, v)$, $H(\Gamma, k, v)$, $\{\Gamma, k, v\}$, $\{\Gamma, k, v\}_0$ are vector spaces over the field of complex numbers. For if S denotes any one of these families, and f and $g \in S$, then $\alpha f + \beta g \in S$ for all complex α, β. We prove in the next theorem that the last two families have finite dimension.

Let

$$\xi = \xi(\Gamma, v) := \kappa_2 + \kappa_3 + \kappa_\infty. \tag{4.2.1}$$

Here

$$\kappa_2 := \sum_{\mathbb{E}_2(\Gamma)} \varepsilon(\zeta), \quad \kappa_3 := \sum_{\mathbb{E}_3(\Gamma)} \varepsilon(\zeta) \tag{4.2.2}$$

and

$$\kappa_\infty := \sum_{L\infty \in \mathbb{P}} \kappa_L, \tag{4.2.3}$$

the summations being taken over sets of e_2, e_3 and λ incongruent vertices of $\mathbb{E}_2(\Gamma)$, $\mathbb{E}_3(\Gamma)$ and $\mathbb{P}$ (mod Γ), respectively; here $\varepsilon(\zeta)$ is the quantity defined in §4.1.

Theorem 4.2.1. *Let $\hat{\Gamma}$ have index μ in $\hat{\Gamma}(1)$. Then*

$$\dim\{\Gamma, k, v\} \leq \max(0, 1 - \xi + k\mu/12)$$

and

$$\dim\{\Gamma, k, v\}_0 \leq \max(0, 1 - \xi - \lambda' + k\mu/12),$$

where λ' is the number of incongruent cusps $L\infty$ (mod Γ) for which $\kappa_L = 0$.

Proof. Suppose that $\dim\{\Gamma, k, v\} = n \geq 1$. By taking a suitable linear combination of forms, we can find $f \in \{\Gamma, k, v\}$ such that $\operatorname{ord}(f, \infty, \Gamma) \geq \kappa_I + n - 1$. Since $\operatorname{ord}(f, \xi, \Gamma) \geq 0$ for all $\zeta \in \mathbb{H}'$, we obtain, by corollary 4.1.2 and theorem 4.1.4.

$$k\mu/12 \geq n - 1 + \kappa_\infty + \kappa_2 + \kappa_3.$$

For, at each cusp $L\infty$, $\operatorname{ord}(f, L\infty, \Gamma) \geq \kappa_L$. If $f \in \{\Gamma, k, v\}_0$, then $\operatorname{ord}(f, L\infty, \Gamma) \geq \kappa_L + 1$ when $\kappa_L = 0$, and from this the second part of the theorem follows.

Theorem 4.2.1 shows, in particular, that the classes $\{\Gamma, k, v\}$ and $\{\Gamma, k, v\}_0$ are trivial when $k<0$; i.e. they reduce to the zero form.

We consider the special case when $\Gamma=\Gamma(1)$, k is even and $v(T)=1$ for all $T\in\Gamma(1)$ in the following corollary. In this case $\nu(T,z)=\mu(T,z)=(T:z)^k$.

Corollary 4.2.1. (i) *For all even integers $k\geq 0$,*

$$\dim\{\Gamma(1),k,1\}\leq\begin{cases}\left[\dfrac{k}{12}\right]+1 & \text{if } k\not\equiv 2 \pmod{12},\\ \left[\dfrac{k}{12}\right] & \text{if } k\equiv 2 \pmod{12}.\end{cases}$$

In particular, $\{\Gamma(1),2,1\}$ is trivial, and $f\in\{\Gamma(1),0,1\}$ if and only if f is a constant.

(ii) *$\{\Gamma(1),k,1\}_0$ is trivial for all even $k<12$ and for $k=14$. For even $k\geq 12$,*

$$\dim\{\Gamma(1),k,1\}_0\leq\begin{cases}\left[\dfrac{k}{12}\right] & \text{if } k\not\equiv 2 \pmod{12},\\ \left[\dfrac{k}{12}\right]-1 & \text{if } k\equiv 2 \pmod{12}.\end{cases}$$

Proof. By (3.3.3), $\kappa_\infty=\kappa_I=0$, since $v(U)=1$. The values of s_L, m_L, κ_2 and κ_3 are given in table 6.

Table 6

$\frac{1}{2}k \pmod 6$	s_L	m_L	κ_2	κ_3	ξ	$\{k/12\}-\xi$
0	0	0	0	0	0	0
1	1	1	$\frac{1}{2}$	$\frac{2}{3}$	$\frac{7}{6}$	-1
2	0	2	0	$\frac{1}{3}$	$\frac{1}{3}$	0
3	1	0	$\frac{1}{2}$	0	$\frac{1}{2}$	0
4	0	1	0	$\frac{2}{3}$	$\frac{2}{3}$	0
5	1	2	$\frac{1}{2}$	$\frac{1}{3}$	$\frac{5}{6}$	0

In the last column $\{k/12\}$ denotes the fractional part of $k/12$. Accordingly, $k/12-\xi=[k/12]$ or $[k/12]-1$, and the stated inequalities follow from theorem 4.2.1. The statement about $\{\Gamma(1),0,1\}$ follows, since any constant belongs to $\{\Gamma(1),0,1\}$ and the dimension is at most 1, by the previous part.

We shall see later on that, in fact, the inequalities in the corollary can be replaced by equalities.

More generally, for any subgroup Γ, if k is even and $v(T)=1$ for all $T\in\Gamma$, then $\kappa_L=0$ for all $L\in\Gamma(1)$ by (3.3.3), so that $\kappa_\infty=0$. It follows from (4.1.20) and (4.2.2) that

$$\kappa_m=e_m\left\{\frac{k}{2}\left(1-\frac{1}{m}\right)\right\}\quad(m=2,3).\tag{4.2.4}$$

Theorem 4.2.1 then gives, for $k\geq 0$,

$$\left.\begin{aligned}&\dim\{\Gamma,k,1\}\leq\delta_k(\Gamma):=\frac{\mu k}{12}+1-e_2\left\{\frac{k}{4}\right\}-e_3\left\{\frac{k}{3}\right\},\\&\dim\{\Gamma,k,1\}_0\leq\delta_k(\Gamma)-\lambda,\end{aligned}\right\}\tag{4.2.5}$$

and from this we derive table 7.

The genus g of the group is given in the last column of table 7. It is possible to show† by means of the Riemann–Roch theorem that

$$\dim\{\Gamma,k,1\}=\delta_k(\Gamma)-g\quad(k>0,k\text{ even})\tag{4.2.6}$$

Thus the upper bound $\delta_k(\Gamma)$ is attained only when $g=0$.

For the remainder of the section we suppose that k is real, that Γ is a normal subgroup of $\Gamma(1)$ and that Δ is a normal subgroup of Γ with the property that Γ/Δ is abelian. We assume that $-I$ belongs to both Δ and Γ. We write $h:=[\Gamma:\Delta]$ and denote by $\mathfrak{X}(\Gamma,\Delta)$ the group of all characters χ on Γ that are constant on Δ; i.e.

$$\chi(ST)=\chi(S)\chi(T)\quad\text{for all }S,T\in\Gamma,$$

and $\chi(T)=1$ for all $T\in\Delta$ (in particular, $\chi(-I)=1$). Clearly $\mathfrak{X}(\Gamma,\Delta)$ has order h and is isomorphic to Γ/Δ.

Let u be a given MS on Γ of weight k. Then, as χ runs through $\mathfrak{X}(\Gamma,\Delta)$, it is clear that $v=u\chi$ runs through a complete set of multiplier systems v on Γ of weight k for which $v(T)=u(T)$ whenever $T\in\Delta$.

If M is a subspace of $M(\Delta,k,u)$, and $L\in\Gamma(1)$, we write

$$M|L:=\{f|L;f\in M\},$$

so that $M|L$ is a subspace of $M(L^{-1}\Delta L,k,u^L)$, by theorem 4.1.1(i). In particular, for $L\in\Gamma$, $u^L=u$, by §3.1, so that $M|L$ is also a

† See Gunning (1962), §8.

Table 7. k even

$\hat{\Gamma}$	μ	λ	e_2	e_3	ε_2	ε_3	$\delta_k(\Gamma)\ (k \geq 0)$	$\delta_k(\Gamma)-\lambda\ (k \geq 4)$	g
$\hat{\Gamma}(1)$	1	1	1	1	1	1	$\left[\frac{k}{4}\right]+\left[\frac{k}{3}\right]-\frac{k}{2}+1$	$\left[\frac{k}{4}\right]+\left[\frac{k}{3}\right]-\frac{k}{2}$	0
$\hat{\Gamma}^2$	2	1	0	2	1	2	$2\left[\frac{k}{3}\right]-\frac{k}{2}+1$	$2\left[\frac{k}{3}\right]-\frac{k}{2}$	0
$\hat{\Gamma}_U(2), \hat{\Gamma}_V(2), \hat{\Gamma}_W(2)$	3	2	1	0	2	1	$\left[\frac{k}{4}\right]+1$	$\left[\frac{k}{4}\right]-1$	0
$\hat{\Gamma}^3$	3	1	3	0	3	1	$3\left[\frac{k}{4}\right]-\frac{k}{2}+1$	$3\left[\frac{k}{4}\right]-\frac{k}{2}$	0
$\hat{\Gamma}(2)$	6	3	0	0	3	2	$\frac{k}{2}+1$	$\frac{k}{2}-2$	0
$\hat{\Gamma}'(1)$	6	1	0	0	3	2	$\frac{k}{2}+1$	$\frac{k}{2}$	1
$\hat{\Gamma}(3)$	12	4	0	0	6	4	$k+1$	$k-3$	0
$\hat{\Gamma}^{2\prime}$	18	3	0	0	9	6	$\frac{3}{2}k+1$	$\frac{3}{2}k-2$	1
$\hat{\Gamma}^{3\prime}$	24	4	0	0	12	8	$2k+1$	$2k-3$	1
$\hat{\Gamma}(6)$	72	12	0	0	36	24	$6k+1$	$6k-11$	1

subspace of $M(\Delta, k, u)$. We can therefore always extend M so that

$$M|L \subseteq M \quad \text{for all } L \in \Gamma. \tag{4.2.7}$$

We are now in a position to prove†

Theorem 4.2.2. *Let Δ and Γ be subgroups of finite index in $\Gamma(1)$ such that $-I \in \Delta$, Δ is normal in Γ, and Γ/Δ is abelian of order h. Suppose further that M is a subspace of $M(\Delta, k, u)$, where u is a MS of real weight k on Γ, and that (4.2.7) holds. Then M is the direct sum of h subspaces M_v, where $M_v \subseteq M(\Gamma, k, v)$, $v = u\chi$ and $\chi \in \mathfrak{X}(\Gamma, \Delta)$.*

Proof. For any $f \in M$ and $v = u\chi$, where $\chi \in \mathfrak{X}(\Gamma, \Delta)$, write

$$f_v = \frac{1}{h} \sum_{T \in \mathfrak{R}} \bar{v}(T)(f|T),$$

† Rankin (1967*a*), theorem 2.

where the bar denotes the conjugate complex number, and $\Gamma = \Delta \cdot \mathcal{R}$. For $S \in \Delta$, by theorem 4.1.1(ii)

$$\bar{v}(ST)f|ST = \frac{\sigma(S, T)}{v(ST)}(f|S)|T$$

$$= \frac{\sigma(S, T)u(S)}{\chi(ST)u(ST)} f|T$$

$$= \bar{v}(T)f|T,$$

by (3.1.14), since $\chi(S) = 1$. Hence f_v is independent of the choice of the right transversal $\mathcal{R}$. Also $f_v \in M$ since $M|T \subseteq M$. Moreover, for any $S \in \Gamma$,

$$f_v|S = \frac{1}{h} \sum_{T \in \mathcal{R}} \bar{v}(T)(f|T)|S$$

$$= \frac{1}{h} \sum_{T \in \mathcal{R}} \frac{\bar{v}(T)}{\sigma(T, S)} f|TS$$

$$= \frac{v(S)}{h} \sum_{T \in \mathcal{R}} \bar{v}(TS)f|TS$$

$$= v(S)f_v,$$

since $\Delta = \Gamma S = \Delta \cdot \mathcal{R}S$. Hence $f_v \in M(\Gamma, k, v)$.

For each $\chi \in \mathfrak{X}(\Gamma, \Delta)$ and $v = u\chi$, write

$$M_v = \{f_v : f \in M\}.$$

Then clearly M_v is a subspace of M and $M(\Gamma, k, v)$. Further, since

$$f = \sum_v f_v,$$

where the summation is over all $v = u\chi$ for $\chi \in \mathfrak{X}(\Gamma, \Delta)$, and since only the zero form can belong to different subspaces M_v, it follows that M is the direct sum of the h subspaces M_v.

This completes the proof of the theorem. We note that the theorem does not apply to every MS u of weight k on Δ, but only to a MS u that can be extended to a MS on the bigger group Γ.

Corollary 4.2.2. *If, in theorem 4.2.2, $M \subseteq \{\Delta, k, u\}$, then*

$$\dim M = \sum_{\chi \in \mathfrak{X}(\Gamma, \Delta)} \dim M_{u\chi}.$$

In particular

$$\dim\{\Delta, k, u\} = \sum_{\chi \in \mathfrak{X}(\Gamma, \Delta)} \dim\{\Gamma, k, u\chi\}$$

and

$$\dim\{\Delta, k, u\}_0 = \sum_{\chi \in \mathfrak{X}(\Gamma, \Delta)} \dim\{\Gamma, k, u\chi\}_0.$$

For, if $M = \{\Delta, k, u\}$ and $f \in \{\Gamma, k, v\}$ for some $v = u\chi$, then $f \in M$ and, since $f_v = f$, $\{\Gamma, k, v\} \subseteq M_v \subseteq \{\Gamma, k, v\}$. Thus $M_v = \{\Gamma, k, v\}$; similarly for $\{\Gamma, k, v\}_0$.

We can apply theorem 4.2.2 and its corollary to $\Gamma = \Gamma(1)$, $\Delta = \Gamma'(1)$. We deduce that

$$\dim\{\Gamma'(1), k, u\} = \sum_{m=0}^{5} \dim\{\Gamma(1), k, u\chi^m\}, \qquad (4.2.8)$$

where the character χ is defined by (3.4.5, 10). A similar result holds for the corresponding spaces of cusp forms. Applications can also be made with $\Gamma = \Gamma^2$ and Γ^3 and $\Delta = \Gamma^{2'}$ and $\Gamma^{3'}$, respectively.

We give more precise results about dimensions of certain spaces in chapter 6.

The proof of theorem 4.2.2 can also be applied to give

Theorem 4.2.3. *Let v be a MS of real weight k on a subgroup Γ of $\Gamma(1)$, and let Δ be a subgroup of index h in Γ. For any $f \in M(\Delta, k, v)$, write*

$$f_v = \frac{1}{h} \sum_{T \in \mathcal{R}} v(T)(f|T), \qquad (4.2.9)$$

where $\Gamma = \Delta \cdot \mathcal{R}$. Then f_v is independent of the transversal $\mathcal{R}$ and

$$f_v \in M(\Gamma, k, v).$$

If we write

$$\mathrm{Tr}(\Gamma, \Delta, v)f := f_v, \qquad (4.2.10)$$

then $\mathrm{Tr}(\Gamma, \Delta, v)$ is a linear operator from $M(\Delta, k, v)$ into $M(\Gamma, k, v)$ and is called a *trace operator*; see Petersson (1967*b*, 1973). In chapter 8 we shall make use of operators of this kind.

4.3. Relations between modular forms. We have seen in the preceding section that $M(\Gamma, k, v)$ is a vector space. We now investigate multiplication and division of modular forms. Let Γ be a subgroup of $\Gamma(1)$ of finite index and let v_1 and v_2 be multiplier systems on Γ of weights k_1 and k_2, respectively. Then it is clear that v_1v_2 and v_1/v_2 are multiplier systems on Γ of weights k_1+k_2 and k_1-k_2, respectively. We note also that v_1 is a MS on Γ of weight $2m+k$ for every integer m; for, since $w(S, T)$ is an integer for all $S, T \in \Gamma$, it follows from (3.2.1) that $\sigma(S, T)$ is unaltered when k is replaced by $k+2m$, while (3.1.13) shows that $v(-U) = e^{-\pi i(k+2m)}$.

Let $f_1 \in M(\Gamma, k_1, v_1)$ and $f_2 \in M(\Gamma, k_2, v_2)$. Then, clearly,

$$f_1f_2 \in M(\Gamma, k_1+k_2, v_1v_2) \quad \text{and} \quad f_1/f_2 \in M(\Gamma, k_1-k_2, v_1/v_2);$$

for all three conditions (I)–(III) of §4.1 are easily verified.

A modular form of weight zero is called a *modular function.* Suppose that $f \in M(\Gamma, 0, v)$. Then the MS v is a character of the group Γ and so takes the constant value 1 on the commutator subgroup Γ'. Since, for all $T \in \Gamma$,

$$f(Tz) = v(T)f(z) \quad (z \in \mathbb{H}),$$

it follows from (2.1.1) that

$$f'(Tz) = v(T)(T:z)^2 f'(z)$$

and also

$$\frac{f'(Tz)}{f(Tz)} = (T:z)^2 \frac{f'(z)}{f(z)}.$$

Since conditions (I)–(III) are easily verified, we have proved the first part of the following theorem.

Theorem 4.3.1. (i) *If $f \in M(\Gamma, 0, v)$ then $f' \in M(\Gamma, 2, v)$ and $f'/f \in M(\Gamma, 2, 1)$.*

(ii) *Suppose that v is an MS of dimension zero and that, for each $T \in \Gamma$, $v(T)$ is a root of unity.*† *Suppose also that $f \neq 0$ and that $f \in \{\Gamma, 0, v\}$. Then f is a non-zero constant, and $v(T) = 1$ for all $T \in \Gamma$.*

(iii) *If $g \in M(\Gamma, k, v)$, where $k \neq 0$, then*

$$h := kgg'' - (k+1)(g')^2 \in M(\Gamma, 2k+4, v^2).$$

† This holds in particular for $\Gamma = \Gamma(1)$, Γ^2 and Γ^3, since their commutator groups have finite index.

Also, if $g \neq 0$, *then* $h \neq 0$. *Further, if* $g \in \{\Gamma, k, v\}$, *then* $h \in \{\Gamma, 2k+4, v^2\}_0$.

Proof. (ii) If $f \in \{\Gamma, 0, v\}, f \neq 0$ and f is a constant, the relation $f(Tz) = v(T)f(z)$ shows that $v(T) = 1$ for all $T \in \Gamma$. Conversely, let $f \in \{\Gamma, 0, v\}$. Since Γ has a finite number of generators and since $v(ST) = v(S)v(T)$ for all $S, T \in \Gamma$, there exists an $n \in \mathbb{Z}^+$ such that $v^n(T) = 1$ for all $T \in \Gamma$. Hence $f^n \in \{\Gamma, 0, 1\}$. By corollary 4.2.1, f^n and therefore f is constant on $\mathbb{H}$.

Finally (iii), if $h = kgg'' - (k+1)(g')^2$, differentiation two times of the equation

$$g(Tz) = v(T)(T:z)^k g(z)$$

yields

$$h(Tz) = v^2(T)(T:z)^{2k+4} h(z).$$

It is easily shown that the function h can only vanish identically if $g(z) = A(z-B)^{-k}$ for some constants A and B; this is impossible when $A \neq 0$, since $k \neq 0$.

That h is a cusp form when g is entire is clear, since if the order of g (mod Γ) at a cusp is $\kappa + N \geq 0$ then the order of h there is at least $2\kappa + 2N$, when this is positive, and is at least 1 when $\kappa + N = 0$.

We now consider the particular case where the MS is constant and k is an even integer, so that every cusp parameter κ_L is zero. Further, if $k = 0, f \neq 0$ and $f \in M(\Gamma, 0, 1)$, it follows from theorem 4.1.2(ii)(iii) that $\operatorname{ord}(f, \zeta, \Gamma)$ is an integer for all $\zeta \in \mathbb{H}$; the same is true for $\zeta \in \mathbb{P}$ by the previous sentence. We define the *valence* of the modular function f to be

$$V(f, \Gamma) = \sum_{\zeta \in \mathbb{F}}{}' \operatorname{ord}(f, \zeta, \Gamma), \tag{4.3.1}$$

the summation being taken over all points ζ in a proper fundamental region at which $\operatorname{ord}(f, \zeta, \Gamma) > 0$. It follows from theorem 4.1.4 that the corresponding sum of $\zeta \in \mathbb{F}$ for which $\operatorname{ord}(f, \zeta, \Gamma) < 0$ is $-V(f, \Gamma)$.

Also $V(f, \Gamma)$ is a positive integer, when f is not a constant. For if f is not a constant and $V(f, \Gamma) = 0$, then $\operatorname{ord}(f, \zeta, \Gamma) = 0$ for all $\zeta \in \mathbb{H}'$. Hence, if $g(z) = f(z) - f(i)$, $g \in M(\Gamma, 0, 1)$ and $\operatorname{ord}(f, \zeta, \Gamma) \geq 0$ for all $\zeta \in \mathbb{H}'$. By theorem 4.1.4, $\operatorname{ord}(g, \zeta, \Gamma) = 0$ for all $\zeta \in \mathbb{H}'$. This is a contradiction, since $\operatorname{ord}(g, i, \Gamma) > 0$.

For any $n \in \mathbb{Z}$, $f^n \in M(\Gamma, 0, 1)(f \neq 0)$ and it follows that every rational function of f is also in $M(\Gamma, 0, 1)$. Thus $M(\Gamma, 0, 1)$ is a field, and this enables us to discuss algebraic as well as linear dependence of modular functions.

Theorem 4.3.2. *Let $f_1, f_2 \in M(\Gamma, 0, 1)$, neither being constant, and let their valences be q_1 and q_2, respectively. Then there exists an irreducible polynomial $\Phi(x_1, x_2)$ of degrees at most q_2 and q_1 in x_1 and x_2 respectively, and with complex coefficients not all of which are zero, such that $\Phi(f_1, f_2) = 0$ identically.*

Proof. We note that the polynomial Φ cannot have degree zero in either variable, or else f_1 or f_2 would reduce to a constant.

Take any positive integers m_1 and m_2 satisfying

$$(m_1+1)(m_2+1) \geq m_1 q_1 + m_2 q_2 + 2. \tag{4.3.2}$$

The general polynomial $\Phi(x_1, x_2)$ of degrees m_1 and m_2 in x_1 and x_2 contains $(m_1+1)(m_2+1)$ coefficients. For all choices of these coefficients $\Phi(f_1, f_2) \in M(\Gamma, 0, 1)$. By considering the poles of Φ we see that either

$$V(\Phi, \Gamma) \leq m_1 q_1 + m_2 q_2 \tag{4.3.3}$$

or $\Phi(f_1, f_2)$ is identically zero. Now we can choose the coefficients, not all zero, so that $\Phi(f_1, f_2)$ has zeros at $(m_1+1)(m_2+1)-1$ assigned incongruent points of $\mathbb{H} - \mathbb{E}$ at which f_1 and f_2 are holomorphic. Hence, if Φ is not identically zero,

$$0 = \operatorname{ord}(\Phi, \Gamma) \geq (m_1+1)(m_2+1) - 1 - V(\Phi, \Gamma) > 0,$$

which is a contradiction, by (4.3.2, 3). Thus $\Phi(f_1, f_2) = 0$; it is, of course, possible that Φ may be reducible, but we can then take an irreducible factor with the same property.

For each choice of m_1, m_2 satisfying (4.3.2) we get an irreducible polynomial Φ with $\Phi(f_1, f_2)$ identically zero. The inequality (4.3.2) shows that we cannot have $m_1 < q_2$, but we can take $m_1 = q_2$ if we choose $m_2 = (q_1 - 1)q_2 + 1$. Similarly, we cannot have $m_2 < q_1$, but we can take $m_2 = q_1$ if we choose $m_1 = (q_2 - 1)q_1 + 1$. Since there cannot exist two different irreducible polynomials with the property stated, we deduce that $\Phi(f_1, f_2)$ has degrees at most q_2 and q_1 in f_1 and f_2, respectively.

Theorem 4.3.2 is particularly useful when the group Γ has the genus zero, since it can then be shown that there exists a modular function having a single pole in its fundamental region; the following theorem then applies.

Theorem 4.3.3. *Suppose that* $f_1, f_2 \in M(\Gamma, 0, 1)$ *and that*

$$V(f_1, \Gamma) = 1, \quad V(f_2, \Gamma) = q \geq 1.$$

Then

$$f_2(z) = \frac{P(f_1)}{Q(f_1)},$$

where P and Q are polynomials of degrees not exceeding q.

Proof. Here $q_1 = 1$, $q_2 = q$, so that we may take Φ in the form

$$f_2 Q(f_1) - P(f_1) = 0;$$

this gives the desired result, since Q cannot vanish identically.

Corollary 4.3.3. *If* f_2 *has a pole of order q at the pole of* f_1*, then* $f_2 = P(f_1)$*, where P is a polynomial of degree q.*

For clearly Q cannot have positive degree.

Theorem 4.3.4. *Suppose that* $f \in M(\Gamma, 0, 1)$, $g \in M(\Gamma, k, 1)$, *where k is an even integer, and that* $V(f, \Gamma) = 1$. *Then*

$$g = (f')^{\frac{1}{2}k} P(f)/Q(f),$$

where P and Q are polynomials.

This follows from theorem 4.3.3 since $g(f')^{-\frac{1}{2}k} \in M(\Gamma, 0, 1)$.

As an application of theorem 4.3.3 we prove

Theorem 4.3.5. *Let* $\Delta \subseteq \Gamma \subseteq \Gamma(1)$*, where* Δ *is a normal subgroup of finite index in the group* Γ*, and suppose that* $\omega \in M(\Delta, 0, 1)$ *and that* $V(\omega, \Delta) = 1$. *Then, for each* $T \in \Gamma$, $\omega_T = \omega | T \in M(\Delta, 0, 1)$ *and there exists a finite group G of bilinear mappings isomorphic to* Γ/Δ *and such that*

$$\omega_T(z) = R_T\{\omega(z)\} \quad (T \in \Gamma, z \in \mathbb{H}), \tag{4.3.4}$$

where $R_T \in G$*; the coefficients of the bilinear functions need neither be real nor integral. Further,*

$$\omega'(Tz) = (T : z)^2 R'_T\{\omega(z)\}\omega'(z). \tag{4.3.5}$$

Proof. By theorem 4.1.1, $\omega_T \in M(\Delta, 0, 1)$ when $T \in \Gamma$, since Δ is normal in Γ. Also, by theorem 4.3.3, ω_T is expressible in the form (4.3.4), where R_T is a rational function, and R_T is clearly bilinear, since $V(\omega_T, \Delta) = V(\omega, \Delta) = 1$. Since $\omega_{ST} = \omega_T$ when $S \in \Delta$, R_T is defined on the cosets of Δ in Γ, and G is a homomorphic image of Γ/Δ.

This homomorphism is, in fact, an isomorphism. For otherwise, for some $T \in \Gamma$ with $T \notin \Delta$, $R_T(\omega) = \omega$. Let Γ' be the group generated by T and Δ, so that $[\Gamma':\Delta] =: \mu > 1$. But $\omega \in M(\Gamma', 0, 1)$ and

$$1 = V(\omega, \Delta) = \mu V(\omega, \Gamma'),$$

which is a contradiction.

The last part of the theorem follows since, for $T \in \Gamma$,

$$\omega'(z)R'_T\{\omega(z)\} = \frac{d}{dz}\omega(Tz) = (T:z)^{-2}\omega'(Tz).$$

Note that it is always possible to choose the bilinear mappings so that they have determinant unity, but this may be at the expense of some simplicity. We given an illustration of the theorem in §7.2 for the case $\Delta = \Gamma(2)$ and $\Gamma = \Gamma(1)$.

As an illustration of the kind of functional relationship that can hold between two different modular functions belonging to the same group we consider what are called transformations of order n, where n is a positive integer. Write

$$T = \begin{bmatrix} a & b \\ c & d \end{bmatrix},$$

where a, b, c, $d \in \mathbb{Z}$ and $\det T = ad - bc = n$. Then T is called a matrix of order n, and the corresponding bilinear mapping

$$w = \frac{az+b}{cz+d} = Tz$$

is called a *transformation of order n*. When h is an integer dividing a, b, c and d we may put

$$T = hT_1 = h\begin{bmatrix} a_1 & b_1 \\ c_1 & d_1 \end{bmatrix}.$$

Since the bilinear transformation associated with T_1 is the same as the mapping associated with T, we confirm our attention in this section to primitive matrices of order n, i.e. matrices T with

$$\det T = n, \quad (a, b, c, d) = 1.$$

The corresponding transformations are called *primitive transformations of order n.*

We write

$$\Omega_n^* = \left\{ T = \begin{bmatrix} a & b \\ c & d \end{bmatrix} : \det T = n; (a, b, c, d) = 1; a, b, c, d \in \mathbb{Z} \right\} \tag{4.3.6}$$

for the set of all primitive matrices of order n. We also write

$$\Omega^+ = \{T: \det T > 0; a, b, c, d \text{ rational}\}. \tag{4.3.7}$$

Then Ω^+ is a group of matrices containing $\Gamma(1)$ as a subgroup and Ω_n^* as a subset. Ω^+ is therefore a union of double cosets $\Gamma(1)T\Gamma(1)$ $(T \in \Omega^+)$. We are only interested in double cosets of this form for $T \in \Omega_n^*$, and, in particular, in the double coset $\Gamma(1)J_n\Gamma(1)$, where

$$J_n = \begin{bmatrix} 1 & 0 \\ 0 & n \end{bmatrix} \quad (n \in \mathbb{Z}^+). \tag{4.3.8}$$

Theorem 4.3.6. *For any* $n \in \mathbb{Z}^+$,

$$\Omega_n^* = \Gamma(1)J_n\Gamma(1) = \Gamma(1) \cdot \mathbf{T}_n. \tag{4.3.9}$$

The right transversal $\mathbf{T}_n$ *is a finite set of cardinality* $\psi(n)$, *and may be taken to be the set*

$$\mathbf{T}_n^* = \{T \in \Omega_n^*: c = 0, a > 0, ad = n, b \text{ modulo } d\}. \tag{4.3.10}$$

Here 'b modulo d' means that b runs through any complete set of residues modulo d satisfying the condition $(a, b, d) = 1$.

Similar results hold for left transversals of Ω_n^* *with respect to* $\Gamma(1)$, *the condition 'b modulo d' in (4.3.10) being replaced by 'b modulo a'.*

Proof. We call two matrices L_1 and L_2 in Ω_n^* left-equivalent if $L_2 = SL_1$ for some $S \in \Gamma(1)$. This is an equivalence relation, and our object is to partition Ω_n^* into corresponding equivalence classes $\Gamma(1)L$. Clearly L_2 is primitive when L_1 is. Right-equivalence is defined similarly.

If $L \in \Omega_n^*$, then $\Gamma(1)L\Gamma(1) \subseteq \Omega_n^*$, so that Ω_n^* is a union of double cosets in Ω^+ and therefore a union of right cosets, from which the existence of a right transversal $\boldsymbol{T}_n$ for which $\Omega_n^* = \Gamma(1) \cdot \boldsymbol{T}_n$ follows. We now show that $\boldsymbol{T}_n$ may be taken in the form (4.3.10).

Suppose that

$$L = \begin{bmatrix} A & B \\ C & D \end{bmatrix} \in \Omega_n^*$$

and write

$$a = (A, C), \quad A = \alpha a, \quad C = \gamma a,$$

so that $(\alpha, \gamma) = 1$; we can therefore find a matrix $S \in \Gamma(1)$ with these values of α and γ in its first column. Then

$$S^{-1}L = \begin{bmatrix} a & b \\ 0 & d \end{bmatrix}$$

for some integers b and d. Note that a is positive and that it is the highest common factor of the elements in the first column of the original matrix L.

Further, two matrices

$$T_i = \begin{bmatrix} a_i & b_i \\ 0 & d_i \end{bmatrix} \quad (i = 1, 2)$$

are left-equivalent if and only if $a_1 = a_2$, $d_1 = d_2$ and $b_1 \equiv b_2 \pmod{d_1}$. For every matrix left-equivalent to T_1 is of the form

$$ST_1 = \begin{bmatrix} \alpha a_1 & \alpha b_1 + \beta d_1 \\ \gamma a_1 & \gamma b_1 + \delta d_1 \end{bmatrix}$$

for some $S \in \Gamma(1)$. If $T_2 = ST_1$ we must have (since a_1 and a_2 are positive)

$$\gamma = 0, \quad \alpha = \delta = 1, \quad b_2 = b_1 + \beta d_1.$$

Conversely, if these relations hold, then $T_2 = U^\beta T_1$ and so T_1 and T_2 are left-equivalent.

This shows that the right transversal $\boldsymbol{T}_n$ may be chosen to be the set $\boldsymbol{T}_n^*$ of (4.3.10). Clearly $\boldsymbol{T}_n^*$ is a finite set and all other right transversals must have the same number of elements. The canonical form of the members of $\boldsymbol{T}_n^*$ given in (4.3.10) is known as Hermite's normal form; see MacDuffee (1946), theorem 22.1.

To prove that $|\boldsymbol{T}_n^*| = \psi(n)$ we put $h = (b, d)$, so that

$$|\boldsymbol{T}_n^*| = \sum_{d|n} \sum_{\substack{h|d \\ (h,\, n/d)=1}} \varphi(d/h),$$

where $\varphi(d/h)$ is Euler's function defined in §1.4, and where h and d are positive. This shows that

$$|\boldsymbol{T}_{mn}^*| = |\boldsymbol{T}_m^*|\,|\boldsymbol{T}_n^*| \quad \text{when } (m, n) = 1,$$

so that it suffices to take $n = p^r$, where p is prime and $r \in \mathbb{Z}^+$, and prove that $|\boldsymbol{T}_n^*| = \psi(p^r) = p^r + p^{r-1}$.

Now there are p^r and $\varphi(p^{r-s})$ matrices of the types

$$\begin{bmatrix} 1 & b \\ 0 & p^r \end{bmatrix} \quad \text{and} \quad \begin{bmatrix} p^s & b \\ 0 & p^{r-s} \end{bmatrix} \quad (1 \le s \le r),$$

respectively, and it is easily verified that

$$p^r + \sum_{s=1}^{r} \varphi(p^{r-s}) = p^r + p^{r-1}.$$

We shall see later on that it is not accidental that $|\boldsymbol{T}_n|$ is equal to the index of $\Gamma_0(n)$ in $\Gamma(1)$; in §9.1, where we consider a rather more general situation, we shall give a group-theoretic, rather than a number-theoretic, proof of this fact.

We now prove that $\Omega_n^* = \Gamma(1) J_n \Gamma(1)$. This is equivalent to reducing the members of Ω_n^* to Smith's normal form; see MacDuffee (1946), theorem 26.2.

We have already shown that any matrix $L \in \Omega_n^*$ is left-equivalent to one of the form

$$\begin{bmatrix} (A, C) & B_1 \\ 0 & D_1 \end{bmatrix}$$

Similarly, L is right-equivalent to a matrix of the form

$$\begin{bmatrix} (A, B) & 0 \\ C_2 & D_2 \end{bmatrix}.$$

Further, $(A, C) \le |A|$ and $(A, B) \le |A|$. This shows that, after a finite number of steps, as a result of alternate left and right multiplication by elements of $\Gamma(1)$, L can be transformed into a matrix

$$L^* = \begin{bmatrix} A^* & B^* \\ C^* & D^* \end{bmatrix}$$

in which A^* divides both B^* and C^*. Left and right multiplication by suitable powers of W and U then reduce L^* to the form

$$L' = \begin{bmatrix} A' & 0 \\ 0 & D' \end{bmatrix}.$$

We now apply the process again to $L'W$, which is left-equivalent to a matrix with first entry (A', D'). The process clearly terminates when we reach a matrix of the form

$$\begin{bmatrix} a & 0 \\ 0 & d \end{bmatrix}$$

in which $a = (a, d)$, and this happens after a finite number of steps. But, since a must divide d, we may take $a = 1$, since the matrix is primitive.

It is clear that similar results for right-equivalence can be proved by similar methods. This completes the proof of the theorem.

We now take Γ to be any subgroup of $\Gamma(1)$ of finite index and put

$$\Gamma(1) = \Gamma \cdot \boldsymbol{R}, \tag{4.3.11}$$

so that, by (4.3.9),

$$\Omega_n^* = \Gamma(1) J_n \Gamma(1) = \Gamma \cdot \boldsymbol{R} \cdot \boldsymbol{T}_n. \tag{4.3.12}$$

This double coset is a union of double cosets $\Gamma L \Gamma$, where $L \in \Omega_n^*$. We may therefore put, for any $L \in \Omega_n^*$,

$$\Gamma L \Gamma = \Gamma \cdot \boldsymbol{T}_n^L, \quad \text{where } \boldsymbol{T}_n^L \subseteq \boldsymbol{R}\boldsymbol{T}_n \tag{4.3.13}$$

and we also write

$$\Gamma_L := L^{-1} \Gamma L \cap \Gamma \quad (L \in \Omega_n^*). \tag{4.3.14}$$

Γ_L is called a *transformation group of order n*.

Theorem 4.3.7. *Let $L \in \Omega_n^*$ and let Γ be a subgroup of $\Gamma(1)$ of finite index μ. Then, for each $T \in \boldsymbol{T}_n^L$, the group Γ_T is a conjugate subgroup of Γ_L in Γ. Conversely, each subgroup of Γ conjugate to Γ_L is of the form Γ_T for some $T \in \boldsymbol{T}_n^L$. Moreover,*

$$\mu_L(\Gamma) := [\Gamma : \Gamma_L] \le \mu\psi(n). \tag{4.3.15}$$

In particular, when $\Gamma = \Gamma(1)$, Γ_L is conjugate to $\Gamma_0(n)$ for each $L \in \Omega_n^*$ and so $\mu_L(\Gamma(1)) = \psi(n)$.

Let $F \in M(\Gamma, 0, 1)$ and define F_T on $\mathbb{H}$ by

$$F_T(z) = F(Tz) \quad (T \in \Gamma L \Gamma). \tag{4.3.16}$$

Then $F_T \in M(\Gamma_T, 0, 1)$ and there exists an irreducible polynomial Φ_L such that

$$\Phi_L(F, F_T) = 0 \quad \textit{for all } T \in \boldsymbol{T}_n^L.$$

Proof. If $T \in \boldsymbol{T}_n^L$, then

$$T = S_1^{-1} L S_2 \quad \text{for some } S_1, S_2 \text{ in } \Gamma.$$

Hence, by (4.3.14),

$$\Gamma_T = S_2^{-1} L^{-1} \Gamma L S_2 \cap \Gamma = S_2^{-1} \Gamma_L S_2.$$

Conversely, if $S_2 \in \Gamma$, then $LS_2 = S_1 T$ for some $S_1 \in \Gamma$ and $T \in \boldsymbol{T}_n^L$, and so $S_2^{-1} \Gamma_L S_2 = \Gamma_T$.

If $\Gamma = \Gamma(1)$, the double coset $\Gamma L \Gamma$ is just Ω_n^* and, for $S \in \Gamma(1)$,

$$J_n S J_n^{-1} = \begin{bmatrix} \alpha & \beta/n \\ \gamma n & \delta \end{bmatrix}. \tag{4.3.17}$$

It follows that $\Gamma_{J_n} = \Gamma^0(n)$ and that the different transformation groups Γ_T, for $T \in \Omega_n^*$, are just the conjugates of $\Gamma^0(n)$ (and therefore of $\Gamma_0(n)$). Accordingly,

$$\mu_L(\Gamma(1)) = \psi(n).$$

We now return to the general case and prove (4.3.15). We use the fact that, if G_1 and G_2 are subgroups of finite index in a group G then

$$[G_1 : G_1 \cap G_2] \leq [G : G_2];$$

see Hall (1959), theorem 1.5.5. Write $\Gamma_L(1) := L^{-1}\Gamma(1)L \cap \Gamma(1)$. Then

$$[\Gamma : \Gamma \cap \Gamma_L(1)] \leq [\Gamma(1) : \Gamma_L(1)] = \psi(n)$$

and

$$\begin{aligned}[\Gamma \cap \Gamma_L(1) : \Gamma_L] &= [\Gamma \cap L^{-1}\Gamma(1)L : (\Gamma \cap L^{-1}\Gamma(1)L) \cap L^{-1}\Gamma L] \\ &\leq [L^{-1}\Gamma(1)L : L^{-1}\Gamma L] = [\Gamma(1) : \Gamma] = \mu.\end{aligned}$$

Hence

$$\mu_L(\Gamma) = [\Gamma : \Gamma_L] \leq \mu \psi(n).$$

Now take any $S \in \Gamma_T$ and write $S^* = TST^{-1}$, so that $S^* \in \Gamma$ and therefore

$$F_T(Sz) = F(TSz) = F(S^*Tz) = F(Tz) = F_T(z).$$

Thus $F_T \in M'(\Gamma_T, 0, 1)$, since F_T is a meromorphic function.

We must therefore consider the behaviour of $F_T(z)$ at the cusp $S_1\infty$, where $S_1 \in \Gamma(1)$. Choose $S_2 \in \Gamma(1)$ so that $TS_1\infty = S_2\infty$ and put

$$S_2^{-1}TS_1 = L = \begin{bmatrix} A & B \\ C & D \end{bmatrix}.$$

Then, since $L\infty = \infty$, $C = 0$ and $F_T(S_1 z)$, which is

$$F_{S_2}\{(Az + B)/D\},$$

has an expansion of the required form at $z = \infty$. Hence $F_T \in M(\Gamma_T, 0, 1)$.

The existence of the irreducible polynomial Φ_L for which

$$\Phi_L(F, F_L) = 0$$

follows from theorem 4.3.2. For any $T \in \boldsymbol{T}_n^L$ we take S_1 and S_2 as above and find that

$$0 = \Phi_L(F, F_L)|S_2 = \Phi_L(F|S_2, F_L|S_2) = \Phi_L(F, F_{LS_2})$$
$$= \Phi_L(F, F_{S_1T}) = \Phi_L(F, F_T).$$

The irreducible equation $\Phi_L(F, F_T) = 0$ is called a *modular* (or *transformation*) equation.

The following theorem enables us to obtain further information about the polynomial Φ_L.

Theorem 4.3.8. *Under the same assumptions as in theorem 4.3.7 we have, for* $L \in \Omega_n^*$ *and* $T \in \Gamma L \Gamma$,

$$V(F, \Gamma_T) = V(F_T, \Gamma_T) = \mu_L(\Gamma) V(F, \Gamma). \qquad (4.3.18)$$

It follows that the irreducible polynomial Φ_L *has degree not exceeding* $\mu_L(\Gamma)V(F, \Gamma)$ *in each of its two variables.*

Proof. Choose a finite constant C differing from the finite set of values of $F(z)$ and $F_T(z)$ at points $z \in \mathbb{E} \cup \mathbb{P}$, and put $G(z) = F(z) - C$. It follows that, when computing the valences $V(G, \Gamma)$,

$V(G, \Gamma_T)$ and $V(G_T, \Gamma_T)$, we need only add up the orders of G or G_T at its zeros in a fundamental region. We deduce at once that

$$V(G, \Gamma_T) = \mu_L(\Gamma) V(G, \Gamma) \quad \text{and} \quad V(G, \Gamma_T) = V(G_T, \Gamma_T),$$

since there is a one-to-one correspondence between zeros of G and G_T in a fundamental region for $\hat{\Gamma}_T$. For if $Tz_1 \equiv Tz_2 \pmod{\Gamma_T}$ then, for some $S \in \Gamma$, $Tz_1 = STz_2$ and so $z_1 = T^{-1}STz_2$. But $T^{-1}ST \in \Gamma_T$ and therefore $z_1 \equiv z_2 \pmod{\Gamma_T}$.

Equation (4.3.18) follows since F and G have the same valences for Γ and for Γ_T. The last part follows immediately from theorem 4.3.2.

We now consider the effect of transformations of order n on modular forms. For this purpose we need to extend the notation $T: z$, first defined in (2.2.1), from $T \in \Theta$ to any non-singular matrix

$$T = \begin{bmatrix} a & b \\ c & d \end{bmatrix}$$

with real entries. As before, we put

$$T: z = cz + d \tag{4.3.19}$$

and we easily verify that (2.2.2) continues to hold; i.e.

$$ST: z = (S: Tz)(T: z) \tag{4.3.20}$$

when S and T are non-singular and $z \neq \infty$, $z \neq T^{-1}\infty$.

Now let Γ be a subgroup of $\Gamma(1)$ of finite index, let v be a MS of weight $k \in \mathbb{R}$ on Γ and suppose that $f \in M(\Gamma, k, v)$. We extend the definition of the stroke operator $f|L$, given in §4.1, by putting

$$f_L(z) = f(z)|L = (\det L)^{\frac{1}{2}k} \{\mu(L, z)\}^{-1} f(Lz) \tag{4.3.21}$$

for any non-singular matrix L with real entries, where, as usual,

$$\mu(L, z) = (L: z)^k.$$

When $L \in \Gamma(1)$ this agrees with the previous notation. Observe that, if $L = hL_1$, where $h > 0$, then $f|L = f|L_1$. Accordingly, when L is a matrix of order $n \in \mathbb{Z}^+$ we may, without loss of generality, assume that L is primitive. Note also that (4.3.21) agrees with (4.3.16) in the particular case when $F = f$, $k = 0$ and $v = 1$.

Now suppose that $L \in \Omega_n^*$, where $n \in \mathbb{Z}^+$. Then we can show, exactly as in §3.1, that an AF ν^L and MS v^L may be defined as in

(3.1.17, 20) not only on Γ_L but on the larger group

$$\Gamma_L^* = L^{-1}\Gamma L \cap \Gamma(1). \tag{4.3.22}$$

In fact, they can be defined on $L^{-1}\Gamma L$, but this group need not be contained in $\Gamma(1)$. For this purpose we can apply §3.1 directly, with L replaced by $L_1 := n^{-\frac{1}{2}}L$, since then $L_1 \in \Omega$.

Moreover, for non-singular matrices S and T with real entries, we may define $\sigma(S, T)$ as in (3.1.15) and deduce, with the help of (4.3.20) that $|\sigma(S, T)| = 1$; $w(S, T)$ may also be defined as in §3.2 and has similar properties.

We now prove an extension of theorem 4.1.1.

Theorem 4.3.9. *Suppose that $f \in M(\Gamma, k, v)$ and that L, L_1 and L_2 belong to $\Omega^*(n)$, $\Omega^*(n_1)$ and $\Omega^*(n_2)$, respectively, where n, n_1 and n_2 are positive integers. Then we have:*

(i) $f_L \in M(\Gamma_L^*, k, v^L)$.

(ii) $f|L_1L_2 = \sigma(L_1, L_2)(f|L_1)|L_2$.

(iii) *If* $T \in \Gamma$, $f_{TL} = \sigma(T, L)v(T)f_L$ *and* $f_{-L} = e^{\pm\pi ik}f_L$.

(iv) $f_L/f \in M(\Gamma_L, 0, v^L/v)$.

(v) *If f is a holomorphic, entire or cusp form, so is f_L.*

Proof. (i) Any member of Γ_L^* can be expressed as $L^{-1}TL$ where $T \in \Gamma$. Then, as in the proof of theorem 4.1.1(i),

$$\begin{aligned} f_L(L^{-1}TLz) &= n^{\frac{1}{2}k}\{\mu(L, L^{-1}TLz)\}^{-1}f(TLz) \\ &= n^{\frac{1}{2}k}\{\mu(L, L^{-1}TLz)\}^{-1}\nu(T, Lz)f(Lz) \\ &= \nu^L(L^{-1}TL, z)f_L(z) \end{aligned}$$

so that $f_L \in M'(\Gamma_L^*, k, v^L)$.

Parts (ii) and (iii) are proved by arguments similar to those used in theorem 4.1.1. Note that L_1L_2 need not be primitive when L_1 and L_2 are.

We now have to examine the behaviour of f_L at cusps. Take any $s \in \Gamma(1)$ and consider the cusp $S\infty$. Then $LS\infty$ is a cusp and so $LS\infty = S_1\infty$ for some $S_1 \in \Gamma(1)$. Put $T = S_1^{-1}LS$ so that $T \in \Omega_n^*$ and

$T\infty=\infty$. Accordingly $c=0$, so that $\mu(T,z)=d^k$ and therefore

$$\begin{aligned} f_L(z)|S &= \{\sigma(L,S)\}^{-1} f_{LS}(z) \\ &= n^{\frac{1}{2}k}\{\mu(LS,z)\sigma(L,S)\}^{-1} f(LSz) \\ &= \frac{n^{\frac{1}{2}k}\mu(S_1,Tz)}{\mu(S_1,Tz)\sigma(L,S)} f_{S_1}(Tz) \\ &= \frac{n^{\frac{1}{2}k}\sigma(S_1,T)}{\mu(T,z)\sigma(L,S)} f_{S_1}(Tz) \\ &= \left(\frac{a}{d}\right)^{\frac{1}{2}k} \frac{\sigma(S_1,T)}{\sigma(L,S)} f_{S_1}\left(\frac{az+b}{d}\right). \end{aligned}$$

From this it is clear that f_L has an expansion of the required form at $S\infty$, and therefore $f \in M(\Gamma_L^*, k, v^L)$.

Part (v) now follows and part (iv) is obvious.

In theorem 4.3.7, 8 we have been concerned with a single double coset $\Gamma L \Gamma$ and this gave rise to a subset $\boldsymbol{T}_n^L$ of $\boldsymbol{RT}_n$; see (4.3.13). Each double coset $\Gamma L \Gamma$ in Ω_n^* determines a different subset of $\boldsymbol{RT}_n$, and these subsets are disjoint. There therefore exists a finite set $\boldsymbol{S}$ of elements of Ω_n^* such that

$$\boldsymbol{RT}_n = \bigcup_{L\in \boldsymbol{S}} \boldsymbol{T}_n^L \quad \text{(disjoint union)}, \tag{4.3.23}$$

and we may clearly take $\boldsymbol{S} \subseteq \boldsymbol{RT}_n$ if we wish. The number of elements of $\boldsymbol{S}$ is just the number of different double cosets $\Gamma L \Gamma$ in Ω_n^*. For each $L \in \boldsymbol{S}$ we derive an irreducible polynomial Φ_L satisfying the conditions of theorems 4.3.7, 8. These polynomials will, in general, be different.

We have seen that the transformation groups Γ_L have finite index in Γ. We observe also that, when Γ is a congruence group, so is Γ_L. For, if $\Gamma(q) \subseteq \Gamma$ for some $q \in \mathbb{Z}^+$ then, for any $L \in \Omega_n^*$, $L^{-1}\Gamma(nq)L \subseteq \Gamma(q)$. It follows that $\Gamma(nq) \subseteq \Gamma_L$.

Finally, take $\Gamma = \Gamma(1)$ and suppose that $F \in M(\Gamma(1), 0, 1)$. Then, as T runs through $\boldsymbol{T}_n$, we obtain $\psi(n)$ functions F_T, each satisfying the same modular equation

$$\Phi(F, F_T) = 0.$$

Since $\Gamma(n) \subseteq \Gamma_0(n)$, we note that $F_T \in M(\Gamma(n), 0, 1)$ for all $T \in \Omega_n^*$.

We shall consider explicit examples of this theory in §§6.5 and 7.2.

4.4. Modular forms of weight 2. We know from theorem 4.3.1(i) that the derivative of a modular function is a modular form of weight 2 but there may exist modular forms of weight 2 that are not derivatives of modular functions. In this section we prove a theorem about the residues of modular forms of weight 2 and restrict our attention to constant multiplier systems.

For any function f meromorphic at a point $\zeta \in \mathbb{C}$ we write $\operatorname{res}(f, \zeta)$ for the residue of f at ζ; $\operatorname{res}(f, \zeta)$ is the coefficient of $(z-\zeta)^{-1}$ in the Laurent series expansion of $f(z)$ about the point ζ.

Now suppose that $f \in M(\Gamma, 2, 1)$, where Γ is a subgroup of $\Gamma(1)$ of finite index and let $\zeta \in \mathbb{H}$. We define the residue of f at ζ (mod Γ) to be

$$\operatorname{res}(f, \zeta, \Gamma) := \frac{\operatorname{res}(f, \zeta)}{|\hat{\Gamma}_\zeta|}, \tag{4.4.1}$$

where Γ_ζ is the stabilizer of ζ (mod Γ); this definition is analogous to that of $\operatorname{ord}(f, \zeta, \Gamma)$ in (4.1.13).

Now suppose that $\zeta = L\infty$, where $L \in \Gamma(1)$; since the MS v is constant, $\kappa_L = 0$ and we have, by (4.1.6, 8, 9, 10),

$$f_L(z) = \sum_{m=N_L}^{\infty} a_m(f_L)\, e^{2\pi i m z/n_L},$$

where

$$n_L = n(\zeta, \Gamma) = [\hat{\Gamma}_\zeta(1) : \hat{\Gamma}_\zeta], \tag{4.4.2}$$

and where we have written $a_m(f_L)$ in place of $a_m(L)$. We define

$$\operatorname{res}(f, \zeta, \Gamma) := n_L a_0(f_L). \tag{4.4.3}$$

Observe that, by (4.1.5), $\operatorname{res}(f, \zeta, \Gamma)$ depends only on ζ and not on the choice of L.

Theorem 4.4.1. *Suppose that $f \in M(\Gamma, 2, 1)$ and that $L \in \Gamma(1)$. Then, for all $\xi \in \mathbb{H}'$,*

$$\operatorname{res}(f, L\zeta, \Gamma) = \operatorname{res}(f_L, \zeta, L^{-1}\Gamma L). \tag{4.4.4}$$

In particular, if $\zeta_1 \equiv \zeta_2$ (mod Γ), then

$$\operatorname{res}(f, \zeta_1, \Gamma) = \operatorname{res}(f, \zeta_2, \Gamma). \tag{4.4.5}$$

Proof. Let l be any piecewise differentiable arc in $\mathbb{H}$ with endpoints z_1 and z_2; in applications l will be a circular arc or a straight line segment. Put $w = Lz$ for $z \in l$. Then, as z goes from z_1 to z_2 along l, w goes from Lz_1 to Lz_2 along Ll. Since

$$f(w) = (L:z)^2 f_L(z), \quad \frac{dw}{dz} = (L:z)^{-2}, \tag{4.4.6}$$

it follows that

$$\int_{Ll} f(w)\,dw = \int_l f_L(z)\,dz, \tag{4.4.7}$$

when f is holomorphic on l. In particular, we deduce from Cauchy's theorem that

$$\operatorname{res}(f, L\zeta) = \operatorname{res}(f_L, \zeta) \tag{4.4.8}$$

for any $\zeta \in \mathbb{H}$.

Now, by (2.2.22),

$$L^{-1}\hat{\Gamma}_{L\zeta}L = (L^{-1}\hat{\Gamma}L)_\zeta,$$

so that $\hat{\Gamma}_{L\zeta}$ and $(L^{-1}\hat{\Gamma}L)_\zeta$ have the same order. From this and (4.4.1, 8) we obtain (4.4.4) in the case when $\zeta \in \mathbb{H}$.

If $\zeta = S\infty$, for $S \in \Gamma(1)$, we have, by (2.2.25) and theorem 4.1.1(i),

$$\begin{aligned}\operatorname{res}(f, L\zeta, \Gamma) &= n(LS\infty, \Gamma)a_0(f_{LS}) \\ &= n(S\infty, L^{-1}\Gamma L)a_0(f_L|S) \\ &= \operatorname{res}(f_L, \zeta, L^{-1}\Gamma L).\end{aligned}$$

This completes the proof of the theorem.

We are now in a position to state and prove

Theorem 4.4.2. *Let $f \in M(\Gamma, 2, 1)$ and let $\mathbb{F}$ be a proper fundamental region for $\hat{\Gamma}$. Then*

$$\sum_{\zeta \in \mathbb{F}} \operatorname{res}(f, \zeta, \Gamma) = 0. \tag{4.4.9}$$

Proof. We note that the summation on the left of (4.4.9) is over a finite number of points ζ at which f has singularities. By (4.4.5), the sum is independent of the choice of $\mathbb{F}$.

We assume, in the first place, that $\Gamma = \Gamma(1)$. In this case the theorem is easy to prove; we have merely to adapt the proof of theorem 4.1.4 in an obvious way. We consider the integral

$$\frac{1}{2\pi i}\int f(z)\,dz \tag{4.4.10}$$

taken around the boundary $\partial\mathbb{F}'_\varepsilon$ of a compact subset $\mathbb{F}'_\varepsilon$ of a fundamental region. Here $\mathbb{F}'_\varepsilon$ is chosen, as in theorem 4.1.4, so that the fixed points ∞, ρ, $-\rho^2$ and i are excluded by the straight line segment λ_1 and the indentations $\lambda_2, \lambda_3, \lambda_4$. Other indentations are made round poles of f that would otherwise be on the contour; see fig. 7. Then, for sufficiently small $\varepsilon > 0$, the integral (4.4.10) is equal to

$$\sum{}' \operatorname{res}(f, \zeta) = \sum{}' \operatorname{res}(f, \zeta, \Gamma(1)),$$

where the dash denotes that the summation is over all points ζ of $\mathbb{F}_I$ with the exception of ∞, ρ, $-\rho^2$ and i.

On the other hand, it follows from (4.4.7) and theorem 2.4.4(iv) that the integrals over the sides l_U, l'_U, l_V, l'_V contribute zero. From the remaining arcs λ_1, λ_2, λ_3 and λ_4 we obtain the contributions

$$\frac{1}{2\pi i}\int_{\lambda_1} f(z)\,dz = -\operatorname{res}(f, \infty, \Gamma(1)), \tag{4.4.11}$$

$$\frac{1}{2\pi i}\left\{\int_{\lambda_2} f(z)\,dz + \int_{\lambda_3} f(z)\,dz\right\} = -\operatorname{res}(f, \rho, \Gamma(1)), \tag{4.4.12}$$

and

$$\frac{1}{2\pi i}\int_{\lambda_4} f(z)\,dz = -\operatorname{res}(f, i, \Gamma(1)). \tag{4.4.13}$$

Since (4.4.11) is obvious, it suffices to prove (4.4.12); the proof of (4.4.13) is similar, but simpler. By (4.4.7),

$$\int_{\lambda_3} f(z)\,dz = \int_{U^{-1}\lambda_3} f(z)\,dz,$$

so that the left-hand side of (4.4.12) is expressible as

$$\frac{1}{2\pi i}\int_{\gamma} f(z)\,dz,$$

where γ consists of the arc $U^{-1}\lambda_3$ followed by λ_2. Let z_1 be the initial point of γ; then its final point is Pz_1. Further the arcs γ, $P\gamma$

and $P^2\gamma$ form a simple closed curve γ^* encircling ρ once in a clockwise direction. Hence, by (4.4.7),

$$\frac{1}{2\pi i}\int_\gamma f(z)\,dz = \frac{1}{6\pi i}\int_{\gamma^*} f(z)\,dz = -\tfrac{1}{3}\operatorname{res}(f,\rho)$$

$$= -\operatorname{res}(f,\rho,\Gamma(1)),$$

so that (4.4.12) holds.

Combining all these results, we obtain

$$\sum_{\zeta\in F_I} \operatorname{res}(f,\zeta,\Gamma(1)) = 0,$$

so that the theorem is true for $\Gamma(1)$.

We now return to the general case and suppose that $f\in M(\Gamma,2,1)$ and write

$$\hat\Gamma(1) = \hat\Gamma\cdot\mathcal{R}.$$

Define

$$F(z) := \sum_{L\in\mathcal{R}} f_L(z). \tag{4.4.14}$$

Clearly $F(z)$ is independent of the choice of the right transversal $\mathcal{R}$.

Take any $S\in\Gamma(1)$. By theorem 1.1.1, $\mathcal{R}S$ is also a right transversal for $\hat\Gamma$ so that

$$f(z)|S = \sum_{L\in\mathcal{R}} f_{LS}(z) = F(z).$$

It follows that $F\in M(\Gamma(1),2,1)$, and so

$$\sum_{\zeta\in F_I} \operatorname{res}(F,\zeta,\Gamma(1)) = 0. \tag{4.4.15}$$

Now take any $z\in\mathbb{H}$ and suppose that S generates the stabilizer $\hat\Gamma_\zeta$ of ζ (mod Γ). We apply theorem 1.1.2 with $\Gamma_1 = \hat\Gamma(1)$ and $\Gamma_2 = \hat\Gamma$. We may take

$$\mathcal{R} = \bigcup_{i=1}^m \mathcal{S}_i.$$

By (4.4.8),

$$\operatorname{res}(f_{L_iS^k},\zeta) = \operatorname{res}(f, L_i\zeta),$$

so that, by (4.4.14),

$$\operatorname{res}(F,\zeta) = \sum_{i=1}^m \sigma_i \operatorname{res}(f, L_i\zeta),$$

where

$$\sigma_i = [\hat{\Gamma}_{L_i\zeta}(1):\hat{\Gamma}_{L_i\zeta}] = [\hat{\Gamma}_\zeta(1):\hat{\Gamma}_{L_i\zeta}] \quad (1 \leq i \leq m),$$

by (1.1.14). On division by $|\hat{\Gamma}_\zeta(1)|$ we find that

$$\operatorname{res}(F, \zeta, \Gamma(1)) = \sum_{i=1}^{m} \operatorname{res}(f, L_i\zeta, \Gamma). \tag{4.4.16}$$

We obtain a similar result when $\zeta = \infty$. Take $S = U$ in theorem 1.1.2 and note that, by (2.2.25),

$$\sigma_i = n(L_i\infty, \Gamma)$$

so that

$$\operatorname{res}(F, \infty, \Gamma(1)) = a_0(F) = \sum_{i=1}^{m} \sigma_i a_0(f_{L_i})$$

$$= \sum_{i=1}^{m} \operatorname{res}(f, L_i\zeta, \Gamma).$$

It follows that (4.4.16) holds for each $\zeta \in \mathbb{F}_I$; the number m depends, of course, on the choice of ζ. Now each proper fundamental region $\mathbb{F}$ for $\hat{\Gamma}$ contains exactly m incongruent points $L_i\zeta$ $(1 \leq i \leq m)$ belonging to the orbit $[\zeta]$. From this and (4.4.15), we obtain (4.4.9).

This completes the proof of theorem 4.4.2. The theorem could have been proved in the general case by an extension of the method used for the case when $\Gamma = \Gamma(1)$. In principle this extension is straightforward, but a careful treatment of all the details is somewhat troublesome. It may be noted that the theorem is a consequence of a general theorem that asserts that the sum of the residues of a meromorphic differential (in this case $f(z)\,dz$) on a Riemann surface is zero; see, for example, Springer (1957), theorem 6-10.

We conclude by considering briefly the 'abelian integral'

$$F(z) = \int_{z_0}^{z} f(u)\,du \quad (z \in \mathbb{H}), \tag{4.4.17}$$

where $f \in H(\Gamma, 2, 1)$ and z_0 is a given point of $\mathbb{H}$. Since f is holomorphic on $\mathbb{H}$, $F(z)$ is independent of the path from z_0 to z. Write

$$\pi(T) = \pi_f(T, z_0) := F(Tz_0). \tag{4.4.18}$$

We call $\pi(T)$ a *period* of f. Then, for $T \in \hat{\Gamma}$,

$$F(Tz) = \int_{Tz_0}^{Tz} f(u)\,du + \int_{z_0}^{Tz_0} f(u)\,du$$
$$= F(z) + \pi(T) \tag{4.4.19}$$

for all $z \in \mathbb{H}$, by (4.4.7, 18).

It follows at once from (4.4.19) that

$$\pi(ST) = \pi(S) + \pi(T) \quad (S, T \in \hat{\Gamma}), \tag{4.4.20}$$

so that the map $\pi : \hat{\Gamma} \to \mathbb{C}$ is a homomorphism of $\hat{\Gamma}$ into the additive group of complex numbers. In particular, we deduce from (4.4.20) that

$$\pi(T) = 0 \quad \text{for all } T \in \hat{\Gamma}'.$$

We also deduce immediately from (4.4.20) that $\pi(T) = 0$ for every elliptic transformation $T \in \hat{\Gamma}$.

Now suppose that T is a parabolic member of $\hat{\Gamma}$ with fixed point $\zeta = L\infty$ $(L \in \Gamma(1))$. The stabilizer of ζ (mod Γ) is generated by

$$S = LU^n L^{-1}, \tag{4.4.21}$$

where $n = n_L = n(\zeta, \Gamma)$. Hence

$$T = S^m \quad \text{for some } m \in \mathbb{Z}. \tag{4.4.22}$$

We show that

$$\pi(T) = m\pi(S) = m \operatorname{res}(f, \zeta, \Gamma). \tag{4.4.23}$$

For, by (4.4.18, 7, 21),

$$\pi(S) = \int_{z_0}^{Sz_0} f(u)\,du = \int_{L^{-1}z_0}^{L^{-1}Sz_0} f_L(u)\,du$$
$$= \int_{L^{-1}z_0}^{n+L^{-1}z_0} f_L(u)\,du = na_0(f_L),$$

and (4.4.23) follows from this and (4.4.3, 22).

It is possible to evaluate $\pi(S)$ for a hyperbolic transformation S in terms of the constant term in the expansion of $(z-\zeta_1)(z-\zeta_2)f'(z)$ in powers of an appropriate variable; here ζ_1 and ζ_2 are the fixed points of S. See Hecke (1925), pp. 214–15.

We note, in conclusion that, if $f \in \{\Gamma, 2, 1\}_0$, it follows that $\pi(T) = 0$ for every elliptic and parabolic transformation $T \in \hat{\Gamma}$. If $\hat{\Gamma}$

is generated by elliptic and parabolic transformations, we can conclude that f must be the zero form. For then, by (4.4.19),

$$F(Tz)=F(z) \quad (T\in\hat{\Gamma}, z\in\mathbb{H})$$

and it is easily shown that F satisfies condition (III) of §4.1, since f is a cusp form. Further, $F\in M(\Gamma, 0, 1)$ and so, by Theorem 4.3.1(ii), F is a constant. It follows that $f=0$.

4.5. The order of magnitude of entire forms and of their Fourier coefficients. In this section we shall consider a given entire form f, and the constants C, C_1, etc. that we introduce will depend on f but not on the value of $z\in\mathbb{H}$ considered. We write $y=\operatorname{Im} z$ throughout.

Theorem 4.5.1. *Suppose that $f\in\{\Gamma, k, v\}$, where $k>0$. Then, for some positive constant C and all $z\in\mathbb{H}$,*

$$|f(z)|\le C \quad (y\ge 1), \quad |f(z)|\le Cy^{-k} \quad (0<y\le 1). \tag{4.5.1}$$

If, however, f is a cusp form then, for some positive constant C,

$$|f(z)|\le Cy^{-\frac{1}{2}k} \quad (y>0). \tag{4.5.2}$$

Proof. As pointed out in §4.1, there are essentially only λ different transforms f_T of f. It follows from (4.1.11) that there exists a constant C such that

$$|f_T(\zeta)|\le Ce^{-2\pi\eta(\kappa_T+N_T)/n_T} \tag{4.5.3}$$

for all $T\in\Gamma(1)$, and all $\zeta\in\mathbb{H}$ for which $\eta=\operatorname{Im}\zeta\ge\frac{1}{2}$. Here $\kappa_T+N_T=\operatorname{ord}(f, T\infty, \Gamma)$, as usual, and n_T is the width of the cusp $T\infty$. Let δ be the minimum of the λ different quantities $2\pi(\kappa_T+N_T)/n_T$, so that $\delta\ge 0$, and $\delta>0$ if $f\in\{\Gamma, k, v\}_0$.

If $y\ge\frac{1}{2}$, it follows that $|f(z)|\le C$, from (4.5.3) with $\zeta=z$ and $T=I$. If $0<y\le\frac{1}{2}$ we take $T\in\Gamma(1)$ so that $\zeta=T^{-1}z\in\mathbb{F}_I$. Then $c\ne 0$ and, by (4.5.3),

$$|f(z)|=|f(T\zeta)|=|T:\zeta|^k|f_T(\zeta)|\le C|cz-a|^{-k}\le Cy^{-k}.$$

From these results the inequalities (4.5.1) follow.

Finally, if $f\in\{\Gamma, k, v\}_0$, take $T\in\Gamma(1)$ so that $\zeta=T^{-1}z\in\mathbb{F}_I$, where z is any point of $\mathbb{H}$. Then

$$|y^{\frac{1}{2}k}f(z)|=\eta^{\frac{1}{2}k}|T:\zeta|^{-k}|f(T\zeta)|=\eta^{\frac{1}{2}k}|f_T(\zeta)|\le C\eta^{\frac{1}{2}k}e^{-\delta\eta}\le C_1,$$

since $\delta>0$ and so $\eta^{\frac{1}{2}k}e^{-\delta\eta}$ is bounded for all $\eta>0$.

Theorem 4.5.2. *Let $f \in \{\Gamma, k, v\}_0$ $(k>0)$, so that*

$$f(z) = e^{2\pi i\kappa z/n} \sum_{m=N}^{\infty} a_m e^{2\pi i m z/n} \quad (N+\kappa>0)$$

for all $z \in \mathbb{H}$. Then, for large m,

$$\text{(i) } a_m = O(m^{\frac{1}{2}k}), \quad \text{(ii) } \sum_{\mu=N}^{m} a_\mu = O(m^{\frac{1}{2}k} \log m),$$

$$\text{(iii) } \sum_{\mu=N}^{m} |a_\mu|^2 = O(m^k), \quad \text{(iv) } \sum_{\mu=N}^{m} |a_\mu| = O(m^{\frac{1}{2}(k+1)}).$$

Similar estimates hold for the Fourier coefficients of the transformed functions f_L.

Proof. Suppose that $m \geq 2$ and put $z = x + i/m$. Then, by uniform convergence,

$$a_m = \frac{1}{n} \int_0^n f(z)\, e^{-2\pi i(m+\kappa)z/n}\, dx,$$

and so, by theorem 4.5.1,

$$|a_m| \leq C m^{\frac{1}{2}k} e^{-2\pi(m+\kappa)/mn} \leq C_1 m^{\frac{1}{2}k},$$

which proves (i). Also

$$\sum_{\mu=N}^{m} a_\mu = \frac{1}{n} \int_0^n f(z)\, e^{-2\pi i\kappa z/n} \frac{e^{-2\pi i z N/n} - e^{-2\pi i(m+1)z/n}}{1 - e^{-2\pi i z/n}}\, dx.$$

Hence

$$\left| \sum_{\mu=N}^{m} a_\mu \right| \leq \frac{C}{n} m^{\frac{1}{2}k} \int_0^n \frac{2\, dx}{|1 - e^{-2\pi i z/n}|}.$$

Put $2\pi x = n\theta$, $\alpha = e^{2\pi y/n}$ and observe that

$$\begin{aligned} |1 - e^{-2\pi i z/n}|^2 &= |1 - \alpha\, e^{-i\theta}|^2 \\ &= (1+\alpha)^2 \sin^2(\tfrac{1}{2}\theta) + (1-\alpha)^2 \cos^2(\tfrac{1}{2}\theta) \\ &= : \psi(\theta). \end{aligned}$$

Hence

$$\frac{2}{n} \int_0^n \frac{dx}{1 - e^{-2\pi i z/n}} = \frac{1}{\pi} \int_0^{2\pi} \frac{d\theta}{\{\psi(\theta)\}^{\frac{1}{2}}}.$$

For large n, $\alpha-1$ is small and positive and less than $\frac{1}{2}\pi$. Hence the integral on the right is equal to

$$\frac{2}{\pi}\int_0^{\alpha-1}\frac{d\theta}{\{\psi(\theta)\}^{\frac{1}{2}}}+\frac{2}{\pi}\int_{\alpha-1}^{\pi}\frac{d\theta}{\{\psi(\theta)\}^{\frac{1}{2}}}$$
$$\leq\frac{2\sqrt{2}}{\pi}\int_0^{\alpha-1}\frac{d\theta}{\alpha-1}+\frac{2}{\pi}\int_{\alpha-1}^{\pi}\frac{d\theta}{(\alpha+1)\sin\frac{1}{2}\theta}$$
$$\leq\frac{2\sqrt{2}}{\pi}+\frac{2}{\alpha+1}\int_{\alpha-1}^{\pi}\frac{d\theta}{\theta}\leq\frac{2\sqrt{2}}{\pi}+2\log\pi-2\log(\alpha-1)$$
$$\leq C_2\log m,$$

since $\alpha=1+O(1/m)$. From this (ii) follows.

Also

$$\sum_{\mu=N}^{\infty}|a_\mu|^2 e^{-4\pi(\kappa+\mu)/(mn)}=\frac{1}{n}\int_0^n|f(z)|^2\,dx\leq Cm^k.$$

If $\mu\leq m$,

$$e^{-4\pi(\kappa+\mu)/(mn)}\geq e^{-4\pi(\kappa+m)/(mn)}\geq e^{-8\pi/n}=C_3>0,$$

so that

$$\sum_{\mu=N}^{m}|a_\mu|^2\geq(C/C_3)m^k,$$

and (iv) follows by Cauchy's inequality. Note that (i) follows immediately from (iii).

A similar argument to that of part (i) of theorem 4.5.2 gives

Theorem 4.5.3. *Let $f\in\{\Gamma,k,v\}$ $(k>0)$. Then, in the same notation as theorem 4.5.2,*

$$a_m=O(m^k).$$

4.6. Further results. Theorem 4.3.1(ii) states that an entire modular function is a constant, when each multiplier $v(T)$ is a root of unity. This latter restriction can be removed; see Knopp, Lehner and Newman (1965).

The modular form h defined in theorem 4.3.1(iii) is only one of the many forms of different weights that can be expressed as polynomials in the derivatives of a given form g. It is possible to define a sequence of polynomials $\psi_2 = h, \psi_3, \psi_4, \ldots$ with the property that, if P is a polynomial in g and its derivatives $g', g'', \ldots, g^{(n)}$, and is a modular form for Γ, then

$$P = g^t Q(\psi_2, \psi_3, \ldots, \psi_n)$$

where $t \in \mathbb{Z}$ and Q is a polynomial. See Rankin (1956, 1957). A basis of a different type has been given by Resnikoff (1966), and other variations are possible; see also Petersson (1967*a*).

We mention one further result of this type, which is due to Bol (1949) and has been used by Petersson (1950). Suppose that $k \in \mathbb{Z}$ and that

$$N := 1 - k > 0.$$

Then, if $h \in M(\Gamma, k, v)$, it follows that

$$h^{(N)} \in M(\Gamma, 2-k, v).$$

The argument given in the last paragraph of §4.4 can be used to show that a subgroup $\hat{\Gamma}$ of finite index in $\hat{\Gamma}(1)$ possesses a system of generators consisting entirely of parabolic and elliptic elements if and only if its genus is zero; see Lehner (1963).

The results in Theorem 4.5.2 are taken from Hecke (1937); see also Wilton (1929) and Walfisz (1933). These results are capable of considerable improvement. Let $f \in \{\Gamma, k, v\}_0$ $(k > 0)$ and put

$$f(x) = e^{2\pi i \kappa z/n} \sum_{m=N}^{\infty} a_m e^{2\pi i m z/n},$$

where n is the width of the cusp ∞ and κ is its parameter. When Γ is a congruence subgroup of $\Gamma(1)$, it can be shown that

$$\sum_{m \le x} |a_m|^2 = \alpha x^k + O(x^{k-(2/5)}) \tag{4.6.1}$$

for large x, where α is a certain positive constant. We shall see later on that $\{\Gamma, k, v\}_0$ is a Hilbert space. If the norm of f in this space is denoted by $\|f\|$, then α is proportional to $\|f\|^2$.

To prove (4.6.1) one considers the Dirichlet series

$$\sum_{m=N}^{\infty} \frac{|a_m|^2}{(m+\kappa)^s} = F(s),$$

say. If follows from theorem 4.5.2(iii) that the series is absolutely convergent for $\sigma = \operatorname{Re} s > k$, and so defines $F(s)$ as a holomorphic function of s in this half-plane. In fact, it can be shown that $F(s)$ can be continued analytically over the whole s-plane as a meromorphic function having simple poles at $s=k$ and $s=k-1$. $F(s)$ also satisfies a functional equation from which it can be deduced, by standard methods of analytical number theory, that (4.6.1) holds. In the case when $\kappa=0$ this is proved in Rankin (1939*b*); see also Selberg (1940) and, for the case $\Gamma=\Gamma(1)$, Hardy (1940) chapter 10. For the more general case the proof is similar; see Selberg (1965).

From (4.6.1) it follows immediately that

$$a_m = O(m^{k/2-1/5}), \tag{4.6.2}$$

and, by introducing a theory of vectorial modular forms, Selberg (1965) has shown that this remains true when Γ is not a congruence group, provided that it is of finite index in $\Gamma(1)$. If Γ is not a congruence group, (4.6.1) need not hold, as Selberg has shown.

The estimate (4.6.2) is an improvement on theorem 4.5.2(i). A still further improvement is possible when $k \geq 2$ and when the MS v is of a particular form. By means of an estimate of Weil (1948) for the order of magnitude of Kloosterman sums (see §§5.3, 5.8), it can be shown that

$$a_m = O(m^{\frac{1}{2}k-\frac{1}{4}+\varepsilon}) \tag{4.6.3}$$

for arbitrary $\varepsilon > 0$. In particular, (4.6.3) holds for $f \in \{\Gamma(1), k, 1\}_0$.

From (4.6.1) it is clear that the 'average' order of a_m is $O(m^{\frac{1}{2}(k-1)})$. The estimate

$$a_m = O(m^{\frac{1}{2}(k-1)+\varepsilon}) \tag{4.6.4}$$

is a weak form of what is known as the Ramanujan–Petersson conjecture; in its stronger form this conjecture relates to coefficients of cusp forms that are eigenforms of the Hecke operators studied in chapter 9. It has been shown by Deligne (1969, 1974) that this conjecture holds (in its stronger form) for functions belonging to $\{\Gamma(1), k, 1\}_0$. For $k \geq 2$ his method extends to the case when Γ is a congruence group; for $k=1$, see Deligne and Serre (1974). It may be noted that this conjecture was proved earlier, in the particular case when $k=2$, by Eichler (1954) and Igusa (1959).

We note that theorem 4.5.2(iii) can be improved to yield

$$\sum_{m \leq x} a_m = O(x^{k/2-1/10}),$$

when $f \in \{\Gamma(N), k, 1\}_0$, by making use of (4.6.1); see Rankin (1940). However, in view of the proof of the Ramanujan–Petersson conjecture, the right-hand side can now be replaced by $O(x^{k/2-1/6+\varepsilon})$, by using the method of Walfisz (1933).

Various other 'Ω-results' for coefficients of modular forms have been obtained; see Rankin (1973*a*) for references and Joris (1975).

An illuminating discussion of these and other related topics will be found in Selberg (1965).

5: The construction of modular forms

5.1. Poincaré series. Our object is to construct a modular form belonging to $M(\Gamma, k, v)$ as a sum of an infinite series. We recall that an infinite series $\sum u_n(z)$ is said to be *absolutely uniformly convergent* on a set $\mathbb{D}$ if and only if $\sum |u_n(z)|$ is uniformly convergent for $z \in \mathbb{D}$. When the terms of an absolutely uniformly convergent series are rearranged, the new series remains absolutely uniformly convergent and has the same sum.

Theorem 5.1.1. *Let A be a non-negative constant, k a real number greater than 2, and suppose that, for each pair of integers μ, ν with $\mu \neq 0$, a function $f_{\mu,\nu}$ is defined on $\mathbb{H}$ and that*

$$|f_{\mu,\nu}(z)| \leq \exp\{Ay/|\mu z + \nu|^2\} \tag{5.1.1}$$

for all $z \in \mathbb{H}$, where $y = \operatorname{Im} z$. Then the double series

$$\sum_{\substack{\mu=-\infty \\ \mu \neq 0}}^{\infty} \sum_{\nu=-\infty}^{\infty} \frac{f_{\mu,\nu}(z)}{|\mu z + \nu|^k} \tag{5.1.2}$$

is absolutely convergent for all $z \in \mathbb{H}$ and absolutely uniformly convergent on every compact subset of $\mathbb{H}$. Further, for every $\varepsilon > 0$, there exists a positive number B, depending only on A, k and ε, such that, if $F(z)$ is the sum of the series (5.1.2), *then*

$$|F(z)| \leq B\, e^{A/|z|}(|z|^{-k} + |z|^{-\frac{1}{2}k}) \tag{5.1.3}$$

for all $z \in \mathbb{A}_\varepsilon$, where

$$\mathbb{A}_\varepsilon := \{z \in \mathbb{H} : \varepsilon \leq \operatorname{ph} z \leq \pi - \varepsilon\}.$$

Proof. It suffices to show that

$$\sum_{\mu=1}^{\infty} \sum_{\nu=-\infty}^{\infty} \frac{\exp\{Ay|\mu z + \nu|^{-2}\}}{|\mu z + \nu|^k} \tag{5.1.4}$$

is uniformly convergent on every compact subset of $\mathbb{H}$ and to obtain an inequality of the form (5.1.3) for its sum $G(z)$ on $\mathbb{A}_\varepsilon$.

Write $z = re^{i\theta}$, where $r>0$ and $\theta = \text{ph}\, z$. Take any positive $\varepsilon < \frac{1}{2}\pi$ and suppose that $z \in \mathbb{A}_\varepsilon$, so that $\varepsilon \leq \theta \leq \pi - \varepsilon$. Then, for $\mu > 0$, $\nu \neq 0$,

$$\begin{aligned}|\mu z + \nu|^2 &= \mu^2 r^2 + 2\mu\nu r \cos\theta + \nu^2 \\ &= (\mu r + \nu)^2 \cos^2 \tfrac{1}{2}\theta + (\mu r - \nu)^2 \sin^2 \tfrac{1}{2}\theta \\ &\geq (\mu r + |\nu|)^2 \sin^2 \tfrac{1}{2}\varepsilon \geq 4\mu|\nu| r \sin^2 \tfrac{1}{2}\varepsilon.\end{aligned}$$

Hence, putting $C = 4\sin^2 \frac{1}{2}\varepsilon$ we have, since $y \leq r$,

$$\begin{aligned}\sum_{\mu=1}^{N} \sum_{\nu=-N}^{N} \frac{\exp\{Ay/|\mu z + \nu|^2\}}{|\mu z + \nu|^k} &\leq \frac{e^{A/r}}{r^k} \sum_{\mu=1}^{N} \mu^{-k} + 2 \sum_{\mu=1}^{N} \sum_{\nu=1}^{N} \frac{e^{A/C}}{(C\mu\nu r)^{\frac{1}{2}k}} \\ &\leq e^{A/r}\zeta(k) r^{-k} + 2e^{A/C}\zeta^2(\tfrac{1}{2}k)(Cr)^{-\frac{1}{2}k}\end{aligned}$$

for every integer $N \geq 1$, where

$$\zeta(s) = \sum_{n=1}^{\infty} n^{-s} \quad (\text{Re}\, s > 1).$$

From this the convergence of (5.1.4) on $\mathbb{A}_\varepsilon$ follows. Further, if

$$B = \max\{\zeta(k), 2\, e^{A/C}\zeta^2(\tfrac{1}{2}k)C^{-\frac{1}{2}k}\},$$

then

$$G(z) \leq B\, e^{A/|z|}\{|z|^{-k} + |z|^{-\frac{1}{2}k}\} \quad (z \in \mathbb{A}_\varepsilon).$$

Finally, if $\mathbb{B}$ is a compact subset of $\mathbb{H}$, there exists a positive $\varepsilon < \frac{1}{2}\pi$ such that

$$\mathbb{B} \subseteq \mathbb{B}_\varepsilon = \mathbb{A}_\varepsilon \cap \{z : y \geq \varepsilon\}.$$

It is enough to prove that (5.1.4) is uniformly convergent on $\mathbb{B}_\varepsilon$. This follows from Weierstrass's M-test by taking $r \geq \varepsilon$ in the above analysis.

Now suppose that Γ is a subgroup of finite index in $\Gamma(1)$, and let ν be an AF of weight k for Γ. We suppose, as usual, that $-I \in \Gamma$ and write v for the associated MS. Subject to the assumptions that $k > 2$ and that ν exists, we shall construct a form belonging to $H(\Gamma, k, v)$.

Let $\zeta = L^{-1}\infty$ be any point of $\mathbb{P}$, where $L \in \Gamma(1)$. For convenience put

$$M = L^{-1} \tag{5.1.5}$$

and write

$$n:=n(L^{-1}\infty,\Gamma)=n_M,\quad \kappa:=\kappa(L^{-1}\infty,\Gamma,v)=\kappa_M \tag{5.1.6}$$

for the width of the cusp ζ and its parameter, respectively. Then $L^{-1}U^nL\in\Gamma$ and, by (3.3.2.3),

$$e^{2\pi i\kappa}=\nu^M(U^n,z)=v(L^{-1}U^nL) \tag{5.1.7}$$

for any $z\in\mathbb{H}$. The positive integer n is the index of the stabilizer $\hat{\Gamma}_\zeta$ in $\hat{\Gamma}$, and we put

$$\hat{\Gamma}=\hat{\Gamma}_\zeta\cdot\mathcal{R}_L. \tag{5.1.8}$$

We now define, for any $m\in\mathbb{Z}$, $L\in\Gamma(1)$ and $k>2$, the *Poincaré series*

$$G_L(z;m;\Gamma,k,v)=G_L(z,m):=\sum_{T\in\mathcal{R}_L}\frac{\exp\left\{\dfrac{2\pi i(m+\kappa)}{n}LTz\right\}}{\mu(L,Tz)\nu(T,z)}, \tag{5.1.9}$$

where $\mu(L,Tz)$ is defined by (3.1.8). We prove the following theorem.

Theorem 5.1.2. *The series (5.1.9) defines $G_L(z,m)$ as a holomorphic function on $\mathbb{H}$, when $k>2$. The series is absolutely convergent on $\mathbb{H}$ and absolutely uniformly convergent on every compact subset of $\mathbb{H}$. Its sum $G_L(z,m)$ does not depend upon the choice of the transversal $\mathcal{R}_L$, and $G_L(z,m)\in H(\Gamma,k,v)$. More generally, for any $S\in\Gamma(1)$,*

$$G_L(z;m;\Gamma,k,v)|S=\{\sigma(L,S)\}^{-1}G_{LS}(z;m;\Gamma^S,k,v^S). \tag{5.1.10}$$

If (a) $m+\kappa>0$, then $G_L\in\{\Gamma,k,v\}_0$ and may vanish identically; here κ is defined by (5.1.6). If (b) $m+\kappa=0$ (so that $m=\kappa=0$), $G_L\in\{\Gamma,k,v\}$ and is called an Eisenstein series; it does not vanish identically and $\operatorname{ord}(G_L.\,L^{-1}\infty,\Gamma)=0$. *If (c) $m+\kappa<0$ (so that $m\le-1$), G_L does not vanish identically and* $\operatorname{ord}(G_L,L^{-1}\infty,\Gamma)=m+\kappa$. *In both cases (b) and (c)* $\operatorname{ord}(G_L,\zeta,\Gamma)>0$ *at every cusp* $\zeta\not\equiv L^{-1}\infty \pmod{\Gamma}$.

Proof. We divide the proof into three parts. We first examine the convergence of the series (5.1.9). Next we show that the definition of $G_L(z,m)$ does not depend upon the particular choice of $\mathcal{R}_L$, and finally that G_L is a modular form with the required properties.

We note that there is at most one term in the series (5.1.9) for which the matrix LT has a given second row. For if $LT' = U^s LT$ for $T, T' \in \mathcal{R}_L$ and $s \in \mathbb{Z}$ then $T'T^{-1} = L^{-1}U^sL$; since this belongs to $\hat{\Gamma}$, it must belong to $\hat{\Gamma}_\zeta$ and this, by the definition of $\mathcal{R}_L$, implies that $T' = T$.

In particular, there is at most one term in (5.1.9) for which $LT \in \hat{\Gamma}_U$. If this term, when it exists, is removed, the remaining series is of the form (5.1.2), since

$$\left|\frac{\exp\left\{\dfrac{2\pi i(m+\kappa)}{n}LTz\right\}}{\mu(L, Tz)\nu(T, z)}\right| = \frac{\exp\left\{\dfrac{-2\pi(m+\kappa)y}{n|LT:z|^2}\right\}}{|LT:z|^k}$$

and $LT: z = \mu z + \nu$, where $\mu \neq 0$. Thus $f_{\mu,\nu}(z) = 0$ for $(\mu, \nu) \neq 1$ and (5.1.1) holds with

$$A = \max\{0, -2\pi(m+\kappa)/n\}.$$

We therefore conclude that the series (5.1.9) is absolutely convergent on $\mathbb{H}$, and therefore its terms may be taken in any order. Further, it is uniformly convergent on every compact subset of $\mathbb{H}$; since each term of the series is holomorphic on $\mathbb{H}$, the sum $G_L(z, m)$ of the series is therefore holomorphic on $\mathbb{H}$.

To show that the definition of G_L does not depend upon the particular choice of $\mathcal{R}_L$ it is enough to show that, in any term, T may be replaced by RT, where $R = L^{-1}U^nL$. Now

$$\exp\left\{\frac{2\pi i(m+\kappa)}{n}LRTz\right\} = \exp\left\{\frac{2\pi i(m+\kappa)}{n}U^nLTz\right\}$$
$$= e^{2\pi i\kappa}\exp\left\{\frac{2\pi i(m+\kappa)}{n}LTz\right\}.$$

Further

$$\mu(L, RTz)\nu(RT, z) = v(RT)\mu(L, RTz)\mu(RT, z) \quad \text{by (3.1.10)},$$
$$= v(R)v(T)\sigma(L, RT)\sigma(R, T)\mu(U^nLT, z) \quad \text{by (3.1.14, 15)},$$
$$= v(R)v(T)\sigma(L, T)\mu(LT, z) \quad \text{by (3.2.23)},$$

and similarly,

$$\mu(L, Tz)\nu(T, z) = v(T)\sigma(L, T)\mu(LT, z).$$

The required result follows since $v(R)=e^{2\pi i\kappa}$ by (5.1.7). We note also that each term on the right of (5.1.9) is unaltered when T is replaced by $-T$.

We have seen that $LT\in\hat{\Gamma}_U$ for at most one term of the series. If $LT=U^s$ for some $T\in\mathcal{R}_L$ and $s\in\mathbb{Z}$, then the corresponding term in the series is

$$\frac{\exp\{2\pi i(m+\kappa)(z+s)/n\}}{\mu(L,L^{-1}U^sz)\nu(L^{-1}U^s,z)}=:\delta_L\exp\{2\pi i(m+\kappa)z/n\}.$$

It follows from this and (3.1.15) and (3.2.15) that

$$\delta_L=\delta_L(\Gamma,m,v)=\frac{\exp\{2\pi is(m+\kappa)/n\}}{v(L^{-1}U^s)\sigma(L,L^{-1})},\tag{5.1.11}$$

and this can be further simplified by using (3.2.19); we note that $|\delta_L|=1$. We now define δ_L to be zero, when $LT\notin\hat{\Gamma}_U$ for all $T\in\hat{\Gamma}$. Then δ_L is defined for all $L\in\Gamma(1)$ and does not depend upon the choice of $\mathcal{R}_L$ as the earlier analysis shows. Also, by (5.1.3),

$$|G_L(m;z;\Gamma,k,v)-\delta_L\exp\{2\pi i(m+\kappa)z/n\}|\\ \leq Be^{A/|z|}\{|z|^{-k}+|z|^{-\frac{1}{2}k}\}\tag{5.1.12}$$

for all $z\in\mathbb{H}$ for which $0<\varepsilon\leq\text{ph}\,z\leq\pi-\varepsilon$, where B is a non-negative number depending on ε, k and m. We shall need this inequality when we examine the behaviour of G_L at points of $\mathbb{P}$.

Now let $S\in\Gamma(1)$ and consider the transform

$$G_L(z;m;\Gamma,k,v)|S=\{\mu(S,z)\}^{-1}G_L(Sz;m;\Gamma,k,v)$$
$$=\sum_{T\in\mathcal{R}_L}\frac{\exp\left\{\frac{2\pi i}{n}(m+\kappa)LTSz\right\}}{\mu(S,z)\mu(L,TSz)\nu(T,Sz)}.\tag{5.1.13}$$

By (5.1.8) and (2.2.22),

$$\hat{\Gamma}^S:=S^{-1}\Gamma S=S^{-1}\hat{\Gamma}_\zeta S\cdot\mathcal{R}'_{LS}=\hat{\Gamma}^S_{\zeta'}\cdot\mathcal{R}'_{LS},\tag{5.1.14}$$

where $\zeta':=S^{-1}\zeta=S^{-1}L^{-1}\infty$ and $\mathcal{R}'_{LS}:=S^{-1}\mathcal{R}_LS$. As T runs through $\mathcal{R}_L$, $T'=S^{-1}TS$ runs through $\mathcal{R}'_{LS}$. Note that $TS=ST'$. Accordingly, we have from (5.1.13), writing T in place of T',

$$G_L(z;m;\Gamma,k,v)|S=\sum_{T\in\mathcal{R}'_{LS}}\frac{\exp\left\{\frac{2\pi i}{n}(m+\kappa)LSTz\right\}}{\mu(S,z)\mu(L,STz)\nu(STS^{-1},Sz)}.\tag{5.1.15}$$

Note that, since $\zeta' = S^{-1}L^{-1}\infty$, $n(\zeta', \Gamma^S) = n$, by (2.2.25). Also, if

$$\kappa' := \kappa(\zeta', \Gamma^S, v^S) = \kappa(S^{-1}M\infty, \Gamma^S, v^S),$$

then

$$e^{2\pi i\kappa'} = v^{S.S^{-1}M}(U^n, z) = v^M(U^n, z) = e^{2\pi i\kappa},$$

by (3.3.2), (3.1.18) and (5.1.7), so that $\kappa' = \kappa$. Further, by (3.1.17, 15),

$$\frac{\mu(S, z)\mu(L, STz)v(STS^{-1}, Sz)}{\mu(LS, Tz)v^S(T, z)} = \frac{\mu(L, STz)\mu(S, Tz)}{\mu(LS, Tz)} = \sigma(L, S)$$

and hence

$$\begin{aligned} G_L(z; m; \Gamma, k, v)|S &= \{\sigma(L, S)\}^{-1} \sum_{T \in \mathcal{R}'_{LS}} \frac{\exp\left\{\frac{2\pi i}{n}(m+\kappa)LSTz\right\}}{\mu(LS, Tz)v^S(T, z)} \\ &= \{\sigma(L, S)\}^{-1} G_{LS}(z; m; \Gamma^S, k, v^S), \end{aligned}$$

which is (5.1.10).

In particular, when $S \in \Gamma$ we deduce from (3.1.5) that

$$\begin{aligned} \mu(S, z)\mu(L, STz)v(STS^{-1}, Sz) &= \mu(L, STz)v(ST, z)\mu(S, z)/v(S, z) \\ &= \bar{v}(S)\mu(L, STz)v(ST, z). \end{aligned}$$

It follows from (5.1.5) that, when $S \in \Gamma$,

$$\begin{aligned} G_L(z; m; \Gamma, k, v)|S &= v(S) \sum_{T \in \mathcal{R}'_{LS}} \frac{\exp\left\{\frac{2\pi i}{n}(m+\kappa)LSTz\right\}}{\mu(L, STz)v(ST, z)} \\ &= v(S)G_L(z; m; \Gamma, k, v), \end{aligned}$$

since $\hat{\Gamma} = \hat{\Gamma}^S_{\zeta'} \cdot \mathcal{R}'_{LS} = \hat{\Gamma}_\zeta \cdot \mathcal{R}'_{LS}$.

Thus G_L is an unrestricted modular form. To complete the proof that $G_L \in M(\Gamma, k, v)$ we examine the behaviour of $G_L|S$ at ∞. By (5.1.10) it follows that we must consider $G_{LS}(z; m; \Gamma^S, k, v^S)$ in the neighbourhood of ∞. As usual, we write

$$n_S = n(S\infty, \Gamma) \quad \text{and} \quad \kappa_S = \kappa(S\infty, \Gamma, v).$$

Let $\delta'_{LS} = \delta_{LS}(\Gamma^S, m, v^S)$; see (5.1.11). Then $\delta'_{LS} = 0$ except when $LST \in \hat{\Gamma}_U$ for some $T \in \hat{\Gamma}^S$ and then $|\delta'_{LS}| = 1$. Hence $\delta'_{LS} \neq 0$ if and only if $LTS \in \hat{\Gamma}_U$ for some $T \in \hat{\Gamma}$, i.e. if and only if $S\infty \equiv L^{-1}\infty$ (mod Γ). This shows that, when $\delta'_{LS} \neq 0$, $n_S = n = n(\zeta', \Gamma^S)$ and

$\kappa_S = \kappa = \kappa'$. However, when $\delta'_{LS} = 0$, n_S need not equal n nor need we have $\kappa_S = \kappa$.

By theorem 4.1.1(iv) (or (4.1.6, 7)), $G_{LS}(z; m) \exp(-2\pi i \kappa_S z/n_S)$ is periodic with period n_S for any $S \in T(1)$. When examining the behaviour of $G_{LS}(z; m)$ at ∞ we may therefore assume that $0 \geq \operatorname{Re} z \leq n_S$ and that $|z| \geq 1$. We can then choose $\varepsilon > 0$ so that $\varepsilon \leq \operatorname{ph} z \leq \pi - \varepsilon$ for these values of z. We then have, by (5.1.12),

$$|G_{LS}(m; z; \Gamma^S, k, v^S) - \delta'_{LS} \exp\{2\pi i(m+\kappa_S)z/n_S\}| \leq B_\varepsilon |z|^{-\frac{1}{2}k}. \tag{5.1.16}$$

Put $t = e^{2\pi i z/n_S}$, $t^{\kappa_S} = e^{2\pi i \kappa_S z/n_S}$. Then $G_L(z; m)t^{-\kappa_S}$, and therefore $G_L(z; m)t^{-\kappa_S} - \delta'_{LS} t^m$, is expressible as a Laurent series in powers of t for $0 < |t| < 1$. Since $t \to 0$ as $|z| \to \infty$ it follows from (5.1.16) that

$$G_{LS}(z; m; \Gamma^S, k, v^S) = t^{\kappa_S}\left\{\delta'_{LS} t^m + \sum_{j=0}^{\infty} g_j t^j\right\}, \tag{5.1.17}$$

where $g_0 = 0$ if $\kappa_S = 0$. From this we deduce that $G_L \in M(\Gamma, k, v)$, and in fact $G_L \in H(\Gamma, k, v)$. Also we obtain the following information about $\operatorname{ord}(G_L, \zeta, \Gamma)$ for any cusp $\zeta = S\infty$:

$$\operatorname{ord}(G_L, \zeta, \Gamma) \begin{cases} = m + \kappa_S, & \text{when } m + \kappa_S \leq 0 \text{ and } \zeta \equiv L^{-1}\infty \pmod{\Gamma}, \\ \geq 1 - \{1 - \kappa_S\}, & \text{when } m + \kappa_S > 0 \text{ and } \zeta \equiv L^{-1}\infty \pmod{\Gamma}, \\ \geq 1 - \{1 - \kappa_S\}, & \text{when } m \in \mathbb{Z} \text{ and } \zeta \not\equiv L^{-1}\infty \pmod{\Gamma}. \end{cases} \tag{5.1.18}$$

Here $\{1 - \kappa_S\}$ is the fractional part of $1 - \kappa_S$; see the footnote on p. 94.

From this all the results of the theorem follow.

Corollary 5.1.2. *Under the conditions of theorem 5.1.2, the Poincaré series is absolutely uniformly convergent on* $\mathbb{F}_I$.

Proof. $\mathbb{F}_I \subseteq \mathbb{B}_\varepsilon$ (see the proof of theorem 5.1.1) for some $\varepsilon > 0$.

We note that, if we put $L = I$ in (5.1.11), we may take $s = 0$ and get

$$\delta_I = \delta_I(\Gamma, m, v) = 1. \tag{5.1.19}$$

If $L \in \Gamma(1)$, we say that L is a *regular* matrix if

$$L \neq -U^r \quad \text{for all } r \in \mathbb{Z}. \tag{5.1.20}$$

It follows that, if $L \in \Gamma(1)$, then either L or $-L$ is regular. The reason for this definition is that, when L is regular, it follows from (3.2.19) that $\sigma(L, L^{-1}) = 1$, and so the formula (5.1.11) gives a simple expression for δ_L when $\delta_L \neq 0$.

We now prove a decomposition theorem, which is due in its simplest form to Hecke (1927); see also Petersson (1932*b*). The theorem asserts that a holomorphic modular form can be expressed as the sum of a cusp form and a linear combination of Poincaré series $G_L(z; m)$ $(m \leq 0)$.

Theorem 5.1.3. *Let $f \in H(\Gamma, k, v)$, where $k > 2$ and let $\mathfrak{S}$ be a set of $\lambda = \lambda[\Gamma]$ regular matrices such that the λ cusps $L\infty$ $(L \in \mathfrak{S})$ are incongruent modulo Γ. Further, suppose that, for each $L \in \mathfrak{S}$,*

$$f_L(z) = e^{2\pi i \kappa_L z/n_L} \sum_{m=-\infty}^{\infty} a_m(L)\, e^{2\pi i m z/n_L} \quad (z \in \mathbb{H});$$

only a finite number of the coefficients $a_m(L)$ for $m \leq 0$ are, of course, non-zero. Let

$$H(z) := f(z) - \sum_{S \in \mathfrak{S}} \sum_{m+\kappa_S \leq 0} a_m(S) G_{S^{-1}}(z; m; \Gamma, k, v). \tag{5.1.21}$$

Then $H \in \{\Gamma, k, v\}_0$. In particular, if $f \in \{\Gamma, k, v\}$, then

$$H(z) = f(z) - \sum_{\substack{S \in \mathfrak{S} \\ \kappa_S = 0}} a_0(S) G_{S^{-1}}(z; 0; \Gamma, k, v) \tag{5.1.22}$$

and $H \in \{\Gamma, k, v\}_0$.

Proof. By (5.1.10, 17, 19), for $L \in \mathfrak{S}$, $S \in \mathfrak{S}$, with $t = e^{2\pi i z/n_L}$,

$$\begin{aligned} G_{S^{-1}}(z; m; \Gamma, k, v)|L &= \{\sigma(S^{-1}, L)\}^{-1} G_{S^{-1}L}(z; m; \Gamma^L, k, v^L) \\ &= \sum_{j+\kappa_L > 0} c_j(S, L) t^{j+\kappa_L}, \end{aligned}$$

when $S \neq L$, while

$$\begin{aligned} G_{L^{-1}}(z; m; \Gamma, k, v)|L &= G_I(z; m; \Gamma^L, k, v^L) \\ &= t^{m+\kappa_L} + \sum_{j+\kappa_L > 0} c_j(L^{-1}, L) t^{j+\kappa_L}. \end{aligned}$$

Hence $H \in M(\Gamma, k, v)$ and

$$H_L(z) = \sum_{j+\kappa_L>0} h_j(L) t^{j+\kappa_L}$$

for each $L \in \mathfrak{S}$, so that H is a cusp form.

In the notation of theorem 4.2.1, let λ' be the number of cusps $S\infty$ $(S \in \mathfrak{S})$ for which $\kappa_S = 0$. The corresponding λ' Eisenstein series $G_{S^{-1}}(z; 0)$ are clearly linearly independent, since at each of the λ' cusps exactly one of them does not vanish. The remaining $\lambda - \lambda'$ forms are cusp forms. Hence we have

Theorem 5.1.4. *If $k > 2$, $\dim\{\Gamma, k, v\}_0 = \dim\{\Gamma, k, v\} - \lambda'$. In particular, if k is an even integer and $v(T) = 1$ for all $T \in \Gamma$, $\dim\{\Gamma, k, 1\}_0 = \dim\{\Gamma, k, 1\} - \lambda$.*

Because of convergence, the argument we have used to construct the Poincaré series G_L only works when $k > 2$. However, we can produce modular forms of other dimensions by division. For example,

$$G_L(z; -1; \Gamma, k, v_1)/G_L(z; -2; \Gamma, k, v_2)$$

is clearly a non-constant meromorphic function belonging to $M(\Gamma, 0, v_1/v_2)$. The function may, of course, have poles at points in $\mathbb{H}$, whereas the functions G_L are holomorphic on $\mathbb{H}$.

The following theorem gives information about the relationship between Poincaré series on a group and on one of its normal subgroups.

Theorem 5.1.5. *Suppose that $k > 2$ and that $-I \in \Delta \subseteq \Gamma$, where Δ is normal in Γ, and let $\mu = [\hat{\Gamma} : \hat{\Delta}]$. Let v be a MS on Γ (and therefore on Δ) of weight k. Define n and κ by (5.1.6), where $L \in \Gamma(1)$ and $\zeta = L^{-1}\infty$, and put*

$$n' = n(\zeta, \Delta), \quad \kappa' = \kappa(\zeta, \Delta, v). \tag{5.1.23}$$

Then $n' = ln$ and $\kappa' = \{l\kappa\}$ (fractional part), where l is a positive integral divisor of μ.

Let $\hat{\Delta} = \hat{\Delta}_\zeta \cdot \mathcal{R}$, where $\hat{\Delta}_\zeta$ is the stabilizer of ζ modulo Δ. Then there exists a set $\mathcal{L}$ of μ/l matrices L_j $(1 \le j \le \mu/l)$ in Γ such that

$$\hat{\Gamma} = \hat{\Gamma}_\zeta \cdot \mathcal{R} \cdot \mathcal{L}, \tag{5.1.24}$$

and, for any $m \in \mathbb{Z}$,

$$G_L(z; m; \Gamma, k, v) = \sum_{j=1}^{\mu/l} \frac{G_{LL_j}(z; lm+[l\kappa]; \Delta, k, v)}{v(L_j)\sigma(L, L_j)}. \tag{5.1.25}$$

Proof. That $n' = nl$, where l divides μ is obvious from the definitions of n and n'. By (3.3.4) (with $m = l$),

$$e^{2\pi il\kappa} = v(L^{-1}U^{nl}L) = v(L^{-1}U^{n'}L) = e^{2\pi i\kappa'},$$

so that, since $0 \leq \kappa' < 1$, $\kappa' = \{l\kappa\}$.

We deduce the existence of $\mathscr{L}$ and (5.1.24) from theorem 1.1.3, with $\Gamma_1 = \hat{\Gamma}, \Gamma_2 = \hat{\Delta}, S = L^{-1}U^nL$ and $\sigma = l$. From it and (5.1.9) we have

$$G_L(z; m; \Gamma, k, v) = \sum_{j=1}^{\mu/l} \sum_{T \in \mathscr{R}} \frac{\exp\left\{\dfrac{2\pi i(m+\kappa)}{n} LTL_jz\right\}}{\mu(L, TL_jz)\nu(TL_j, z)}. \tag{5.1.26}$$

Now, by (3.1.5, 10),

$$\mu(L, TL_jz)\nu(TL_j, z) = \mu(L, TL_jz)\nu(T, L_jz)v(L_j)\mu(L_j, z),$$

while

$$\frac{m+\kappa}{n} = \frac{lm+l\kappa}{n'} = \frac{lm+[l\kappa]+\{l\kappa\}}{n'},$$

so that, by (5.1.10, 26),

$$\begin{aligned} G_L(z; m; \Gamma, k, v) &= \sum_{j=1}^{\mu/l} \frac{G_L(L_jz; lm+[l\kappa]; \Delta, k, v)}{v(L_j)\mu(L_j, z)} \\ &= \sum_{j=1}^{\mu/l} \bar{v}(L_j) G_L(z; lm+[l\kappa]; \Delta, k, v)|L_j \\ &= \sum_{j=1}^{\mu/l} \frac{G_{LL_j}(z; lm+[l\kappa]; \Delta, k, v)}{v(L_j)\sigma(L, L_j)}. \end{aligned}$$

For, since Δ is normal in Γ and $L_j \in \Gamma$, $\Delta^{L_j} = \Delta$ and $v^{L_j} = v$.

We conclude by observing that the Poincaré series G_L is not essentially changed when L is replaced by $\pm U^qL$. In fact, when $k > 2$ and $q \in \mathbb{Z}$,

$$G_{U^qL}(z; m; \Gamma, k, v) = \exp\{2\pi i(m+\kappa)q/n\} G_L(z; m; \Gamma, k, v). \tag{5.1.27}$$

This follows directly from (5.1.9), since we may take $\mathcal{R}_{U^qL}$ to be $\mathcal{R}_L$. Also, if

$$L=\begin{bmatrix} A & B \\ C & D \end{bmatrix}, \tag{5.1.28}$$

$$G_{-L}(z;m;\Gamma,k,v)=e^{\pm\pi ik}G_L(z;m;\Gamma,k,v), \tag{5.1.29}$$

where the plus sign is taken when $C>0$ or $C=0$, $D<0$ and the minus sign otherwise.

5.2 The Hilbert space of cusp forms. Theorem 5.1.3 shows that the construction of a basis for the vector space $\{\Gamma, k, v\}$ can be reduced to the same problem for $\{\Gamma, k, v\}_0$ and our object is to show that such a basis can be chosen from the Poincaré series $G_L(z;m)$. In order to do this we first define an inner product on $\{\Gamma, k, v\}_0$, which makes it a Hilbert space with certain rather convenient properties.

As usual we assume that $-I\in\Gamma$ and that Γ has finite index

$$\mu=\mu(\Gamma)=[\Gamma(1):\Gamma]=[\hat{\Gamma}(1):\hat{\Gamma}] \tag{5.2.1}$$

in $\Gamma(1)$. To begin with we assume that $k\geq 0$. Later, when we consider Poincaré series we shall require k to be greater than 2. We write $\Gamma(1)=\Gamma\cdot\mathcal{R}$. Then

$$\mathbb{F}=\bigcup_{L\in\mathcal{R}}\mathbb{F}_L$$

is a fundamental region for $\hat{\Gamma}$. Now suppose that f and g belong to $H(\Gamma, k, v)$ and that their product fg, which belongs to $H(\Gamma, 2k, v^2)$, belongs to $\{\Gamma, 2k, v^2\}_0$. The behaviour of f and g in $\mathbb{F}_L$ is determined by that of f_L and g_L in $\mathbb{F}_I$. Write $z=x+iy$ (x, y real). Then, by (4.1.12), for some $\kappa'_L>0$,

$$f_L(z)g_L(z)=\exp(4\pi i\kappa'_L z/n_L)\sum_{j=0}^{\infty}c_j(L)\exp(2\pi ijz/n_L)$$

for all $z\in\mathbb{H}$. Since $y\geq\frac{1}{2}\sqrt{3}$ when $z\in\mathbb{F}_I$, it follows that there exists a positive constant A_L such that

$$|f_L(z)g_L(z)|\leq A_L\exp(-4\pi\kappa'_L y/n_L). \tag{5.2.2}$$

Now suppose that $\mathbb{D}$ is any subset of $\mathbb{H}$ whose frontier consists of a finite number of arcs or circles, or segments of straight lines, orthogonal to $\mathbb{R}$, and suppose that the integral

$$\iint_{\mathbb{D}} f(w)\overline{g(w)}v^{k-2}\,du\,dv$$

is absolutely convergent; here $w = u + iv$, where u and v are real, and the bar denotes the complex conjugate. Write, for any $L \in \Omega$, $w = Lz$, so that $z \in \mathbb{H}$ and

$$v = \frac{y}{|L:z|^2}, \quad \frac{\partial(x, y)}{\partial(u, v)} = \left|\frac{dz}{dw}\right|^2 = |L:z|^4$$

and therefore, by (4.3.21),

$$f(w)\overline{g(w)} = f_L(z)\overline{g_L(z)}|L:z|^{2k}.$$

It follows that

$$\iint_{\mathbb{D}} f(w)\overline{g(w)}v^{k-2}\,du\,dv = \iint_{L^{-1}\mathbb{D}} f_L(z)\overline{g_L(z)}y^{k-2}\,dx\,dy \tag{5.2.3}$$

and that the second integral is also absolutely convergent. In particular, for any $L \in \Gamma$, by theorem 4.1.1(iii),

$$\iint_{\mathbb{D}} f(z)\overline{g(z)}y^{k-2}\,dx\,dy = \iint_{L^{-1}\mathbb{D}} f(z)\overline{g(z)}y^{k-2}\,dx\,dy. \tag{5.2.4}$$

Conversely, since, for any $L \in \Omega$, $L^{-1}\mathbb{D}$ is a region of the same type, it follows that, if the right-hand side of (5.2.3) is an absolutely convergent integral, so is the left-hand side.

We apply these results to $\mathbb{D} = L\mathbb{F}_I = \mathbb{F}_L$, where $L \in \mathcal{R}$. By (5.2.1),

$$\iint_{\mathbb{F}_I} f_L(z)\overline{g_L(z)}y^{k-2}\,dx\,dy$$

is absolutely convergent, since, by (5.2.2), it is majorized by the integral

$$\iint_{\mathbb{F}_I} A_L y^{k-2}\exp(-4\pi\kappa'_L y/n_L)\,dx\,dy$$

$$\leq A_L \int_{\frac{1}{2}\sqrt{3}}^{\infty} y^{k-2}\exp(-4\pi\kappa'_L y/n_L)\,dy =: A'_L.$$

It follows that

$$\iint_{\mathbb{F}_I} f_L(z)\overline{g_L(z)}y^{k-2}\,dx\,dy = \iint_{\mathbb{F}_L} f(z)\overline{g(z)}y^{k-2}\,dx\,dy, \tag{5.2.5}$$

both integrals being absolutely convergent.

We now define the *inner product* (or *scalar product*) (f, g) of f and g to be

$$(f, g) = (f, g; \Gamma) := \frac{1}{\mu(\Gamma)} \iint_{\mathbb{F}} f(z)\overline{g(z)}y^{k-2}\,dx\,dy. \tag{5.2.6}$$

This integral is absolutely convergent and, by (5.2.3), it is equal to

$$\frac{1}{\mu} \sum_{L\in\mathcal{R}} \iint_{\mathbb{F}_I} f_L(z)\overline{g_L(z)}u^{k-2}\,dx\,dy.$$

It follows from (5.2.4) (which is valid for $L \in \Gamma$) and theorem 2.3.1 that (f, g) is independent of the choice of fundamental region; we restrict ourselves, of course, to fundamental regions bounded by a finite number of arcs of circles and straight line segments orthogonal to the real axis.

Theorem 5.2.1. *If f and g belong to $H(\Gamma, k, v)$ and $fg \in \{\Gamma, 2k, v^2\}_0$, where $k \geq 0$, then the inner product (f, g), as defined by (5.2.6), is independent of the choice of the fundamental region $\mathbb{F}$ and the double integral is absolutely convergent. Further the inner product has the following properties*:

(i) $\overline{(f, g)} = (g, f)$.

(ii) *Let $f \in \{\Gamma, k, v\}_0$. Then $(f, f) \geq 0$, and $(f, f) = 0$ if and only if $f = 0$.*

(iii) *If $f_i, g_j \in H(\Gamma, k, v)$, $f_i g_j \in \{\Gamma, 2k, v^2\}_0$ and $\alpha_i, \beta_j \in \mathbb{C}$ for $1 \leq i \leq m$, $1 \leq i \leq n$, then*

$$\left(\sum_{i=1}^{m} \alpha_i f_i, \sum_{j=1}^{n} \beta_j f_j\right) = \sum_{i=1}^{m} \sum_{j=1}^{n} \alpha_i \overline{\beta_j} (f_i, g_j).$$

(iv) *The inner product $(f, g; \Gamma)$ is independent of the group Γ; i.e., if the conditions on f and g are valid for two different subgroups*

Γ_1 and Γ_2 of $\Gamma(1)$, then

$$(f, g; \Gamma_1) = (f, g; \Gamma_2). \tag{5.2.7}$$

(v) *Let $L \in \Omega_n^*$, where $n \in \mathbb{Z}^+$, and put $\Gamma_n = \Gamma \cap \bar{\Gamma}(n)$. Then*

$$(f, g; \Gamma) = (f, g; \Gamma_n) = (f|L, g|L; L^{-1}\Gamma_n L). \tag{5.2.8}$$

Proof. Parts (i), (ii) and (iii) follow immediately from the definition of the inner product. To prove (iv), let $\Gamma = \Gamma_1 \cap \Gamma_2$. A fundamental region for $\hat{\Gamma}$ is the union of $[\hat{\Gamma}_i : \hat{\Gamma}]$ fundamental regions $L\mathbb{F}_i$ for $\hat{\Gamma}_i$, where $L \in \mathcal{R}_i$ and $\hat{\Gamma}_i = \hat{\Gamma} \cdot \mathcal{R}_i$ $(i = 1, 2)$. From (5.2.4) with $\mathbb{D} = L\mathbb{F}_i$ we obtain

$$\iint_{\mathbb{F}} f(z)\overline{g(z)}y^{k-2}\,dx\,dy = [\hat{\Gamma}_i : \hat{\Gamma}] \iint_{\mathbb{F}_i} f(z)\overline{g(z)}y^{k-2}\,dx\,dy$$

and from this (5.2.7) follows since

$$[\hat{\Gamma}(1) : \hat{\Gamma}] = [\hat{\Gamma}(1) : \hat{\Gamma}_i][\hat{\Gamma}_i : \hat{\Gamma}].$$

(v) If J_n is defined as in (4.3.8), then

$$J_n^{-1} S J_n = \begin{bmatrix} \alpha & \beta n \\ \gamma/n & \delta \end{bmatrix}$$

and so

$$J_n^{-1}\bar{\Gamma}(n)J_n \subseteq J_n^{-1}\Gamma_0(n, n)J_n = \Gamma_0(1, n^2).$$

Now, by theorem 4.3.6, there exist S and T in $\Gamma(1)$ such that $L = SJ_nT$ and so, since $\Gamma_n \subseteq \bar{\Gamma}(n)$,

$$L^{-1}\Gamma_n L \subseteq T^{-1}J_n^{-1}\bar{\Gamma}(n)J_nT \subseteq T^{-1}\Gamma_0(1, n^2)T.$$

It follows that $L^{-1}\Gamma_n L \subseteq \Gamma(1)$ and

$$\begin{aligned}
[\Gamma(1) : L^{-1}\Gamma_n L] &= [\Gamma(1) : T^{-1}\Gamma_0(1, n^2)T][T^{-1}\Gamma_0(1, n^2)T : L^{-1}\Gamma_n L] \\
&= [\Gamma(1) : \Gamma_0(1, n^2)][\Gamma_0(1, n^2) : J_n^{-1}S^{-1}\Gamma_n SJ_n] \\
&= [\Gamma(1) : \Gamma_0(1, n^2)][\Gamma_0(n, n) : S^{-1}\Gamma_n S] \\
&= [\Gamma(1) : \Gamma_0(n, n)][\Gamma_0(n, n) : \bar{\Gamma}(n)][\bar{\Gamma}(n) : S^{-1}\Gamma_n S] \\
&= [\Gamma(1) : \bar{\Gamma}(n)][\bar{\Gamma}(n) : \Gamma_n] \\
&= [\Gamma(1) : \Gamma_n].
\end{aligned} \tag{5.2.9}$$

Now, exactly as in the proof of theorem 2.3.2(i), we may show that, if $\mathbb{D}$ is a fundamental region for $\hat{\Gamma}_n$, then $L^{-1}\mathbb{D}$ is a fundamental region for $L^{-1}\hat{\Gamma}_n L$. It now follows from (5.2.3, 6, 9) that (5.2.8) holds.

It may be remarked that part (v) holds if L is any matrix with positive determinant and rational entries. For we can then find positive integers q and n such that $qL=:L'\in\Omega_n^*$ and can then apply what has been proved to L' rather than L.

When we use the notation (f, g) we shall usually assume that both f and g are cusp forms. Then $\{\Gamma, k, v\}_0$ becomes a finite-dimensional Hilbert space over the field $\mathbb{C}$, with inner product (f, g). If $f\in\{\Gamma, k, v\}_0$, we define the *norm* of f to be

$$\|f\|:=(f,f)^{\frac{1}{2}}=\left\{\frac{1}{\mu}\iint_{\mathbb{F}}|f(z)|^2y^{k-2}\,dx\,dy\right\}^{\frac{1}{2}}.$$

The metric associated with this norm is given by

$$d(f,g):=\|f-g\|=(f-g,f-g).$$

We can obtain an orthogonal basis for $\{\Gamma, k, v\}_0$ in the usual way by the Gram–Schmidt orthogonalization process; see, for example, Berberian (1961), §2.6. For this purpose we assume that

$$\dim\{\Gamma,k,v\}_0=N>0. \qquad (5.2.10)$$

Then there exist N forms $\varphi_i\in\{\Gamma, k, v\}_0$ $(i=1, 2, \ldots, N)$ forming a basis for $\{\Gamma, k, v\}_0$ with the property

$$(\varphi_i,\varphi_j)=\delta_{ij}\quad(i\neq j), \qquad (5.2.11)$$

when $1\le i\le N$, $1\le j\le N$. We may also, if we wish, assume that the basis is orthonormal, so that (5.2.11) holds in addition when $i=j$.

Now let $\mathbb{F}$ be a proper fundamental region for $\hat{\Gamma}$ and take any $L\in\Gamma(1)$. Let $\zeta=L^{-1}\infty$ and define M, n, κ and $\mathcal{R}_L$ as in (5.1.5, 6, 8). Put

$$\mathbb{S}=\{z:-\tfrac{1}{2}n\le x=\operatorname{Re} z<\tfrac{1}{2}n,\ y=\operatorname{Im} z>0\} \qquad (5.2.12)$$

and, for each $T\in\mathcal{R}_L$, let $\mathbb{F}^T$ be the part of $\mathbb{S}$ congruent to $LT\mathbb{F}$ modulo Γ_{U^n}; i.e.

$$\mathbb{F}^T=\{z\in\mathbb{S}:z\equiv z'\ (\text{mod}\ \Gamma_{U^n})\text{ for some }z'\in(LT\mathbb{F})\cap\mathbb{H}\}. \qquad (5.2.13)$$

Then the regions $\mathbb{F}^T$, for $T \in \mathcal{R}_L$, cover $\mathbb{S}$ without overlapping, except possibly at elliptic fixed points. For if $z \in \mathbb{S}$ then, for some $q \in \mathbb{Z}$ and $T \in \mathcal{R}_L$, $L^{-1}z \in L^{-1}U^{nq}LT\mathbb{F}$. If we take $z' = z - qn$ we see that $z' \in LT\mathbb{F}$ and $z \equiv z' \pmod{\Gamma_{U^n}}$, so that $z \in \mathbb{F}^T$. Further, for each $z \in \mathbb{S} - \mathbb{E}$, the transformation T is uniquely determined. It follows that

$$\iint_{\mathbb{S}} F(x, y)\, dx\, dy = \sum_{T \in \mathcal{R}_L} \iint_{\mathbb{F}_T} F(x, y)\, dx\, dy$$

whenever one side exists with $F(x, y)$ replaced by $|F(x, y)|$.

Theorem 5.2.2. *Suppose that $k > 2$, $f \in \{\Gamma, k, v\}_0$, $L = M^{-1} \in \Gamma(1)$ and that*

$$f_M(z) = \sum_{r+\kappa>0} b_r \exp\{2\pi i(r+\kappa)z/n\},$$

where $n = n_M$, $\kappa = \kappa_M$, and $L \neq -U^q$ $(q \in \mathbb{Z})$, so that L is regular. Suppose also that $m \geq 0$. Then

$$(f(z), G_L(z; m; \Gamma, k, v)) = \begin{cases} \dfrac{b_m n^k \Gamma(k-1)}{\mu\{4\pi(m+\kappa)\}^{k-1}} & \text{if } m+\kappa > 0, \\ 0 & \text{if } m = \kappa = 0, \end{cases}$$

where $\mu = [\hat{\Gamma}(1) : \hat{\Gamma}]$ and $\Gamma(k-1)$ is the gamma function.

Proof. We note that, since $m \geq 0$, fG_L is a cusp form and therefore (f, G_L) is defined. Let $\mathbb{F}$ be a fundamental region for $\hat{\Gamma}$, where

$$\mathbb{F} = \bigcup_{S \in \mathcal{R}} \mathbb{F}_S$$

and $\Gamma(1) = \hat{\Gamma} \cdot \mathcal{R}$. By theorems 5.1.1 and 5.1.2, the series defining $G_{LS}(w; m; \Gamma^S, k, v^S)$ is absolutely uniformly convergent for $w \in \mathbb{F}_I$. If we put $z = Sw$ for $w \in \mathbb{F}_I$, then $z \in \mathbb{F}_S$ and, by (5.1.10), the terms in $G_{LS}(w; m; \Gamma^S, k, v^S)$ are the same as those in

$$\{\mu(S, w)\}^{-1} G_L(z; m; \Gamma, k, v),$$

apart from a factor $\sigma(L, S)$ of unit modulus. Further, if $y = \operatorname{Im} z$ and $v = \operatorname{Im} w$,

$$|f(z)\{\mu(S, w)\}^{-1} u^{k-2}| = \left| f_S(w) |S : w|^{4-2k} v^{k-2} \right|$$

and this is bounded for $w \in \mathbb{F}_I$, since f_S is a cusp form and $k > 2$. For $|S:w| \geq 1$ for all $w \in \mathbb{F}_I$. It follows that the series for

$$G_L(z; m; \Gamma, k, v)\overline{f(z)}y^{k-2}$$

is absolutely uniformly convergent for $z \in \mathbb{F}_S$ for each $S \in \mathcal{R}$, and so is absolutely uniformly convergent on $\mathbb{F}$. Accordingly, by (5.1.9) and (5.2.6),

$$\mu(G_L, f) = \iint_{\mathbb{F}} \sum_{T \in \mathcal{R}_L} \frac{\exp\left\{\frac{2\pi i(m+\kappa)}{n} LTz\right\}\overline{f(z)}y^{k-2}}{\mu(L, Tz)\nu(T, z)} dx\, dy$$

$$= \sum_{T \in \mathcal{R}_L} \iint_{\mathbb{F}} \frac{\exp\left\{\frac{2\pi i(m+\kappa)}{n} LTz\right\}\overline{f(z)}y^{k-2}}{\mu(L, Tz)\nu(T, z)} dx\, dy.$$

In the double integral we change the variable from z to $w = LTz$, so that, by (3.1.5, 15), since L is regular,

$$\begin{aligned} f(z) = f(T^{-1}Mw) &= \nu(T^{-1}, Mw)f(Mw) \\ &= \{\nu(T, T^{-1}Mw)\}^{-1}\mu(M, w)f_M(w) \\ &= \{\nu(T, z)\mu(L, Tz)\}^{-1}f_M(w). \end{aligned}$$

Thus

$$\overline{f(z)} = |LT:z|^{-2k}\mu(L, Tz)\nu(T, z)\overline{f_M(w)}.$$

Hence, by (5.2.13),

$$\mu(G_L, f) = \sum_{T \in \mathcal{R}_L} \iint_{LT\mathbb{F}} \overline{f_M(w)}v^{k-2} \exp\{2\pi i(m+\kappa)w/n\}\, du\, dv$$

$$= \sum_{T \in \mathcal{R}_L} \iint_{\mathbb{F}^T} \overline{f_M(w)}v^{k-2} \exp\{2\pi i(m+\kappa)w/n\}\, du\, dv,$$

since the integrand is a periodic function of w with period n. Hence

$$\mu(G_L, f) = \iint_{\mathbf{S}} \overline{f_M(w)} v^{k-2} \exp\{2\pi i(m+\kappa)w/n\}\, du\, dv$$

$$= \int_0^\infty v^{k-2}\, dv \int_{-\frac{1}{2}n}^{\frac{1}{2}n} \sum_{r+k>0} \bar{b}_r \exp\Big[\frac{2\pi i}{n}\{(m+\kappa)(u+iv) - (r+\kappa)(u-iv)\}\Big]\, du\, dv.$$

The series in the inner interval is uniformly convergent for $v \geq \delta > 0$ and therefore

$$\int_{-\frac{1}{2}n}^{\frac{1}{2}n} \sum_{r+\kappa>0} \bar{b}_r \exp\Big[\frac{2\pi i}{n}\{(m+\kappa)w - (r+\kappa)\bar{w}\}\Big]\, du$$

$$= \sum_{r+\kappa>0} \bar{b}_r \exp\{-2\pi v(m+r+2\kappa)/n\} \int_{-\frac{1}{2}n}^{\frac{1}{2}n} \exp\{2\pi i n(m-r)/n\}\, du$$

$$= n\bar{b}_m \exp\{-4\pi(m+\kappa)v/n\},$$

this being zero if $m = \kappa = 0$. Hence

$$\mu(G_L, f) = n\bar{b}_m \int_0^\infty v^{k-2} \exp\{-4\pi(m+\kappa)v/n\}\, dv$$

$$= \frac{n^k \bar{b}_m \Gamma(k-1)}{\{4\pi(m+\kappa)\}^{k-1}},$$

from which the theorem follows.

We now suppose as before that (5.2.10) holds, so that, by theorem 4.2.1,

$$0 < N \leq N_0 = 1 - \xi - \lambda' + k\mu/12, \tag{5.2.14}$$

where ξ is defined by (4.2.1) and λ' is the number of incongruent cusps $M\infty \pmod{\Gamma}$ for which $\kappa_M = 0$. Take a fixed $M = L^{-1} \in \Gamma(1)$ and write

$$n = n_M, \quad \kappa = \kappa_M$$

as usual. Then, if $f \in \{\Gamma, k, v\}_0$ and $f \neq 0$, we have

$$\operatorname{ord}(f, M\infty, \Gamma) \neq m + \kappa > 0$$

for some integer m. By the arguments used in the proof of theorem 4.2.1,

$$k\mu/12 \geq m + \kappa - 1 + \xi + \lambda'.$$

This is easily seen by considering the cases $\kappa = 0$, $0 < \kappa < 1$ separately. It follows that

$$0 < m + \kappa \leq N_0. \tag{5.2.15}$$

Note that the number of integers m that satisfy this inequality is N_0.

Now let $f^{(1)}, f^{(2)}, \cdots, f^{(N)}$ be a basis for $\{\Gamma, k, v\}_0$ and write

$$f_M^{(j)}(z) = \sum_{r+\kappa>0} b_{jr}(M) \exp\{2\pi i(r+\kappa)/n\}. \tag{5.2.16}$$

We shall usually write b_{jr} in place of $b_{jr}(M)$. Denote by $\boldsymbol{b}_r$ the vector (column matrix) with components b_{jr} $(j = 1, 2, \ldots, N)$. The vectors $\boldsymbol{b}_r$ for $r + \kappa > 0$ form an infinite matrix B, where the columns $\boldsymbol{b}_r$ are supposed arranged in order of increasing r from left to right. The submatrix B_q of B consists of the q columns $\boldsymbol{b}_r$ for which $0 < r + \kappa \leq q$. Note that there is a one-to-one correspondence between the N rows of B and the N forms $f^{(j)}$ $(1 \leq j \leq N)$.

Theorem 5.2.3. *Let $L = M^{-1} \in \Gamma(1)$. Then, if $q \geq N_0$, the matrix $B_q = B_q(M)$ has rank N. It follows that there exist N integers m_j $(1 \leq j \leq N)$ with*

$$-\kappa < m_1 < m_2 < \cdots < m_N \leq N_0 - \kappa$$

such that the N vectors $\boldsymbol{b}_{m_j}$ $(1 \leq j \leq N)$ are linearly independent.

Further, if $k > 2$, the N forms $G_L(z; m_j; \Gamma, k, v)$ $(1 \leq j \leq N)$ form a basis for $\{\Gamma, k, v\}_0$. Conversely, if N cusp forms $G_L(z; p_j)$ $(p_j + \kappa > 0$, $1 \leq j \leq N)$ are linearly independent, then so are the vectors $\boldsymbol{b}_{p_j}$ $(1 \leq j \leq N)$.

Proof. Take $q \geq N_0$. If B_q has rank less than N, its rows are linearly dependent and there exist complex numbers $x_1, x_2, \ldots, x_N$, not all zero, such that, if

$$f(z) = \sum_{j=1}^{N} x_j f^{(j)}(z),$$

then

$$f_M(z) = \sum_{j=1}^{N} x_j f_M^{(j)}(z) = \sum_{r+\kappa>q} c_r \exp\{2\pi i(r+\kappa)/n\}.$$

It follows that

$$\operatorname{ord}(f, M\infty, \Gamma) > q \geq N_0,$$

and this contradicts (5.2.15).

Secondly, if the N forms $G_L(z; m_j)$ $(1 \leq j \leq N)$ are not linearly independent, then there exist complex numbers $x_1, x_2, \ldots, x_N$ not all zero, such that

$$F(z) := \sum_{j=1}^{N} x_j G_L(z; m_j) = 0$$

for all $z \in \mathbb{H}$. Hence, by theorem 5.2.2 and (5.2.16),

$$0 = (f^{(s)}, F) = \frac{n^k \Gamma(k-1)}{\mu (4\pi)^{k-1}} \sum_{j=1}^{N} \bar{x}_j b_{sm_j} (m_j + \kappa)^{1-k} \quad (1 \leq s \leq N).$$

This states that the vectors $\boldsymbol{b}_{m_j}$ are linearly dependent, which is false.

Finally, if the N forms $G_L(z; p_j)$ $(p_j + \kappa > 0, 1 \leq j \leq N)$ are linearly independent, while the vectors $\boldsymbol{b}_{p_j}$ $(1 \leq j \leq N)$ are linearly dependent, we can find $f \in \{\Gamma, k, v\}_0$ with $f \neq 0$ and Fourier series

$$f_M(z) = \sum_{r+\kappa>0} C_r \exp\{2\pi i (r+\kappa)/n\},$$

where $c_r = 0$ for $r = p_j$ $(1 \leq j \leq N)$. Then, for some $x_1, x_2, \ldots, x_N$ not all zero,

$$f(z) = \sum_{j=1}^{N} x_j G_L(z; p_j)$$

and so

$$(f, f) = \sum_{j=1}^{N} \overline{x_j} (f(z), G_L(z; p_j)).$$

But this is zero, by theorem 5.2.2, and hence $f = 0$, which is a contradiction.

From this theorem and theorem 5.1.3 we immediately deduce the following theorem.

Theorem 5.2.4. *The set of Poincaré series $G_L(z; m; \Gamma, k, v)$ $(m \in \mathbb{Z})$ spans the space $H(\Gamma, k, v)$ $(k > 2)$. The subspace $\{\Gamma, k, v\}_0$ is spanned by the subset of functions $G_L(z; m)$ with $m + \kappa_M > 0$*

($M=L^{-1}$). In particular, if $N_0=N$ in (5.2.14), the first N Poincaré series $G_L(z;m)$ ($m+\kappa_M>0$) form a basis for $\{\Gamma, k, v\}_0$.

Our next theorem gives a symmetric relation between the Fourier coefficients of different Poincaré series.

Theorem 5.2.5. *Suppose that, for $L, S\in\Gamma(1)$ and $k>2$,*

$$G_L(z;m;\Gamma,k,v)|S=\sum_{\kappa_S+r>0} a(r,m,L,S)\exp\{2\pi i(\kappa_S+r)z/n_S\}, \tag{5.2.17}$$

where $m+\kappa_m>0$ and $n_S=n(S\infty,\Gamma)$, $\kappa_S=\kappa(S\infty,\Gamma,v)$. Let L_1 and L_2 be regular matrices in $\Gamma(1)$ and let $n_j=n(L_j^{-1}\infty,\Gamma)$ and $\kappa_j=\kappa(L_j^{-1}\infty,\Gamma,v)$ for $j=1,2$. Then, if $m_j+\kappa_j>0$ ($j=1,2$),

$$\frac{n_2^k a(m_2,m_1,L_1,L_2^{-1})}{(m_2+\kappa_2)^{k-1}}=\frac{n_1^k a(m_1,m_2,L_2,L_1^{-1})}{(m_1+\kappa_1)^{k-1}}.$$

Proof. By theorem 5.2.2,

$$(G_{L_1}(z;m_1),G_{L_2}(z;m_2))=\frac{n_2^k\Gamma(k-1)}{\mu\{4\pi(m_2+\kappa_2)\}^{k-1}}a(m_2,m_1,L_1,L_2^{-1})$$

and a similar result holds for (G_{L_2},G_{L_1}), from which the theorem follows.

We can now deduce two necessary and sufficient conditions for the identical vanishing of a Poincaré series.

Theorem 5.2.6. *Suppose that $G_L(z;m;\Gamma,k,v)\in\{\Gamma,k,v\}_0$, where $k>2$ (so that $m+\kappa_m>0$). Then each of the following two conditions is both necessary and sufficient for the identical vanishing of $G_L(z;m)$:*

(i) *$b_m=0$ for all $f\in\{\Gamma,k,v\}_0$, in the notation of theorem 5.2.2.*
(ii) *$a(m,m;L,L^{-1})=0$ in the notation of (5.2.17).*

Proof. If $G_L=0$, we deduce that $b_m=0$ and that $a(m,m;L,L^{-1})=0$ by applying theorem 5.2.2 to (f,G_L) and (G_L,G_L). The converse result is obtained by taking $f=G_L$ and considering (G_L,G_L).

We conclude by showing that, in certain cases, a Hilbert space of cusp forms can be decomposed into a direct sum of mutually orthogonal spaces of a specially simple kind.

Theorem 5.2.7. *Suppose that M is a subspace of $\{\Delta, k, u\}_0$ satisfying all the conditions of theorem 4.2.2, where $k \geq 0$. Then the subspaces M_v $(v = u\chi, \chi \in \mathfrak{X}(\Gamma, \Delta))$, of which M is the direct sum, are mutually orthogonal.*

Proof. We have to show that, if $f_j \in \{\Gamma, k, u\chi_j\}_0$ $(j = 1, 2)$, where $\chi_1 \neq \chi_2$, then $(f_1, f_2) = 0$.

Let $\Gamma = \Delta \cdot \mathcal{R}$ and suppose that $\mathbb{F}$ is a fundamental region for $\hat{\Gamma}$. Then

$$\bigcup_{T \in \mathcal{R}} T\mathbb{F}$$

is a fundamental region for $\hat{\Delta}$. Hence

$$(f_1, f_2) = \frac{1}{\mu} \sum_{T \in \mathcal{R}} \iint_{T\mathbb{F}} f_1(z)\overline{f_2(z)}y^{k-2}\,dx\,dy$$

$$= \frac{1}{\mu} \sum_{T \in \mathcal{R}} \iint_{\mathbb{F}} \chi_1(T)\overline{\chi_2(T)}f_1(z)\overline{f_2(z)}y^{k-2}\,dx\,dy.$$

This is zero, since

$$\sum_{T \in \mathcal{R}} \chi_1(T)\overline{\chi_2(T)} = 0.$$

For the summation is essentially over the factor group Γ/Δ and $\chi_1\overline{\chi_2}$ is a non-principal character on this abelian group.

5.3. The Fourier coefficients of Poincaré series. In §4.5 we obtained a certain amount of information about the Fourier coefficients of entire modular forms by using general estimates of the magnitude of these forms near the real axis. In the present section we obtain explicit formulae for the Fourier coefficients of the Poincaré series $G_L(z; m)$ $(m \in \mathbb{Z})$. By using theorem 5.2.4 we can therefore obtain information about the Fourier coefficients of any modular form holomorphic on $\mathbb{H}$.

We require certain standard formulae for gamma and Bessel functions. From Hankel's formula for the gamma function† it can be deduced that

$$\int_{-\infty+ic}^{\infty+ic} w^{-k} e^{-2\pi i\mu w}\, dw = (2\pi)^k \frac{\mu^{k-1} e^{-\frac{1}{2}k\pi i}}{\Gamma(k)} \tag{5.3.1}$$

for positive μ and c, and $k>1$. The integral is taken along the line $\operatorname{Im} w = c$. See Whittaker and Watson (1927), §12.2.

The Bessel functions J_{k-1} and I_{k-1} are defined by the absolutely convergent infinite series

$$J_{k-1}(z) = \sum_{m=0}^{\infty} \frac{(-1)^m (\frac{1}{2}z)^{2m+k-1}}{m!\,\Gamma(m+k)} \tag{5.3.2}$$

and

$$I_{k-1}(z) = \sum_{m=0}^{\infty} \frac{(\frac{1}{2}z)^{2m+k-1}}{m!\,\Gamma(m+k)} \tag{5.3.3}$$

for all $z \in \mathbb{C}$. When computing real powers of z, here and elsewhere, it is assumed that $-\pi < \operatorname{ph} z \leq \pi$. We shall only require these functions for real $k>1$. Note that

$$J_{k-1}(-ix) = i\, e^{-\frac{1}{2}k\pi i} I_{k-1}(x) \quad (x>0). \tag{5.3.4}$$

The following integral representation of $J_{k-1}(z)$ is well known.

$$J_{k-1}(z) = \frac{1}{2\pi i} (\tfrac{1}{2}z)^{k-1} \int_{-\infty}^{(0+)} w^{-k} \exp\left(w - \frac{z^2}{4w}\right) dw.$$

See Watson (1922), §6.2, for example. The contour runs from $-\infty$ on the negative real axis, encircles the origin in a counterclockwise direction and returns to $-\infty$. The formula is valid for all non-zero complex z and real k. When $k>1$ the contour can be deformed to yield the following two formulae, after a simple change of variable:

$$\int_{-\infty+ci}^{\infty+ci} w^{-k} \exp\{-2\pi i(\mu_1 w + \mu_2 w^{-1})\}\, dw$$
$$= 2\pi \left(\frac{\mu_1}{\mu_2}\right)^{\frac{1}{2}(k-1)} e^{-\frac{1}{2}k\pi i} J_{k-1}(4\pi \sqrt{(\mu_1\mu_2)}), \tag{5.3.5}$$

† No confusion should arise between the function $\Gamma(k)$ and the group $\Gamma(k)$.

$$\int_{-\infty+ci}^{\infty+ci} w^{-k} \exp\{-2\pi i(\mu_1 w - \mu_2 w^{-1})\}\, dw$$
$$= 2\pi\left(\frac{\mu_1}{\mu_2}\right)^{\frac{1}{2}(k-1)} e^{-\frac{1}{2}k\pi i} I_{k-1}(4\pi\sqrt{(\mu_1\mu_2)}). \tag{5.3.6}$$

In these formulae it is assumed that $k>1$ and that c, μ_1 and μ_2 are positive numbers.

We also have, for large $x>0$,

$$J_{k-1}(x) = \left(\frac{2}{\pi x}\right)^{\frac{1}{2}} \cos(x - \tfrac{1}{2}k\pi + \tfrac{1}{4}\pi) + O(x^{-\frac{3}{2}}) \tag{5.3.7}$$

and

$$I_{k-1}(x) = \frac{e^x}{(2\pi x)^{\frac{1}{2}}}\{1 + O(x^{-1})\}; \tag{5.3.8}$$

see Watson (1922), §§7.21, 7.23.

We use these results in the following theorem.

Theorem 5.3.1. *Suppose that $z \in \mathbb{H}$, $k>1$ and that κ and λ are real numbers. Write*

$$F_\kappa(z, \kappa, \lambda) = \sum_{h=-\infty}^{\infty} (z+h)^{-k} \exp\left\{-2\pi i\left(\kappa h + \frac{\lambda}{z+h}\right)\right\}. \tag{5.3.9}$$

The series is absolutely uniformly convergent on every compact subset of $\mathbb{H}$ and defines $F_\kappa(z, \kappa, \lambda)$ as a holomorphic function on $\mathbb{H}$. Further, $F_\kappa(z, \kappa, \lambda)$ can be expressed as a Fourier series

$$F_\kappa(z, \kappa, \lambda) = \sum_{r+\kappa>0} g_r\, e^{2\pi i(r+\kappa)z} \tag{5.3.10}$$

which is absolutely uniformly convergent on every compact subset of $\mathbb{H}$. The Fourier coefficients g_r are given by the formulae

$$g_r = \frac{(2\pi)^k}{\Gamma(k)} e^{-\frac{1}{2}k\pi i} (r+\kappa)^{k-1} \quad \text{when } \lambda = 0, \tag{5.3.11}$$

$$g_r = 2\pi\, e^{-\frac{1}{2}k\pi i}\left(\frac{r+\kappa}{\lambda}\right)^{\frac{1}{2}(k-1)} J_{k-1}(4\pi\sqrt{\{\lambda(r+\kappa)\}}) \quad \text{when } \lambda>0, \tag{5.3.12}$$

and

$$g_r = 2\pi e^{-\frac{1}{2}k\pi i}\left(\frac{r+\kappa}{|\lambda|}\right)^{\frac{1}{2}(k-1)} I_{k-1}(4\pi\sqrt{\{|\lambda|(r+\kappa)\}}) \quad \textit{when } \lambda<0. \tag{5.3.13}$$

Proof. It is clear that the series in (5.3.9) is absolutely uniformly convergent on every compact subset of $\mathbb{H}$ and so defines $F_k(z, \kappa, \lambda)$ as a holomorphic function of z on $\mathbb{H}$. Now

$$F_k(z+1, \kappa, \lambda) = \sum_{h=-\infty}^{\infty} (z+1+h)^{-k} \exp\left\{-2\pi i\left(\kappa h + \frac{\lambda}{z+1+h}\right)\right\}$$

$$= e^{2\pi i\kappa} F_k(z, \kappa, \lambda),$$

as is seen by summing with respect to $h+1$ instead of h. Put $t = e^{2\pi i z}$ and define

$$G(t) := e^{-2\pi i\kappa z} F_k(z, \kappa, \lambda). \tag{5.3.14}$$

Then, as in §4.1, $G(t)$ is well defined for $0<|t|<1$ and holomorphic on this punctured disc. Hence $G(t)$ can be expressed as a Laurent series

$$G(t) = \sum_{r=-\infty}^{\infty} g_r t^r \quad (0<|t|<1),$$

where

$$g_r = \frac{1}{2\pi i}\int \frac{G(t)\,dt}{t^{r+1}},$$

the integral being taken round the circle $|t|=\rho$, where $0<\rho<1$.

We take $\rho = e^{-2\pi c}$, where $c>0$, so that, by (5.3.14),

$$g_r = \int_{ic}^{1+ic} e^{-2\pi i(r+\kappa)z} F_k(z, \kappa, \lambda)\, dz.$$

Because of the uniform convergence of the series (5.3.9), we have

$$g_r = \sum_{h=-\infty}^{\infty} \int_{ic}^{1+ic} (z+h)^{-k} e^{-2\pi i(r+\kappa)z} \exp\left\{-2\pi i\left(\kappa h + \frac{\lambda}{z+h}\right)\right\} dz$$

$$= \sum_{h=-\infty}^{\infty} \int_{ic+h}^{ic+h+1} z^{-k} \exp[-2\pi i\{(r+\kappa)z + \lambda/z\}]\, dz$$

$$= \int_{-\infty+ic}^{\infty+ic} z^{-k} \exp[-2\pi i\{(r+\kappa)z + \lambda/z\}]\, dz. \tag{5.3.15}$$

If $r+\kappa \leq 0$, then

$$\mathrm{Re}\{-2\pi i(r+\kappa)z\} = 2\pi(r+\kappa)y \leq 0$$

and we can replace the contour of integration by a large semicircle in $\mathbb{H}$. Since $k>1$, we find that $g_r = 0$ when $r+\kappa \leq 0$.

We may therefore assume that $r+\kappa > 0$. Taking $\mu = r+\kappa$ in (5.3.1), we deduce (5.3.11) from (5.3.15) in the case when $\lambda = 0$. When $\lambda \neq 0$, we deduce (5.3.12, 13) from (5.3.15) and (5.3.5, 6) by taking $\mu_1 = r+\kappa$ and $\mu_2 = |\lambda|$. The absolute uniform convergence of the series (5.3.10) on compact sets is immediate, since it is a Laurent series; alternatively the estimates (5.3.7, 8) may be used to verify this.

We now use this theorem to obtain a Fourier expansion of $G_L(z; m; \Gamma, k, v)$, where, as usual, $k>2$ and $L \in \Gamma(1)$. We define M and $\mathcal{R}_L$ as in (5.1.5, 8) but adopt a different notation for the various cusp widths and parameters that arise. We put

$$n_1 := n_I = n(\infty, \Gamma), \quad \kappa_1 := \kappa_I = \kappa(\infty, \Gamma, v), \tag{5.3.16}$$

and

$$n_2 := n_M = n(L^{-1}\infty, \Gamma), \quad \kappa_2 := \kappa_M = \kappa(L^{-1}\infty, \Gamma, v). \tag{5.3.17}$$

We also put

$$\hat{\Gamma}_1 = \hat{\Gamma}_{U^{n_1}}, \quad \hat{\Gamma}_2 = L^{-1}\hat{\Gamma}_{U^{n_2}}L. \tag{5.3.18}$$

Thus $\hat{\Gamma}$ and $\hat{\Gamma}_2$ are the stabilizers of ∞ and $L^{-1}\infty$, respectively, in $\hat{\Gamma}$ and

$$\hat{\Gamma} = \hat{\Gamma}_2 \cdot \mathcal{R}_L. \tag{5.3.19}$$

Our immediate object is to make a particular choice of the right transversal $\mathcal{R}_L$ which will simplify our discussion of $G_L(z; m)$ as given by (5.1.9).

The group $\hat{\Gamma}$ can be expressed as a disjoint union of double cosets $\hat{\Gamma}_2 T \hat{\Gamma}_1$ for $T \in \hat{\Gamma}$, and we denote by $\mathcal{T}$ a transversal of these double cosets. Thus

$$\hat{\Gamma} = \hat{\Gamma}_2 \mathcal{T} \hat{\Gamma}_1$$

and each $S \in \hat{\Gamma}$ determines a unique $T \in \mathcal{T}$ such that $S \in \hat{\Gamma}_2 T \hat{\Gamma}_1$.

Suppose first of all that $L\hat{\Gamma}\cap\hat{\Gamma}_U=\varnothing$; this is equivalent to the statement that $\delta_L\neq 0$, where δ_L is defined by (5.1.11) and the sentence following that equation. In this case we have

$$\hat{\Gamma}=\hat{\Gamma}_2\cdot\mathscr{T}\cdot\hat{\Gamma}_1.$$

For if, for $T, T'\in\mathscr{T}$ and integers p, q, r and s, we have

$$L^{-1}U^{pn_2}LTU^{qn_1}=L^{-1}U^{rn_2}LT'U^{sn_1},$$

then $T'=T$ and

$$U^{(p-r)n_2}LT=LTU^{(s-q)n_1}.$$

Thus $LT\infty$ is a fixed point for $U^{(p-r)n_2}$ and is therefore ∞. Hence $LT\in\hat{\Gamma}_U$, which is a contradiction.

On the other hand, if $L\hat{\Gamma}\cap\hat{\Gamma}_U\neq\varnothing$, then $\delta_L\neq 0$ and $MU^s\in\hat{\Gamma}$ for some $s\in\mathbb{Z}$. We note in passing that $L^{-1}\infty\equiv\infty\ (\mathrm{mod}\ \Gamma)$ in this case, so that $n_2=n_1$ and $\kappa_2=\kappa_1$. We now write

$$\mathscr{T}_L:=L\mathscr{T}-\hat{\Gamma}_U. \tag{5.3.20}$$

The argument given above then shows that

$$\hat{\Gamma}-M\hat{\Gamma}_U=\hat{\Gamma}_2\cdot M\mathscr{T}_L\cdot\hat{\Gamma}_1 \tag{5.3.21}$$

and $\mathscr{T}_L$ may be taken to be any set satisfying (5.3.12). This equation also holds when $\delta_L=0$, since then the left-hand side is $\hat{\Gamma}$ and $\mathscr{T}_L=L\mathscr{T}$.

It is convenient to regard $\mathscr{T}_L$ as a set of matrices, rather than transformations. We note that a matrix and its negative cannot both belong to $\mathscr{T}_L$ and that $\mathscr{T}_L\cap\Gamma_U=\varnothing$. Further, any matrix S in $\mathscr{T}_L$ may be replaced by $U^{pn_2}SU^{qn_1}$ for any $p, q\in\mathbb{Z}$ without destroying the validity of (5.3.21); note also that these changes leave the entry γ in S unaltered. It follows that we may take the set $\mathscr{T}_L$ to be a disjoint union

$$\mathscr{T}_L=\bigcup_{\gamma=1}^{\infty}\mathscr{T}_L(\gamma), \tag{5.3.22}$$

where, for each $\gamma\in\mathbb{Z}^+$, $\mathscr{T}_L(\gamma)$ consists of all matrices $S\in L\Gamma$ for which

$$0\leq\delta<\gamma n_1,\quad 0\leq\alpha<\gamma n_2. \tag{5.3.23}$$

It also follows from (5.3.21) that there is at most one matrix S in $\mathscr{T}_L$ with a given triple of integers γ, δ, α satisfying (5.3.23).

In the particular case when $\Gamma=\bar{\Gamma}(N)$, we have $n_1=n_2=N$ and we put, as usual,

$$L=\begin{bmatrix} A & B \\ C & D \end{bmatrix}. \tag{5.3.24}$$

The set $\mathcal{T}_L(\gamma)$ is empty except when

$$\gamma>0 \quad \text{and} \quad \gamma\equiv\varepsilon C \pmod N, \quad \text{where } \varepsilon=\pm 1. \tag{5.3.25}$$

When (5.3.25) holds, each matrix $S\in\mathcal{T}_L(\gamma)$ has the property that $S\equiv\varepsilon L \pmod N$. From this and (5.3.23) it follows that, when $S\in\mathcal{T}_L(\gamma)$, we have

$$\left.\begin{aligned} 0\le\delta<\gamma N, \quad 0\le\alpha<\gamma N, \\ [\alpha,\delta]\equiv\varepsilon[A,D] \pmod N, \quad \alpha\delta\equiv 1+\varepsilon B\gamma \pmod{N\gamma}. \end{aligned}\right\} \tag{5.3.26}$$

It is easily seen that, conversely, when (5.3.25) holds and α, δ are any integers satisfying (5.3.26), then there exists exactly one matrix S in $\mathcal{T}_L(\gamma)$ having these values of α and δ. When $N=1$ or 2 we may, of course, restrict ε to be 1 in (5.3.25). When $N>2$ and $2C\not\equiv 0 \pmod N$, each positive γ determines at most one value of ε for which (5.3.25) holds; however, if $N>2$, $2C\equiv 0 \pmod N$ and $\gamma\equiv C \pmod N$, then $\mathcal{T}_L(\gamma)$ contains some matrices for which (5.3.26) holds with $\varepsilon=1$ and others for which $\varepsilon=-1$.

We now return to the general case.

For each $\gamma\in\mathbb{Z}^+$, $\mathcal{T}_L(\gamma)$ is a finite set (and may be empty). We define the generalized Kloosterman sum to be

$$W(r,m;\gamma)=\sum_{S\in\mathcal{T}_L(\gamma)} \frac{\exp\left\{\dfrac{2\pi i}{\gamma}\left(\dfrac{(m+\kappa_2)\alpha}{n_2}+\dfrac{(r+\kappa_1)\delta}{n_1}\right)\right\}}{v(MS)\sigma(L,M)}\sigma(M,S). \tag{5.3.27}$$

This depends also on L, Γ, k and the MS v, of course; it is easily checked from (3.2.4, 23) that the right-hand side of (5.3.27) depends only on the residues of α and δ (mod $n_2\gamma$) and (mod $n_1\gamma$), respectively. The sum is zero when $\mathcal{T}_L(\gamma)$ is empty.

By (5.3.23) there are, for each $\gamma>0$, at most $n_1\gamma$ values of δ. Also, since $\alpha\delta\equiv 1 \pmod\gamma$, there are at most n_2 values of α for each pair γ, δ. It follows that $\mathcal{T}_L(\gamma)$ contains at most $n_1n_2\gamma$ matrices

S and so, by (5.3.27),

$$|W(r, m;, \gamma)| \leq n_1 n_2 \gamma. \tag{5.3.28}$$

We shall see that in special cases this estimate can be improved.

We are now ready to prove the following theorem.

Theorem 5.3.2. *Let $L \in \Gamma(1)$, $m \in \mathbb{Z}$, $k > 2$ and put $M = L^{-1}$. Then*

$$G_L(z; m; \Gamma, k, v) = \delta_L \exp\{2\pi i(m+\kappa_1)z/n_1\}$$
$$+ \sum_{r+\kappa_1>0} a(r, m; L) \exp\{2\pi i(r+\kappa_2)z/n_2\}. \tag{5.3.29}$$

Here n_1, n_2, κ_1 and κ_2 are defined by (5.3.16, 17) and $\delta_L = 0$, except when $MU^s \in \Gamma$ for some $s \in \mathbb{Z}$, in which case $n_1 = n_2$, $\kappa_1 = \kappa_2$ and

$$\delta_L = \exp\{2\pi is(m+\kappa_1)/n_1\}/\{v(MU^s)\sigma(L, M)\}. \tag{5.3.30}$$

The coefficients $a(r, m; L)$ are given by the following formulae for $r+\kappa_1>0$:

$$a(r, m; L) = \frac{(2\pi)^k}{\Gamma(k)} e^{-\frac{1}{2}k\pi i}(r+\kappa_1)^{k-1} \sum_{\gamma=1}^{\infty} \frac{W(r, 0; \gamma)}{(n_1\gamma)^k} \quad (m = \kappa_2 = 0), \tag{5.3.31}$$

$$a(r, m; L) = 2\pi e^{-\frac{1}{2}k\pi i} \frac{n_2^{\frac{1}{2}(k-1)}}{n_1^{\frac{1}{2}(k+1)}} \left(\frac{r+\kappa_1}{m+\kappa_2}\right)^{\frac{1}{2}(k-1)}$$
$$\times \sum_{\gamma=1}^{\infty} \frac{W(r, m; \gamma)}{\gamma} J_{k-1}\left(\frac{4\pi}{\gamma}\sqrt{\frac{(r+\kappa_1)(m+\kappa_2)}{n_1 n_2}}\right) \tag{5.3.32}$$

when $m+\kappa_2>0$, and

$$a(r, m; L) = 2\pi e^{-\frac{1}{2}k\pi i} \frac{n_2^{\frac{1}{2}(k-1)}}{n_1^{\frac{1}{2}(k+1)}} \left|\frac{r+\kappa_1}{m+\kappa_2}\right|^{\frac{1}{2}(k-1)}$$
$$\times \sum_{\gamma=1}^{\infty} \frac{W(r, m; \gamma)}{\gamma} I_{k-1}\left(\frac{4\pi}{\gamma}\sqrt{\frac{(r+\kappa_1)|m+\kappa_2|}{n_1 n_2}}\right) \tag{5.3.33}$$

when $m+\kappa_2<0$.

Proof. By (5.1.9), we have

$$G_L(z; m) = \sum_{T \in \mathcal{R}_L} \frac{\exp\left\{\frac{2\pi i(m+\kappa_2)}{n_2} LTz\right\}}{\mu(L, Tz)\gamma(T, z)}.$$

We choose $\mathcal{R}_L$ as described previously, so that we have

$$G_L(z;m)=\delta_L\exp\{2\pi i(m+\kappa_1)z/n_1\}$$
$$+\sum_{LT\in\mathcal{T}_L\hat{\Gamma}_1}\frac{\exp\left\{\dfrac{2\pi i(m+\kappa_2)}{n_2}LTz\right\}}{\mu(L,Tz)\nu(T,z)} \tag{5.3.34}$$

where, by (5.1.11),

$$\delta_L=\frac{\exp\{2\pi is(m+\kappa_1)/n_1\}}{v(MU^s)\sigma(L,M)}$$

if $MU^s\in\Gamma$ and is otherwise zero; when $\delta_L\neq 0$, $n_2=n_1$ and $\kappa_2=\kappa_1$, as has already been remarked.

The mappings belonging to $\mathcal{T}_L\hat{\Gamma}_1$ are defined by the matrices

$$LT=SU^{hn_1}\quad(S\in\mathcal{T}_L,\ h\in\mathbb{Z}).$$

We take $\mathcal{T}_L$ as in (5.3.22) and write

$$\zeta_S:=\left(z+\frac{\delta}{\gamma}\right)\Big/n_1. \tag{5.3.35}$$

Then, by (3.1.15, 5, 10) and (3.3.4),

$$\begin{aligned}\mu(L,Tz)\nu(T,z)&=\mu(L,MSU^{hn_1}z)\nu(MSU^{hn_1},z)\\&=e^{2\pi ih\kappa_1}v(MS)\sigma(L,MS)\mu(S,U^{hn_1}z)\\&=(\gamma n_1)^k\,e^{2\pi ih\kappa_1}v(MS)\sigma(L,M)\sigma^{-1}(M,S),\end{aligned}$$

by (3.2.1, 4). Also

$$\begin{aligned}SU^{hn_1}z=S(z+hn_1)&=\frac{\alpha(z+hn_1)+\beta}{\gamma(z+hn_1)+\delta}\\&=\frac{\alpha}{\gamma}-\frac{1}{\gamma^2n_1(\zeta_S+h)},\end{aligned}$$

so that, by (5.3.9, 34),

$$\begin{aligned}G_L(z;m)&=\delta_L\exp\{2\pi i(m+\kappa_1)z/n_1\}\\&\quad+\sum_{\gamma=1}^{\infty}\sum_{S\in\mathcal{T}_L(\gamma)}\frac{\sigma(M,S)\exp\left\{\dfrac{2\pi i}{n_2\gamma}(m+\kappa_2)\alpha\right\}}{(\gamma n_1)^k v(MS)\sigma(L,M)}\\&\quad\times F_k\left(\zeta_S,\kappa_1,\frac{m+\kappa_2}{n_1n_2\gamma^2}\right)\\&=\delta_L\exp\{2\pi i(m+\kappa_1)z/n_1\}\\&\quad+\sum_{r+\kappa_1>0}a(r,m;L)\exp\{2\pi i(r+\kappa_1)z/n_1\},\end{aligned} \tag{5.3.36}$$

where, by theorem 5.3.1 and (5.3.27), for $r+\kappa_1>0$,

$$a(r,m;L)=\sum_{\gamma=1}^{\infty}\frac{W(r,m;\gamma)}{(\gamma n_1)^k}g_r.$$

Here g_r is given by (5.3.11–13) with $\kappa=\kappa_1$ and $\lambda=(m+\kappa_2)/(n_1n_2\gamma^2)$. The inversion of the order of summation is justified by absolute convergence.

From (5.3.31–33) we deduce the following estimates for the coefficients $a(r,m;L)$, with the help of (5.3.2, 3, 7, 8, 28).

Theorem 5.3.3. *In the notation of theorem* 5.3.2, *we have, for large* r,

$$a(r,0;L)=O(r^{k-1})\quad(\kappa_2=0),\tag{5.3.37}$$

$$a(r,m;L)=O\left(\frac{r^{\frac{1}{2}k}}{(m+\kappa_2)^{\frac{1}{2}k-1}}\right)\quad(m+\kappa_2>0),\tag{5.3.38}$$

$$a(r,m;L)=O\left\{\frac{r^{\frac{1}{2}k}}{|m+\kappa_2|^{\frac{1}{2}k-1}}\exp\left(4\pi\left|\frac{(r+\kappa_1)(m+\kappa_2)}{n_1n_2}\right|^{\frac{1}{2}}\right)\right\}\quad(m+\kappa_2<0).\tag{5.3.39}$$

The constants implied by the O-notation do not depend on r or m.

It follows from (5.3.37) and theorems 5.1.3 and 5.2.4 that the estimate for a_m given in theorem 4.5.3 can be improved to give $a_m=O(m^{k-1})$ when $k>2$. On the other hand, the estimate of the order of magnitude of the Fourier coefficients of a cusp form given by theorem 4.5.2(i) is not improved by (5.3.38). A smaller bound than (5.3.28) for the Kloosterman sum $W(r,m;\gamma)$ would, however, yield a corresponding improvement in theorem 5.3.3 and theorem 4.4.2(i). In certain cases it is possible to improve on (5.3.28), as we show in the next section.

5.4. Kloosterman sums. In this section we obtain estimates of the magnitude of the Kloosterman sum $W(r,m;\gamma)$ defined in (5.3.27) in certain special cases. The estimates we give are not the best possible, but are sufficient for our purpose; they are obtained by number-theoretic arguments that are elementary, but which cannot easily be found in any one place in the literature.

We restrict our attention to the case when $\Gamma = \bar{\Gamma}(N)$ and k is an integer. When $N > 2$ we assume that $v(T) = 1$ for all $T \in \Gamma(N)$, so that $v(T) = (-1)^k$ for $T \in \bar{\Gamma}(N) - \Gamma(N)$; for odd k this MS is the MS that was introduced in §3.3. When $N = 1$ or 2, we assume that k is even and that $v(T) = 1$ for all $T \in \Gamma(N)$. In all cases, therefore, $v(T) = 1$ when $T \in \Gamma(N)$. We deduce that

$$n(M\infty, \Gamma) = N \quad \text{for all } M \in \Gamma(1) \tag{5.4.1}$$

and

$$\kappa_m = \kappa(M\infty, \Gamma, v) = 0 \quad \text{for all } M \in \Gamma(1). \tag{5.4.2}$$

As usual we write $M = L^{-1}$ and take L to be given by (5.3.24).

In (5.3.27) the Kloosterman sum $W(r, m; \gamma)$ is defined in terms of a set $\mathcal{T}_L(\gamma)$ of matrices S. This set is empty except when γ satisfies (5.3.25), in which case the entries α and δ must satisfy (5.3.26), and this determines S uniquely. Further, since $S \equiv \varepsilon L \pmod N$ when $S \in \mathcal{T}_L(\gamma)$, where $\varepsilon = \pm 1$, it follows that $v(MS) = \varepsilon^k$, and we therefore have

$$W(r, m; \gamma) = \sum \varepsilon^k \exp\left\{\frac{2\pi i}{N\gamma}(m\alpha + r\delta)\right\}, \tag{5.4.3}$$

where the summation is over all α and δ satisfying (5.3.26). The ε^k is taken inside the summation since, when $N > 2$ and $2C \equiv 0 \pmod N$, both values $\varepsilon = 1$ and -1 have to be summed over; we refer to these conditions as the *awkward case.*

We now write

$$N = N_1 N_2, \quad \gamma = \gamma_1 \gamma_2, \tag{5.4.4}$$

where N_1 is the largest factor of N prime to γ and γ_1 is the largest factor of γ prime to N. Thus N_2 and γ_2 are either both unity or are composed of positive powers of the same prime numbers.

Any pair of numbers α and δ modulo N may be expressed uniquely in the forms

$$\alpha = \alpha_1 N\gamma_2 + \alpha_2 \gamma_1, \quad \delta = \delta_1 N\gamma_2 + \delta_2 \gamma_1, \tag{5.4.5}$$

where α_1 and δ_1 are taken modulo γ_1 and α_2 and δ_2 are taken modulo $N\gamma_2$.The conditions (5.3.26) are then equivalent to the conditions

$$\alpha_1 \delta_1 (N\gamma_2)^2 \equiv 1 \pmod{\gamma_1} \tag{5.4.6}$$

and

$$\alpha_1\gamma_1 \equiv \varepsilon A \pmod{N}, \quad \delta_2\gamma_1 = \varepsilon D \pmod{N},$$
$$\alpha_2\delta_2\gamma_1^2 \equiv 1 + \varepsilon B\gamma_1\gamma_2 \pmod{N\gamma_2}. \tag{5.4.7}$$

We observe that δ_2 is necessarily prime to γ_2. Choose γ_1' modulo N so that

$$\gamma_1\gamma_1' \equiv 1 \pmod{N}.$$

Then if δ_2 is chosen so that

$$\delta_2 \equiv \varepsilon\gamma_1' D \pmod{N},$$

we can find exactly one residue $\alpha_2 \pmod{N\gamma_2}$ to satisfy (5.4.7). For if we put $\alpha_2 = \varepsilon\gamma_1' A + nN$, we have to choose $n \pmod{\gamma_2}$ to satisfy

$$nN\gamma_1^2\delta_1 \equiv 1 + \varepsilon B\gamma_1\gamma_2 - A\gamma_1\delta_2 \pmod{N\gamma_2}.$$

Since $(N\gamma_1^2\delta_2, N\gamma_2) = N$ and since

$$1 + \varepsilon B\gamma_1\gamma_2 - A\gamma_1\delta_2 \equiv 1 + BC - AD \equiv 0 \pmod{N},$$

this reduces to a congruence modulo γ_2 that has a unique solution. Accordingly, the congruence (5.4.6) has $\varphi(\gamma_1)$ different pairs of solutions α_1, δ_1, while (5.4.7) has γ_2 different pairs of solutions α_2, δ_2 for each admissible value of ε; here $\varphi(\gamma_1)$ is Euler's function. Hence, by (5.4.3, 5),

$$\begin{aligned} W(r, m; \gamma) &= \sum_{\alpha_1,\delta_1} \exp\{2\pi i(m\alpha_1 + r\delta_1)/\gamma_1\} \\ &\quad \times \sum_{\alpha_2,\delta_2} \varepsilon^k \exp\{2\pi i(m\alpha_2 + r\delta_2)/(N\gamma_2)\} \\ &= S(r, lm; \gamma_1)T(r, m; N\gamma_2), \end{aligned} \tag{5.4.8}$$

say, where l is chosen modulo γ_1 so that

$$l(N\gamma_2)^2 \equiv 1 \pmod{\gamma_1} \tag{5.4.9}$$

and, for any integers u, v, q,

$$S(u, v; q) = \sum_{h=1}^{q} \exp\{2\pi i(uh + vh')/q\}, \tag{5.4.10}$$

where $hh' \equiv 1 \pmod{q}$. In particular, we have

$$|W(r, m; \gamma)| \leq 2\gamma_2 |S(r, lm; \gamma_1)| \tag{5.4.11}$$

and

$$|W(0, 0; \gamma)| = \rho\gamma_2\varphi(\gamma_1) \tag{5.4.12}$$

where ρ is ε^k, except in the awkward case, when $\rho = 1 + (-1)^k$.

We now suppose that $m=0$ and take, in the first place, $N=1$. Then $\gamma=\gamma_1$, $\gamma_2=N=l=1$ and

$$W(r, 0; \gamma) = \sum_{\substack{h=1\\(h,\gamma)=1}}^{\gamma} \exp(2\pi i r h/\gamma) \quad (N=1).$$

This is just the well known Ramanujan's sum $c_\gamma(r)$ and

$$W(r, 0; \gamma) = c_\gamma(r) = \sum_{d/(r,y)} d\mu(\gamma/d) \quad (N=1); \qquad (5.4.13)$$

see Hardy and Wright (1938), §16.6. Here $\mu(\gamma/d)$ is the Möbius function and the summation is over all positive divisors of r and γ.

In the general case when $N>1$ it is also possible to evaluate $W(r, 0; \gamma)$ explicitly. Thus it can be shown that

$$W(r, 0; \gamma) = \varepsilon^k \gamma_2 \exp(2\pi i r \gamma_1' \varepsilon D/(\gamma_2 N)) c_{\gamma_1}(r/\gamma_2), \qquad (5.4.14)$$

where $c_{\gamma_1}(r/\gamma_2)$ is defined to be zero when γ_2 does not divide r; in the awkward case the right-hand side of (5.4.14) must be summed over both values of ε. We do not prove (5.4.14) since all we require is the upper bound

$$|W(r, 0; \gamma)| \leq 2\gamma_2 \sum_{d|(\gamma,r)} d \leq 2\gamma_2 \sum_{d|r} d =: 2\gamma_2\sigma_1(|r|), \qquad (5.4.15)$$

which is easily obtained from (5.4.11), since

$$S(r, 0; \gamma) = \sum_{\substack{h=1\\(h,q)=1}}^{q} \exp(2\pi i r h/\gamma) = c_\gamma(r).$$

It is clear that we also have

$$|W(0, m; \gamma)| \leq 2\gamma_2\sigma_1(|m|). \qquad (5.4.16)$$

For the remainder of the section we shall be mainly concerned with the simple Kloosterman sum $S(u, v; q)$. We note, in the first place that

$$S(u, v; q) = S(ua, va'; q) \qquad (5.4.17)$$

for any a prime to q, where $aa' \equiv 1 \pmod q$. Also, if $q=q_1q_2$, where $(q_1, q_2)=1$, we can write $h=h_1q_2+h_2q_1$, where h_1, h_2 run through reduced residue systems modulo q_1 and modulo q_2, respectively. Then

$$h' \equiv h_1'(q_2')^2 q_2 + h_2'(q_1')^2 q_1 \pmod{q_1 q_2}$$

where $h_1 h_1' \equiv q_2 q_2' \equiv 1 \pmod{q_1}$ and $h_2 h_2' \equiv q_1 q_1' \equiv 1 \pmod{q_2}$. Hence

$$S(u, v; q) = S(u, vq_2'^2; q_1) S(u, vq_1'^2; q_2). \tag{5.4.18}$$

By repeated applications of this formula it is clear that the evaluation of $S(u, v; q)$ can be reduced to that of evaluating $S(u, v; p^n)$ for all u and v, where p is a prime and n is a positive integer.

We consider first the case when $n = 1$, so that q is a prime number p, and we use a method due to Mordell (1932). Since

$$S(0, 0; p) = p - 1, \quad S(u, 0; p) = S(0, u; p) = -1 \quad (p \nmid u), \tag{5.4.19}$$

we have, by (5.4.17),

$$\sum_{u=1}^{p} \sum_{v=1}^{p} |S(u, v; p)|^4 = (p-1)^4 + 2(p-1) + (p-1) \sum_{m=1}^{p-1} |S(1, m; p)|^4. \tag{5.4.20}$$

On the other hand, the left-hand side of (5.4.20) is equal to

$$\sum_{u=1}^{p} \sum_{v=1}^{p} \sum_{x_1, x_2=1}^{p-1} \sum_{y_1, y_2=1}^{p-1} \exp\left\{\frac{2\pi i}{p}[u(x_1 + x_2 - y_1 - y_2) + v(x_1' + x_2' - y_1' - y_2')]\right\} = p^2 s, \tag{5.4.21}$$

where s is the number of solutions of the congruences

$$x_1 + x_2 \equiv y_1 + y_2, \quad x_1' + x_2' \equiv y_1' + y_2' \pmod{p}. \tag{5.4.22}$$

We note that, if $x_1 + x_2 \not\equiv 0 \pmod{p}$, it follows that $x_1 x_2 \equiv y_1 y_2 \pmod{p}$. Hence s is equal to the number of solutions of the congruences

$$x_1 + x_2 \equiv y_1 + y_2 \equiv 0, \quad x_1 x_2 \equiv y_1 y_2 \pmod{p} \tag{5.4.23}$$

plus the number of solutions of the congruences

$$x_1 + x_2 \equiv y_1 + y_2 \equiv 0 \not\equiv x_1^2 \not\equiv y_1^2 \not\equiv 0 \pmod{p}. \tag{5.4.24}$$

It follows from (5.4.23) that either $x_1 = y_1$, $x_2 = y_2$ or $x_1 = y_2$, $x_2 = y_1$, these being different solutions except when $x_1 = x_2$ and $y_1 = y_2$. Hence, if $p \geq 3$,

$$s = \{2(p-1)^2 + (p-1)\} + (p-1)(p-3).$$

For there are $p-1$ possible choices of the pair x_1, x_2 in (5.4.24), leaving $p-3$ choices for y_1, y_2. Hence we have, by (5.4.20, 21), for $1 \leq m < p$, provided that $p \geq 3$,

$$|S(1, m; p)|^4 \leq p^2\{2(p-1)-1+(p-3)\}-(p-1)^3-2$$
$$= 2p^3-3p^2-3p-1.$$

It follows that

$$|S(u, v; p)| \leq 2p^{\frac{3}{4}} \quad ((uv, p)=1), \tag{5.4.25}$$

for all primes p, including $p=2$; for $|S(u, v; 2)|=1$. This result is due to Estermann (1930).

Now suppose that $q=p^n$, where $n \geq 1$ and put

$$p^s = (u, v, p^n),$$

so that $0 \leq s \leq n$. If $s=n$, then $S(u, v; p^n) = S(0, 0; p^n) = p^{n-1}(p-1)$, while

$$S(u, v; p^n) = p^s S(up^{-s}, vp^{-s}; p^{n-s})$$

if $0 \leq s < n$. We show that, in all cases,

$$|S(u, v; p^n)| \leq 4(u, v, p^n)p^{\frac{3}{4}n}. \tag{5.4.26}$$

This is certainly true for $s=n, n-1$ by (5.4.25). To prove it for all s it suffices to show that

$$|S(u, v; p^n)| \leq 4p^{\frac{3}{4}n}$$

when $(u, v, p)=1$ and $n \geq 2$. Take

$$\nu = \left[\frac{n+1}{2}\right],$$

so that $\nu \geq \frac{1}{2}n \geq 1$. In (5.4.10) we write

$$h = l + rp^\nu \quad (0 \leq r < p^{n-\nu}, 0 \leq l < p^\nu, (l, p)=1) \tag{5.4.27}$$

and choose l' (mod p^n) so that $ll' \equiv 1$ (mod p^n). Then we may take

$$h' = l' - r(l')^2 p^\nu$$

so that

$$S(u, v; p^n) = \sum_l \exp\{2\pi i(ul+vl')/q\} \sum_r \exp\{2\pi i(u-vl'^2)p^{\nu-n}\},$$

the summations over l and r being as given in (5.4.27). Hence

$$S(u, v; p^n) = 0 \quad \text{if } l^2u - v \not\equiv 0 \pmod{p^{n-\nu}}$$

has solutions. Since, by (5.4.17), $S(u, v; p^n) = S(1, uv; p^n)$, we have

$$S(u, v; p^n) = p^{n-\nu} \sum \exp\{2\pi i(l + uvl')/q\}$$

where the summation is over all l satisfying

$$(l, p) = 1, \quad 0 \leq l < p^\nu, \quad l^2 \equiv uv \pmod{p^{n-\nu}}.$$

When n is even, $n - \nu = \nu$ and there are at most two (or four when $p = 2$) values of l satisfying these conditions. When n is odd, $n - \nu = \nu - 1$ there are at most $2p$ (or $4p$ when $p = 2$) values of l. Hence we have

$$|S(u, v; p^n)| \leq 4p^\nu \leq 4p^{\frac{3}{4}n}. \tag{5.4.28}$$

This completes the proof of (5.4.26), and we deduce immediately from (5.4.28) that

$$|S(u, v; q)| \leq (u, v, q)q^{\frac{3}{4}}4^{\omega(q)}, \tag{5.4.29}$$

where $\omega(q)$ is the number of different prime factors of q.

We note that, since $2^{\omega(q)} \leq d(q)$, where $d(q)$ is the number of positive divisors of q,

$$4^{\omega(q)} \leq \{d(q)\}^2 = O(q^{\varepsilon_1}), \tag{5.4.30}$$

for large q and arbitrary $\varepsilon_1 > 0$; see Hardy and Wright (1938), §§18.1, 22.15. Accordingly we have, by (5.4.11, 29),

$$|W(r, m; \gamma)| \leq 4^{\omega(\gamma)}(r, m, \gamma_1)\gamma_1^{\frac{3}{4}}(2\gamma_2) \tag{5.4.31}$$

$$\leq A(r, m, \gamma_1)\gamma_1^{4/5}\gamma_2 \tag{5.4.32}$$

where A is a sufficiently large positive constant.

In §5.7 we shall need to consider various infinite series containing Kloosterman sums and it is convenient to discuss these here. These series are of the type

$$\sum_\gamma \frac{W(r, m; \gamma)}{\gamma^{2+2s}} \tag{5.4.33}$$

where $s \geq 0$ and the summation is carried out over all positive γ that are congruent to $\pm C \pmod{N}$. When both r and m are not zero we need only show that the series is absolutely convergent;

when $r=m=0$ the series diverges when $s=0$ and we need to know how the sum behaves as $s \to 0+$.

Suppose, in the first place, that $m \neq 0$. Then, by (5.4.32), the series (5.4.33) is majorized by the series

$$A|m| \sum_{\gamma} \frac{\gamma_1^{4/5} \gamma_2}{\gamma^{2+2s}}$$

in which we may suppose that γ runs through all the positive integers. Thus we have to consider the series

$$A|m| \sum_{\gamma=\gamma_1\gamma_2=1}^{\infty} \gamma_1^{-6/5-2s} \gamma_2^{-1-2s} = \prod_{p \nmid N} (1-p^{-6/5-2s})^{-1} \prod_{p|N} (1-p^{-1-2s})^{-1}$$

$$= \zeta(\tfrac{6}{5}+2s) \prod_{p|N} \left(\frac{1-p^{-6/5-2s}}{1-p^{-1-2s}} \right).$$

Accordingly, when $m \neq 0$, the series (5.4.33) is absolutely convergent, not only for $s \geq 0$ but also for $s > -\frac{1}{10}$. An exactly similar argument shows that, when $r \neq 0$, the series (5.4.33) is absolutely convergent for $s > -\frac{1}{10}$ and therefore for $s \geq 0$.

Now take $r=m=0$, k even, and put

$$g(s) = \sum_{\gamma} \frac{W(0,0;\gamma)}{\gamma^{2+2s}} = \rho \sum_{\gamma} \frac{\gamma_2 \varphi(\gamma_1)}{\gamma^{2+2s}}, \tag{5.4.34}$$

where $s>0$. Note that $\rho=1$, except in the awkward case, when $\rho=2$. We use standard methods of analytic number theory to sum this series. Write

$$(C,D)=d, \quad N=N^*d, \quad C=C^*d, \quad \gamma=\gamma^*d, \quad \gamma_2=\gamma_2^*d, \tag{5.4.35}$$

so that

$$g(s) = \rho d^{-1-2s} \sum \frac{\varphi(\gamma_1)}{\gamma_1^{2+2s}(\gamma_2^*)^{1+2s}}, \tag{5.4.36}$$

the summation being over all $\gamma^* = \gamma_1\gamma_2^* \equiv \pm C^* \pmod{N^*}$. Write

$$\delta_N^* = 1 \ (N=1,2), \quad \delta_N^* = 2 \ (N>2), \tag{5.4.37}$$

and let χ be any Dirichlet character modulo N^* and

$$L(2s+1,\chi) = \sum_{n=1}^{\infty} \chi(n) n^{-1-2s}$$

the associated Dirichlet series. Put

$$g_\chi(s) = d^{-1-2s} \sum_{\gamma^*=1}^{\infty} \frac{\chi(\gamma^*)\varphi(\gamma_1)}{\gamma_1^{2+2s}(\gamma_2^*)^{1+2s}}. \tag{5.4.38}$$

Then, because of the multiplicative properties of the summand, it is easily seen that

$$g_\chi(s) = d^{-1-2s}\frac{L(2s+1,\chi)}{L(2s+2,\chi)} \prod_{p|N} (1-\chi(p)p^{-2-2s})^{-1}. \tag{5.4.39}$$

In fact, when χ is not the principal character $\chi_0 \pmod{N^*}$, the right-hand side of (5.4.39) is a holomorphic function of the complex variable s over the half-plane $\text{Re}\, s \geq -\frac{1}{2}$. On the other hand, since

$$L(2s+1,\chi_0) = \zeta(2s+1) \prod_{p|N^*} (1-p^{-1-2s}),$$

$$g_{\chi_0}(s) = d^{-1-2s}\frac{\zeta(2s+1)}{\zeta(2s+2)} \prod_{p|N^*} (1-p^{-2s-1}) \prod_{p|N} (1-p^{-2s-2})^{-1}. \tag{5.4.40}$$

This function is therefore holomorphic for $\text{Re}\, s \geq 0$, with the exception of a simple pole at $s=0$. Accordingly,

$$\lim_{s\to 0+} s g_{\chi_0}(s) = \frac{3}{\pi^2 d} \prod_{p|N^*} (1-p^{-1}) \prod_{p|N} (1-p^{-2})^{-1}. \tag{5.4.41}$$

Now

$$g(s) = \frac{\rho}{\varphi(N^*)} \sum_{\chi} \{\bar{\chi}(C^*) + (\delta_N^* - 1)\bar{\chi}(-C^*)\}\, g_\chi(s) \tag{5.4.42}$$

except in the awkward case, when the term $(\delta_N^* - 1)\bar{\chi}(-C^*)$ must be omitted; for then $-C^* \equiv C^* \pmod{N^*}$. Observe that the coefficient of $g_{\chi_0}(s)$ on the right of (5.4.42) is $\delta_N^*/\varphi(N^*)$ in all cases. Thus the series for $g(s)$ is absolutely convergent for $s>0$ and

$$\lim_{s\to 0+} s g(s) = \frac{\delta_N^*}{\varphi(N^*)} \lim_{s\to 0+} s g_{\chi_0}(k) = \frac{3\delta_N^*}{\pi^2 N} \prod_{p|N} (1-p^{-2})^{-1}. \tag{5.4.43}$$

5.5. Poincaré series belonging to $\Gamma(N)$. In this section we apply the results of §5.3 to the case when $\Gamma = \bar{\Gamma}(N)$ $(N \geq 1)$. We make the same assumptions about the weight and multiplier system as are described in the second paragraph of §5.4, but require in addition that $k>2$ in order to ensure convergence of the Poincaré series. We shall consider the case $k=2$ in §5.7.

For the purpose of evaluating the Fourier coefficients of the Poincaré series considered in §5.3 we found it convenient to take a transversal $\mathcal{R}_L$ (see (5.1.8)) with special properties and introduced the set $\mathcal{T}_L$ for this purpose. We shall make use of the particular transversal in our further consideration of Fourier coefficients in the present section, but remark that a slightly different choice of $\mathcal{R}_L$ simplifies the definition (5.1.9) of the series $G_L(z;m;\bar{\Gamma}(N),k,v)$. We may take the matrices in $\mathcal{R}_L$ to belong to $\Gamma(N)$ and this implies that the set $L\mathcal{R}_L$ may be taken to consist of all matrices $S=LT$ with the properties that (i) $S\equiv L \pmod N$, and (ii) two different matrices S_1 and S_2 in $L\mathcal{R}_L$ have second rows $[\gamma_1,\delta_1],[\gamma_2,\delta_2]$ satisfying $[\gamma_1,\delta_1]\neq\pm[\gamma_2,\delta_2]$. This set $L\mathcal{R}_L$ is maximal with respect to these two properties.

Then, by (5.1.9) and (5.4.1, 2), for any $m\in\mathbb{Z}$ and $L\in\Gamma(1)$,

$$G_L(z;m;\bar{\Gamma}(N),k,v)=\sum_{S\equiv L \pmod N}{}' (S:z)^{-k}\exp\left\{\frac{2\pi i m}{N}Sz\right\}, \tag{5.5.1}$$

where the prime signifies that the summation is subject to the conditions (i) and (ii) just stated.

We now restate theorem 5.3.2 in the particular case we are considering. Then $\delta_L=0$ except when $L\equiv\varepsilon U^s \pmod N$ for some $s\in\mathbb{Z}$ and $\varepsilon=\pm1$, in which case $\delta_L=\varepsilon^k e^{2\pi i s m/N}$. In terms of the entries B, C and D of L this states that

$$\delta_L=\begin{cases} e^{2\pi i m B/N} & \text{if } C\equiv0, D\equiv1 \pmod N,\\ (-1)^k e^{-2\pi i m B/N} & \text{if } C\equiv0, D\equiv-1 \pmod N,\\ 0 & \text{otherwise.}\end{cases} \tag{5.5.2}$$

Theorem 5.5.1. *Let $L\in\Gamma(1)$, $m\in\mathbb{Z}$ and let the MS v and weight k be subject to the restrictions imposed at the beginning of this section. Then*

$$G_L(z;m;\bar{\Gamma}(N),k,v)=\delta_L e^{2\pi i m z/N}+\sum_{r=1}^{\infty}a(r,m;L)e^{2\pi i r z/N}, \tag{5.5.3}$$

where, for $r \geq 1$,

$$a(r, 0; L) = \left(\frac{2\pi r}{Ni}\right)^k \frac{1}{r\Gamma(k)} \sum_{\substack{\gamma=1 \\ \gamma \equiv \pm C \pmod N}}^{\infty} \gamma^{-1} W(r, 0; \gamma), \tag{5.5.4}$$

$$a(r, m; L) = \frac{2\pi}{Ni^k}\left(\frac{r}{m}\right)^{\frac{1}{2}(k-1)} \times \sum_{\substack{\gamma=1 \\ \gamma \equiv \pm C \pmod N}}^{\infty} \gamma^{-1} W(r, m; \gamma) J_{k-1}\{4\pi\sqrt{(rm)}/(\gamma N)\}, \tag{5.5.5}$$

for $m > 0$ *and*

$$a(r, m; L) = \frac{2\pi}{Ni^k}\left(\frac{r}{|m|}\right)^{\frac{1}{2}(k-1)} \times \sum_{\substack{\gamma=1 \\ \gamma \equiv \pm C \pmod N}}^{\infty} \gamma^{-1} W(r, m; \gamma) I_{k-1}\{4\pi\sqrt{(r|m|)}/(\gamma N)\}, \tag{5.5.6}$$

when $m < 0$. *Here* δ_L *is defined by (5.5.2) and* $W(r, m; \gamma)$ *by (5.4.3).*

For the remainder of this section we take $m = 0$, so that G_L is an Eisenstein series, and we investigate its Fourier coefficients, as given by (5.5.4), in greater detail. It is clear from (5.5.1) that G_L depends only on the second row of the matrix L and we therefore write

$$E_k(z; C, D; N) := G_L(z; 0; \bar{\Gamma}(N), k, v) = \sum_{\substack{\gamma \equiv C, \delta \equiv D \pmod N \\ (\gamma, \delta) = 1}} (\gamma z + \delta)^{-k}. \tag{5.5.7}$$

This summation is subject to the condition (ii) stated previously.

The numbers C and D are relatively prime; however, since they only arise in congruences modulo N, it is only necessary to assume that

$$(C, D, N) = 1, \tag{5.5.8}$$

i.e. that C, D is a primitive pair modulo N, as defined in §1.4. We also note that, for any $T \in \Gamma(1)$, by (5.1.10),

$$E_k(z; C, D; N)|S = E_k(z; Ca + Dc, Cb + Da; N). \tag{5.5.9}$$

The number of different Eisenstein series is $\hat{\lambda}(N)$ (see (2.4.17)), one for each cusp of a fundamental region.

In (5.5.7) we also write

$$a(r; C, D; N) = a(r, 0; L) \tag{5.5.10}$$

so that

$$E_k(z; C, D; N) = \delta_L + \sum_{r=1}^{\infty} a(r; C, D; N)\, e^{2\pi i r z/N} \tag{5.5.11}$$

where δ_L is given by (5.5.2) with $m = 0$.

It is clear from (5.5.4) that the coefficients $a(r; C, D; N)$ can be expressed in terms of the Kloosterman sums $W(r, 0; \gamma)$ for which an explicit formula is given in (5.4.14). However, it is rather more convenient to obtain the Fourier coefficients of a related modular form $E_k^*(z; C, D; N)$ and then express the Eisenstein series (5.5.7) in terms of these new functions. We write, for any integers C and D

$$E_k^*(z; C, D; N) := \sum_{m \equiv C, n \equiv D \pmod N}{}' (mz+n)^{-k}, \tag{5.5.12}$$

where the prime indicates that the term with $m = n = 0$ is omitted when $C \equiv D \equiv 0 \pmod N$. As usual, $z \in \mathbb{H}$ and k is an integer greater than 2. (Note that, if k is odd, E_k^* vanishes identically when $N = 1$ or 2.)

We may assume that C, D is a primitive pair modulo N; i.e. that (5.5.8) holds. For if $(C, D, N) = h$, it follows easily that

$$E_k^*(z; C, D; N) = h^{-k} E_k^*(z; C/h, D/h; N/h) \tag{5.5.13}$$

and C/h, D/h is a primitive pair modulo N/h.

The following theorem shows the relationship between the functions E_k and E_k^*.

Theorem 5.5.2. *Let C, D be a primitive pair modulo N and k an integer greater than 2. Then, if $N > 2$,*

$$E_k^*(z; C, D; N) = \sum_{\substack{h=1 \\ (h,N)=1}}^{N} \Bigg\{ \sum_{\substack{m=1 \\ mh \equiv 1 \pmod N}}^{\infty} m^{-k} \Bigg\} E_k(z; hC, hD; N) \tag{5.5.14}$$

and

$$E_k(z; C, D: N) = \sum_{\substack{h=1 \\ (h,N)=1}}^{N} \Bigg\{ \sum_{\substack{m=1 \\ mh \equiv 1 \pmod N}}^{\infty} \frac{\mu(m)}{m^k} \Bigg\} E_k^*(z; hC, hD; N). \tag{5.5.15}$$

Further,

$$E_k^*(z) := E_k(z; C, D; 1) = 2\zeta(k) E_k(z; C, D; 1) =: 2\zeta(k) E_k(z), \tag{5.5.16}$$

and

$$E_k^*(z; C, D; 2) = 2(1-2^{-k})\zeta(k) E_k(z; C, D; 2). \tag{5.5.17}$$

Finally, for all $N \geq 1$ and all $S \in \Gamma(1)$,

$$E_k^*(z, C, D; N)|S = E_k^*(z; C\alpha + D\gamma, C\beta + D\delta; N). \tag{5.5.18}$$

Proof. In the summation on the right of (5.5.12), write $(m, n) = d$. Then, since $(C, D, N) = 1$, d is prime to N. Write $m = d\gamma$, $n = d\delta$ so that we have

$$E^*(z; C, D; N) = \sum_{\substack{d=1 \\ (d,N)=1}}^{\infty} d^{-k} \sum_{\substack{d\gamma \equiv C \\ d\delta \equiv D}}' (\gamma z + \delta)^{-k},$$

the congruences being modulo N. Choose h so that $hd \equiv 1 \pmod N$ and (5.5.14) follows (with m in place of d).

To derive (5.5.15) we use the fact that

$$\sum_{d|n} \mu(d) = \begin{cases} 1 & \text{if } n = 1, \\ 0 & \text{if } n > 1. \end{cases}$$

Hence we have

$$\begin{aligned} E_k(z; C, D; N) &= \sum_{\substack{\gamma \equiv C, \gamma \equiv D \\ (\gamma,\delta)=1}} (\gamma z + \delta)^{-k} \\ &= \sum_{\gamma \equiv C, \delta \equiv D} (\gamma z + \delta)^{-k} \sum_{d|(\gamma,\delta)} \mu(d) \\ &= \sum_{\substack{d=1 \\ (d,N)=1}}^{\infty} \mu(d) d^{-k} \sum_{\substack{\gamma' d \equiv C \\ \delta' d \equiv D}} (\gamma' z + \delta')^{-k} \end{aligned}$$

where we have put $\gamma = \gamma' d$, $\delta = \delta' d$ and have used the fact that d must be prime to N, since $(C, D, N) = 1$. Take h so that $hd \equiv 1 \pmod N$ and (5.5.15) follows (with m in place of d).

When $N=1$ or 2 the situation is simpler. For (5.5.16) is obvious and so is (5.5.17), since

$$\sum_{\substack{d=1\\ d\ \text{odd}}}^{\infty} d^{-k}=(1-2^{-k})\zeta(k);$$

the factor 2 arises in each case because of condition (ii) at the beginning of this section. Finally (5.5.18) follows from what has been proved and from (5.4.9).

It follows from the theorem that when $(C, D, N)=1$, the functions $E_k^*(z; C, D; N)$ are modular forms and span the same vector space as the Eisenstein series $E_k(z; C, D; N)$. We suppose that they have the Fourier series expansion

$$E_k^*(z; C, D; N)=\delta_L^*+\sum_{r=1}^{\infty} a^*(r; C, D; N)\, e^{2\pi i r z/N}. \tag{5.5.19}$$

We now obtain formulae for the coefficients $a^*(r; C, D; N)$. For this purpose we introduce the functions

$$\sigma_{k-1}(r)=\sum_{d\mid r, d>0} d^{k-1} \tag{5.5.20}$$

$$\sigma_{k-1}(r; C, D; N)=\sum_{\substack{d\mid r\\ \frac{r}{d}\equiv C\ (\mathrm{mod}\ N)}} d^{k-2}|d|\, e^{2\pi i d D/N}. \tag{5.5.21}$$

In this last summation d is allowed to take negative as well as positive values.

Theorem 5.5.3. *Let* $(C, D, N)=1$. *The constant term in the Fourier series expansion of* $E_k^*(z; C, D; N)$ *is*

$$\delta_L^*=\begin{cases}\displaystyle\sum_{r\equiv D\ (\mathrm{mod}\ N)} r^{-k} & \textit{when } C\equiv 0\ (\mathrm{mod}\ N),\\ 0 & \textit{otherwise.}\end{cases} \tag{5.5.22}$$

(*The summation is over all* $r\in\mathbb{Z}$ *for which* $r\equiv D$ (mod N).) *Further, for* $r\geq 1$,

$$a^*(r; C, D; N)=\frac{(2\pi/Ni)^k}{\Gamma(k)}\sigma_{k-1}(r; C, D; N). \tag{5.5.23}$$

Proof. We apply theorem 5.3.1 to (5.5.12, 19), with $\kappa=\lambda=0$, $n=D+hN$ $(h\in\mathbb{Z})$ and either $(mz+D)/N$ or $-(mz+D)/N$ in place of z according as $m>0$ or $m<0$; for the imaginary part is required to be positive. We can have $m=0$ only when $C\equiv 0 \pmod N$ and this produces the constant term δ_L^* given by (5.5.22). Hence we have, in the notation of theorem 5.3.1,

$$\begin{aligned} E_k^*(z;C,D;N)-\delta_L^* &= \sum_{\substack{m\equiv C\\ m>0}} N^{-k}F_k\{(mz+D)/N,0,0\} \\ &\quad +(-1)^k \sum_{\substack{m\equiv C\\ m<0}} N^{-k}F_k\{-(mz+D)/N,0,0\} \\ &= \frac{(2\pi/Ni)^k}{\Gamma(k)}\left\{\sum_{\substack{m\equiv C\\ m>0}}\sum_{n=1}^{\infty} n^{k-1}\,e^{2\pi in(mz+D)/N}\right. \\ &\quad \left.+(-1)^k\sum_{\substack{m\equiv C\\ m<0}}\sum_{n=1}^{\infty} n^{k-1}\,e^{-2\pi in(mz+D)/N}\right\}. \end{aligned}$$

All congruences are modulo N. The series are absolutely convergent so that they may be rearranged in powers of $e^{2\pi irz/N}$, where $r=nm$ and $-nm$, respectively. This gives

$$\begin{aligned} a^*(r;C,D;N)=\frac{(2\pi/Ni)^k}{\Gamma(k)}&\left\{\sum_{\substack{d\mid r,d>0\\ \frac{r}{d}\equiv C}} d^{k-1}\,e^{2\pi idD/N}\right. \\ &\quad \left.-\sum_{\substack{d\mid r,d<0\\ \frac{r}{d}\equiv C}} d^{k-1}\,e^{2\pi idD/N}\right\}, \end{aligned}$$

from which (5.5.23) follows.

We immediately deduce the following theorem.

Theorem 5.5.4. *Let $(C,D,N)=1$ where $N>1$ and k is an integer greater than 2; k is supposed even when $N=2$. Then the Fourier coefficients of the Eisenstein series $E_k(z;C,D;N)$ are given by (5.5.2, 11) with $m=0$ and the following formulae:* (i) *When* $N>2$, $n\geq 1$,

$$a(n;C,D;N)=\frac{(2\pi/Ni)^k}{\Gamma(k)}\sum_{\substack{h=1\\ (h,N)=1}}^{N}\ \sum_{\substack{m=1\\ mh\equiv 1\\ (\text{mod } N)}}^{\infty}\frac{\mu(m)}{m^k}\sigma_{k-1}(n;hC,hD;N). \tag{5.5.24}$$

(ii) *When $N=2$, $n \geq 1$,*

$$a(n; C, D; 2) = \frac{(\pi/i)^k}{2(1-2^{-k})\zeta(k)\Gamma(k)} \sigma_{k-1}(n; C, D; 2). \tag{5.5.25}$$

In case (ii) of the theorem there are only three different choices for the pair C, D namely $0,1; 1,1; 1,0$ corresponding to $L = I, P, P^2$, respectively. The formula (5.5.23, 25) can be simplified in this case. We have

$$\sigma_{k-1}(n; C, D; 2) = 2 \sum_{\substack{\frac{n}{d} \equiv C \pmod 2 \\ d|n, d>0}} d^{k-1}(-1)^{dD}. \tag{5.5.26}$$

The Fourier coefficients for the case when $N=1$ and k is even can also be deduced very simply from theorems 5.5.2, 3 or directly from (5.5.8, 9, 10). We defer this to the next chapter which is mainly concerned with the full modular group ($N=1$).

Similar results can be obtained for Poincaré series $G_L(z; m; \Gamma, k, v)$, where $\bar{\Gamma}(N) \subseteq \Gamma = \bar{\Gamma}$ and v is a MS on Γ that is constant on $\Gamma(N)$. This can either be done directly by using methods similar to those used for $\bar{\Gamma}(N)$, or by applying theorem 5.1.5, with $\Delta = \bar{\Gamma}(N)$, to the results already obtained.

5.6. Poincaré series on $\Gamma(N^2)$. In this section we use theorem 5.1.5 to express a Poincaré series on $\bar{\Gamma}(N)$ in terms of Poincaré series on $\bar{\Gamma}(N^2)$. We take $N>1$ and assume that k is any even or odd integer greater than 2.

We note first that, as is easily verified from (1.1.11), for any S, $T \in \Gamma(N)$,

$$ST \equiv TS \pmod{N^2}.$$

This shows that $\Gamma(N)/\Gamma(N^2)$ is abelian. By theorem 1.4.1, the index of this factor group is N^3.

In fact, $\Gamma(N)/\Gamma(N^2)$ is a product of three cyclic groups of order N. For each of its elements can be represented by a matrix of the form

$$\begin{bmatrix} 1+xN & (u+x)N \\ (w-x)N & 1-xN \end{bmatrix} \pmod{N^2}$$

where u, w and x are integers, and this is congruent (mod N^2), to

$$U^{uN}W^{wN}X^{xN}, \quad \text{where} \quad X = P^{-1}UP.$$

Similar results hold when $N>2$ for $\bar{\Gamma}(N)/\bar{\Gamma}(N^2)$, which is then isomorphic to $\Gamma(N)/\Gamma(N^2)$. However, $\bar{\Gamma}(2)/\bar{\Gamma}(4)$ has order 4 and is a product of two cyclic groups which may be taken to be the groups generated by U^2 and X^2 (mod 4). We note that

$$U^2X^2 = -W^2. \tag{5.6.1}$$

Now take any $L \in \Gamma(1)$ and put

$$U_r = \varepsilon L^{-1}P^{-r}U^NP^rL \quad (r=0, 1, 2), \tag{5.6.2}$$

where

$$\varepsilon = 1\ (N>2), \quad \varepsilon = -1\ (N=2). \tag{5.6.3}$$

Then $U_1 = \varepsilon L^{-1}X^NL$ and $U_2 = \varepsilon L^{-1}W^{-N}L$; further, when $N=2$, $U_0U_1 \equiv U_2$ (mod 4). Then, because of the normality of $\bar{\Gamma}(N)$ in $\Gamma(1)$, $\bar{\Gamma}(N)/\bar{\Gamma}(N^2)$ is a direct product of the cyclic groups generated modulo N^2 by U_0, U_1 and (for $N>2$) U_2. A similar statement holds for the corresponding inhomogeneous groups.

We now choose a fixed MS v_0 on $\bar{\Gamma}(N)$ that is constant on $\Gamma(N^2)$ as follows. For even k we take $v_0(T)=1$ for all $T \in \bar{\Gamma}(N)$. When k is odd and $N>2$ we take $v_0(T)=1$ or -1 according as $T \equiv I$ or $-I$ (mod N). Finally, when k is odd and $N=2$ we take $v_0(T)=1$ when $T \in \tilde{\Gamma}(2)$ and $v_0(T)=-1$ for $T \notin \tilde{\Gamma}(2)$. We recall, from (3.3.18), that $\tilde{\Gamma}(2)$ is normal in $\Gamma(1)$ and is generated by U_0 and U_1; thus $v_0 = \tilde{v}$ in this case.

Then, if v is any MS on $\bar{\Gamma}(N)$ that coincides with v_0 on $\bar{\Gamma}(N^2)$, it is necessarily of the form

$$v(T) = v(\lambda_0, \lambda_1, \lambda_2; T) = \exp\left\{\frac{2\pi i}{N}(\lambda_0 n_0 + \lambda_1 n_1 + \lambda_2 n_2)\right\} v_0(T) \tag{5.6.4}$$

for $N>2$, or of the form

$$v(T) = v(\lambda_0, \lambda_1; T) = (-1)^{\lambda_0 n_0 + \lambda_1 n_1} v_0(T) \tag{5.6.5}$$

for $N=2$. Here, for $r=0, 1, 2$, λ_r and n_r are integers satisfying

$$0 \le \lambda_r < N, \quad 0 \le n_r < N. \tag{5.6.6}$$

The n_r are determined by the relations

$$T \equiv \pm U_0^{n_0}U_1^{n_1}U_2^{n_2}\ (\text{mod}\ N^2) \quad (N>2), \tag{5.6.7}$$

or

$$T \equiv \pm U_0^{n_0} U_1^{n_1} \pmod 4 \quad (N=2). \tag{5.6.8}$$

It is easily checked that, when $N=2$, the MS v is constant on the groups $\tilde{\Gamma}(2), L^{-1}\Gamma^*(2)L, L^{-1}P^{-1}\Gamma^*(2)PL$ and $L^{-1}P^{-2}\Gamma^*(2)P^2L$ in the cases $(\lambda_0, \lambda_1)=(0,0), (1,0), (1,1)$ and $(0,1)$, respectively.

We observe that, if $\zeta_r = L^{-1}P^{-r}\infty$, then

$$\kappa_r = \kappa(\zeta_r, \bar{\Gamma}(N), v),$$

where

$$e^{2\pi i\kappa_r} = v(\varepsilon U_r) = e^{2\pi i\lambda_r/N} v_0(\varepsilon U_r) = \varepsilon^k e^{2\pi i\lambda_r/N},$$

so that

$$\kappa_r = \begin{cases} \dfrac{\lambda_r}{N} & (N>2; \text{ or } N=2, k \text{ even}), \\ \tfrac{1}{2}(1-\lambda_r) & (N=2, k \text{ odd}). \end{cases} \tag{5.6.9}$$

We now apply theorem 5.1.5 with $\Gamma=\bar{\Gamma}(N)$, $\Delta=\bar{\Gamma}(N^2)$. We find that

$$n'=N^2, \quad n=l=N, \quad \kappa'=0, \quad \kappa=\kappa_0.$$

For $\mathscr{L}$ we can take the set of matrices

$$U_1^{n_1}U_2^{n_2} \; (N>2), \quad U_1^{n_1} \; (N=2),$$

where (5.6.6) holds. We then deduce that, for $N>2$,

$$G_L(z; m; \bar{\Gamma}(N), k, v) = \sum_{n_1=0}^{N-1} \sum_{n_2=0}^{N-1} \exp\left\{\frac{-2\pi i}{N}(\lambda_1 n_1 + \lambda_2 n_2)\right\} \times G_{LU_1^{n_1}U_2^{n_2}}(z; mN+\lambda_0; \bar{\Gamma}(N^2), k, v_0), \tag{5.6.10}$$

while, for $N=2$,

$$G_L(z; m; \Gamma(2), k, v) = G_L(z; 2m+2\kappa_0; \bar{\Gamma}(4), k, v_0) + (-1)^{\lambda_1} G_{LU_1}(z; 2m+2\kappa_0; \bar{\Gamma}(4), k, v_0). \tag{5.6.11}$$

These formulae hold for all choices of the MS v, i.e. for all choices of λ_0, λ_1 and (for $N>2$) λ_2 satisfying $0<\lambda_r \le N$. The Fourier coefficients of the Poincaré series on the right-hand sides of (5.6.10, 11) can be found from the theorems in §5.4. It is, accordingly, a straightforward matter to find the Fourier coefficients of the Poincaré series on the left-hand sides of these equations.

Finally, we note that, when $N=2$, $\Gamma(2)$ has three incongruent cusps and there are therefore essentially only three different Poincaré series (for each m) obtained by taking $L=I$, P and P^2. Also G_L is an Eisenstein series if and only if $m=\kappa_0=0$; i.e. if and only if $m=0$ and $v(-U_0)=1$. It is convenient to write

$$U_r^* = -P^{-r}U^2P^r \quad (r \in \mathbb{Z}) \tag{5.6.12}$$

so that $U_r = L^{-1}U_r^*L$, by (5.6.2). Then the four multiplier systems on $\Gamma(2)$ are the multiplier system $v_0=\tilde{v}$ already defined and the three MS $v^{(s)}$ $(s=1, 2, 3)$ defined in table 8; these are constant on the groups listed in the last column. We have, by (3.1.20), that $v^{(s)}=v^{(0)}|P^s$, where we write $v|L$ in place of v^L.

Table 8. Multiplier system on $\Gamma(2)$

T	U_0^*	U_1^*	U_2^*	Γ
$\tilde{v}(T)$	1	1	1	$\tilde{\Gamma}(2)$
$v^{(0)}(T)$	-1	$+1$	-1	$\Gamma^*(2)$
$v^{(1)}(T)$	-1	-1	$+1$	$P^{-1}\Gamma^*(2)P$
$v^{(2)}(T)$	$+1$	-1	-1	$P^{-2}\Gamma^*(2)P^2$

When k is even $G_L(z;0;\Gamma(2), k, v_0) = G_L(z;0;\Gamma(2), k, 1)$ and is an Eisenstein series for all $L \in \Gamma(1)$; hence, by taking $L=P^s$ $(s=0, 1, 2)$ we get three different Eisenstein series. When k is odd, $v_0(-U_r)=-1$ for all r and so there are no Eisenstein series in $\{\Gamma(2), k, v_0\}$.

We now consider the other MS $v^{(s)}$ and note that, for $L=P^r$, $v^{(s)}(-U_0)=v^{(s)}(-U_r^*)=(-1)^k v^{(0)}(U_{r-s}^*)$. Hence

$$v^{(s)}(-U_0)=1 \quad \text{if and only if} \begin{cases} r-s\equiv 1 \pmod 3 & (k \text{ even}), \\ r-s\not\equiv 1 \pmod 3 & (k \text{ odd}). \end{cases} \tag{5.6.13}$$

Hence, for k even, the space $\{\Gamma(2), k, v^{(s)}\}$ contains exactly one

Eisenstein series, namely

$$\begin{aligned} E_k(z; P^{s+1}, v^{(s)}) &:= G_{P^{s+1}}(z; 0; \Gamma(2), k, v^{(s)}) \\ &= G_{P^{s+1}}(z; 0; \bar{\Gamma}(4), k, 1) \\ &\quad - G_{U_1^* P^{s+1}}(z; 0; \bar{\Gamma}(4), k, 1). \end{aligned} \qquad (5.6.14)$$

For odd k, $\{\Gamma(2), k, v^{(s)}\}$ contains two Eisenstein series, namely

$$E_k(z; P^s, v^{(s)}) \quad \text{and} \quad E_k(z; P^{s-1}, v^{(s)}),$$

where

$$\begin{aligned} E_k(z; P^s, v^{(s)}) &= G_{P^s}(s; 0; \Gamma(2), k, v^{(s)}) \\ &= G_{P^s}(z; 0, \bar{\Gamma}(4), k, v_0) + G_{U_1^* P^s}(z; 0; \bar{\Gamma}(4), k, v_0) \end{aligned} \qquad (5.6.15)$$

and

$$\begin{aligned} E_k(z; P^{s-1}, v^{(s)}) &= G_{P^{s-1}}(z; 0; \Gamma(2), k, v^{(s)}) \\ &= G_{P^{s-1}}(z; 0; \bar{\Gamma}(4), k, v_0) \\ &\quad - G_{U_1^* P^{s-1}}(z; 0; \bar{\Gamma}(4), k, v_0). \end{aligned} \qquad (5.6.16)$$

5.7. Modular forms on $\bar{\Gamma}(N)$ of weight 2. In our discussion of Poincaré series we have assumed that $k > 2$, where k is the weight, since this assumption was necessary in order to secure the absolute convergence of the series (5.1.9). We have already seen in §4.4 that forms of weight 2 are of interest, and we therefore use a method of Hecke (1924) to modify the analysis of §§5.1–5.5 in order to construct such modular forms. Our assumptions and notation are the same as in §5.5, except that we take $k = 2$. Thus the group is $\bar{\Gamma}(N)$, $v(T) = 1$ for all $T \in \Gamma(N)$ and (5.4.1, 2) hold. For any $m \in \mathbb{Z}$ we define

$$G_L(z; m, N; s) := \sum_{T \in R_L} \frac{\exp\left\{\dfrac{2\pi i m}{N} LTz\right\}}{(LT: z)^2 |LT: z|^{2s}}. \qquad (5.7.1)$$

This definition is similar to (5.1.9) except for the convergence factor $|LT: z|^{2s}$ in the denominator. Here s is a positive real number and our aim is to consider whether the expression on the right of (5.7.1) tends to a modular form of weight 2 as $s \to 0+$.

We can use theorem 5.1.1, as in the proof of theorem 5.1.2, to show that the series (5.7.1) is absolutely convergent for $s>0$ when $y=\operatorname{Im} z>0$. Moreover, for fixed z, the convergence is uniform for $s\geq s_0>0$ and so $G_L(z; m, N; s)$ is a continuous function of s for $s>0$. As in the proof of theorem 5.1.2, we have, for any $S\in\bar{\Gamma}(N)$,

$$G_L(Sz; m, N; s)=(S:z)^2|S:z|^{2s}G_L(z; m, N; s) \quad (5.7.2)$$

and, more generally, for any $S\in\Gamma(1)$,

$$G_L(Sz;\ m, N; s)=(S:z)^2|S:z|^{2s}G_{LS}(z; m, N; s). \quad (5.7.3)$$

Both (5.7.2) and (5.7.3) hold whenever $y>0$ and $s>0$.

We now adapt the analysis of §5.3 to the study of $G_L(z; m, N; s)$. We find that (5.3.36) now becomes

$$G_L(z; m, N; s)=\delta_L\, e^{2\pi imz/N}+\sum_{\substack{\gamma=1\\ \gamma\equiv\varepsilon C \pmod N}}^{\infty}\ \sum_{\alpha,\delta}\frac{e^{2\pi im\alpha/(N\gamma)}}{(\gamma N)^{2+2s}}F_s(\zeta_S, \lambda), \quad (5.7.4)$$

where δ_L is given by (5.5.2), $\varepsilon=\pm1$ and the inner sum is over all α and δ satisfying (5.3.26). Here

$$\zeta_S=\left(z+\frac{\delta}{\gamma}\right)N^{-1},\quad \lambda=\frac{m}{N^2\gamma^2} \quad (5.7.5)$$

and, for any $\zeta=\xi+i\eta$ with $\eta:=\operatorname{Im}\zeta>0$

$$F_s(\zeta, \lambda)=\sum_{n=-\infty}^{\infty}\frac{\exp\{-2\pi i\lambda/(\zeta+h)\}}{(\zeta+h)^2|\zeta+h|^{2s}}. \quad (5.7.6)$$

To sum this series we make use of the Poisson summation formula (see Mordell (1928))

$$\sum_{h=-\infty}^{\infty} f(h)=\sum_{r=-\infty}^{\infty}\int_{-\infty}^{\infty} f(w)\, e^{-2\pi irw}\, dw. \quad (5.7.7)$$

This is valid for any function f defined on $\mathbb{R}$ for which f'' exists and is continuous, if $f(x)$ and $f'(x)$ tend to zero as $|x|\to\infty$ and $|f|$, $|f'|$ and $|f''|$ are integrable over $\mathbb{R}$. This is certainly true for

$$f(w):=\frac{\exp\{-2\pi i\lambda/(\zeta+w)\}}{(\zeta+w)^2|\zeta+w|^s}=:\frac{e(\zeta, w)}{(\zeta+w)^2|\zeta+w|^{2s}} \quad (5.7.8)$$

for real w and $s>-\frac{1}{2}$. We therefore have, by (5.7.6, 7),

$$F_s(\zeta, \lambda)=\sum_{r=-\infty}^{\infty} e^{2\pi ir\zeta}I_r(\zeta, s), \quad (5.7.9)$$

where

$$I_r(\zeta, s) := \int_{-\infty}^{\infty} \frac{e(\zeta, w)\, e^{-2\pi i r(w+\zeta)}\, dw}{(\zeta+w)^2|\zeta+w|^{2s}}. \tag{5.7.10}$$

In this integral $s > -\frac{1}{2}$ and ζ is any point of $\mathbb{H}$. We shall write

$$I(\zeta, s) = \sum_{r\neq 0} e^{2\pi i r\zeta} I_r(\zeta, s). \tag{5.7.11}$$

We now prove

Theorem 5.7.1. *The functions $F_s(\zeta, \lambda)$ and $I_r(\zeta, s)$ defined by (5.7.7, 11) are continuous functions of s for $s > -\frac{1}{2}$ for each $\zeta \in \mathbb{H}$ and $r \in \mathbb{Z}$. Further*

$$I_r(\zeta, s) = I_r(\xi, s),$$

where $\xi = \mathrm{Re}\,\zeta$. The functions $I_r(\zeta, 0)$ are holomorphic functions of ζ on $\mathbb{H}$ for each r and

$$I_r(\zeta, 0) = 0 \quad (r \leq 0) \tag{5.7.12}$$

while, for $r > 0$

$$I_r(\zeta, 0) = \begin{cases} -4\pi^2 r & (m=0), \quad (5.7.13) \\ -2\pi N\gamma\sqrt{(r/m)}J_1(4\pi\sqrt{(mr)}/(N\gamma)) & (m>0), \quad (5.7.14) \\ -2\pi N\gamma\sqrt{(r/|m|)}I_1(4\pi\sqrt{(|m|r)}/N\gamma)) & (m<0). \quad (5.7.15) \end{cases}$$

Also

$$|e^{2\pi i r\zeta} I_r(\zeta, s)| \leq \frac{2\Gamma(\frac{1}{2})\Gamma(s+\frac{1}{2})}{c^{2s+1}\Gamma(s+1)} e^{-2\pi(|r|c - |\lambda|c)} \tag{5.7.16}$$

for $r \in \mathbb{Z}$, $s > -\frac{1}{2}$ and $\eta = \mathrm{Im}\,\zeta \geq 2c$, where c is any positive number. Finally, when $\lambda = 0$ (i.e. $m = 0$)

$$\lim_{s\to 0+} \frac{I_0(\zeta, s)}{s} = -\frac{\pi}{\eta}. \tag{5.7.17}$$

Proof. Denote by $\mathbb{H}_c$ the half-plane

$$\mathbb{H}_c := \{\zeta \in \mathbb{C} : \mathrm{Im}\,\zeta = \eta \geq 2c\} \tag{5.7.18}$$

for fixed $c > 0$. We note that the denominator of the integral in

(5.7.10) can be expressed as

$$(w+\zeta)^{2+s}(w+\bar{\zeta})^s, \tag{5.7.19}$$

where these powers are evaluated by taking

$$0 \le \mathrm{ph}(w+\zeta) < \pi, \quad -\pi < \mathrm{ph}(w+\bar{\zeta}) < 0.$$

In fact (5.7.19) is a uniquely defined holomorphic function of w for fixed $\zeta \in \mathbb{H}_c$ in the domain obtained by slitting $\mathbb{C}$, with slits parallel to the imaginary axis, from $-\zeta$ to $-\zeta - i\infty$ and from $-\bar{\zeta}$ to $-\bar{\zeta}+i\infty$. In particular, for all $\zeta \in \mathbb{H}_c$, the integrand in (5.7.10) is a holomorphic function of w in the strip $|\mathrm{Im}\, w| < 2c$. If we write, for real u, v, ξ, η,

$$w = u + iv, \quad \zeta = \xi + i\eta,$$

we have

$$|e(\zeta, w)| = \exp\left\{\frac{-2\pi\lambda(v+\eta)}{(u+\xi)^2+(v+\eta)^2}\right\},$$

from which it follows that

$$|e(\zeta, w)| \le e^{2\pi|\lambda|/c} \quad (|v| \le c). \tag{5.7.20}$$

We also have

$$|(w+\zeta)^{2+s}(w+\bar{\zeta})^s| = |w+\zeta|^{2+2\sigma}. \tag{5.7.21}$$

When $r \ne 0$ we replace the line of integration $\mathbb{R}$ in (5.7.10) by the parallel line $v = \varepsilon_1 c$, where $\varepsilon_1 = -1$ or 1 according as $r > 0$ or $r < 0$. A simple application of Cauchy's theorem then shows that

$$I_r(\zeta, s) = \int_{-\infty+ic\varepsilon_1}^{\infty+ic\varepsilon_1} \frac{e(\zeta, w)e^{-2\pi i r(w+\zeta)}\, dw}{(w+\zeta)^{2+s}(w+\bar{\zeta})^s} \quad (r \ne 0). \tag{5.7.22}$$

Since this integral is absolutely convergent for $s > -\frac{1}{2}$ and uniformly convergent for $s \ge -\frac{1}{2}+\delta > -\frac{1}{2}$, it follows that $I_r(\zeta, s)$ is, for fixed $\zeta \in \mathbb{H}_c$ and $r \ne 0$, a continuous function of s when $s > -\frac{1}{2}$. In fact we obtain the upper bound

$$|e^{2\pi i r\zeta} I_r(\zeta, s)| \le e^{2\pi(|\lambda|/c - |r|c)} \int_{-\infty}^{\infty} \frac{dw}{(w^2+c^2)^{1+s}}$$

$$= \frac{2\Gamma(\frac{1}{2})\Gamma(s+\frac{1}{2})}{c^{2s+1}\Gamma(s+1)} e^{-2\pi(|r|c - |\lambda|/c)}.$$

It is clear that this is also an upper bound for $|I_0(\zeta, s)|$ under the same conditions, no change of contour being necessary to establish the result. This proves (5.7.16).

We now examine $I_0(\zeta, s)$ in greater detail in the case when $\lambda = 0$. Then $e(\zeta, w) = 1$ and, by (5.7.10),

$$I_0(\zeta, s) = \eta^{-2s-1} \int_{-\infty}^{\infty} \frac{dx}{(x+i)^2(x^2+1)^s} \quad (\lambda = 0). \quad (5.7.23)$$

The imaginary part of the integrand is an odd function of x so that

$$I_0(\zeta, s) = \frac{2}{\eta^{s+1}} \int_0^{\infty} \frac{(x^2-1)dx}{(x^2+1)^{s+2}} = -\frac{s\Gamma(\frac{1}{2})\Gamma(s+\frac{1}{2})}{\eta^{2s+1}\Gamma(s+2)}. \quad (5.7.24)$$

On letting $s \to 0$ we derive (5.7.17).

It remains to consider the functions $I_r(\zeta, 0)$ $(r \in \mathbb{Z})$. It is clear from (5.7.10) and (5.3.15) that $I_r(\zeta, 0) = g_r$ and g_r was shown to be zero when $r \leq 0$; this proves (5.7.12). For $r > 0$ the results (5.7.13–15) follow from (5.3.11–13), since $k = 2$ and $\lambda = m/(N\gamma)^2$. This completes the proof of theorem 5.7.1.

We now return to (5.7.4) and assume, in the first instance, that $s > 0$. We have, since $I_r(\zeta_s, s) = I_r(z/N, s)$ by (5.7.5) and the theorem,

$$\begin{aligned} &G_L(z; m, N; s) - \delta_L e^{2\pi imz/N} \\ &= \sum_{\gamma} \sum_{\alpha,\delta} \frac{e^{2\pi im\alpha/N\gamma}}{(\gamma N)^{2+2s}} \sum_{r=-\infty}^{\infty} e^{2\pi ir(\gamma z+\delta)/N} I_r(z/N, s) \\ &= \sum_{\gamma} \sum_{r=-\infty}^{\infty} \frac{W(r, m; \gamma)}{(\gamma N)^{2+2s}} I_r(z/N, s)\, e^{2\pi irz/N}. \end{aligned} \quad (5.7.25)$$

We first consider the case when $m \neq 0$. From the discussion of the series (5.4.33) and the upper bound (5.7.16) it is clear that the double series on the right of (5.7.25) is uniformly convergent for $s \geq 0$ and $y = \operatorname{Im} z \geq 2Nc > 0$. Accordingly, its limit as $s \to 0+$ is

$$\sum_{\gamma} \sum_{r=1}^{\infty} \frac{W(r, m; \gamma)}{(\gamma N)^2} I_r(z/N, 0)\, e^{2\pi irz/N},$$

where $I_r(z/N, 0)$ can be evaluated from (5.7.14, 15). Thus we may define

$$G_L(z;m,N):=\lim_{s\to 0+} G_L(z;m,N;s)$$
$$=\delta_L\, e^{2\pi imz/N}+\sum_{\gamma}\sum_{r=1}^{\infty}\frac{W(r,m;\gamma)}{(\gamma N)^2}I_r(z/N,0)\,e^{2\pi irz/N} \tag{5.7.26}$$

and note that it follows, again by uniform convergence, that $G_L(z;m,N)$ is a holomorphic function of z on $\mathbb{H}$: the summation over γ is over positive γ for which $\gamma\equiv\pm C\pmod N$. Further, from (5.7.2, 3), we have

$$G_L(Sz;m,N)=(S:z)^2G_L(z;m,N)\quad (S\in\bar{\Gamma}(N)) \tag{5.7.27}$$

and

$$G_L(Sz;m,N)=(S:z)^2G_{LS}(z;m,N)\quad (S\in\Gamma(1)). \tag{5.7.28}$$

These relations together with the Fourier expansion (5.7.26) show that

$$G_L(z;m,N)\in\{\bar{\Gamma}(N),2,1\}_0\quad (m>0), \tag{5.7.29}$$
$$G_L(z;m,N)\in H(\bar{\Gamma}(N),2,1)\quad (m<0). \tag{5.7.30}$$

When $m=0$ arguments similar to those used above to show that the part of (5.7.25) corresponding to non-zero values of r is continuous for $s\geq 0$, and, when $s=0$, represents a holomorphic function of z. The remaining part is just

$$N^{-2}g(s)I_0(z/N,s),$$

in the notation of (5.4.34). The limit of this as $s\to 0+$ is, by (5.4.43) and (5.7.17)

$$\frac{-3\delta_N^*}{\pi yN^2}\prod_{p|N}(1-p^{-2})^{-1}=\frac{-3}{\pi y\hat{\lambda}(N)}, \tag{5.7.31}$$

by (2.4.17). It will be observed that this is not a holomorphic function of z. In this case we define

$$G_L(z;0,N):=-\frac{3}{\pi y\hat{\lambda}(N)}+\delta_L+\sum_{\gamma}\sum_{r=1}^{\infty}\frac{W(r,0;\gamma)}{(\gamma N)^2}I_r(z/N,0)\,e^{2\pi irz/N} \tag{5.7.32}$$

and (5.7.27, 28) continue to hold. Observe that

$$\delta_L = 1 \Leftrightarrow LU^q \in \bar{\Gamma}(N) \quad \text{for some } q \in \mathbb{Z}. \qquad (5.7.33)$$

In particular, $\delta_I = 1$. Note also that, apart from the non-analytic term in y^{-1}, the expansion (5.7.32), when combined with (5.7.13–15) agrees with that of the expansion of the Poincaré series (5.5.1) when k is not equal to 2 in (5.5.4–6). We return to this remark in §7.2.

We now consider whether, by taking linear combinations of the functions $G_L(z; 0, N)$ for different $L \in \Gamma(1)$, we can remove the term in y^{-1} and so obtain a holomorphic modular form having non-zero constant term. Since $\dim\{\Gamma(1), 2, 1\} = 0$ by corollary 4.2.1 this is not possible when $N = 1$, and in this case there is only one series $G_I(z; 0, N)$.

Accordingly we assume from now on that $N \geq 2$. Let $\lambda = \hat{\lambda}(N)$ be the number of incongruent cusps belonging to the group $\bar{\Gamma}(N)$, so that, by (2.4.17), $\lambda \geq 3$. By applying theorem 1.1.2 we can take a set of λ matrices $L_i \in \Gamma(1)$ $(1 \leq i \leq \lambda)$ such that

$$\Gamma(1) = \Gamma(N) \cdot \bigcup_{i=1}^{\lambda} \mathcal{S}_i,$$

where $\mathcal{S}_i = U\{L_i U^k : 0 \leq k < N\}$. Then the λ cusps $L_i\infty$ are incongruent modulo $\bar{\Gamma}(N)$. For simplicity we take $L_1 = I$. We note that

$$\delta_{L_j^{-1}L_i} = 1 \Leftrightarrow i = j. \qquad (5.7.34)$$

We now define, for $j = 2, 3, \ldots, \lambda$,

$$G_j^*(z; N) := G_I(z; 0, N) - G_{L_j^{-1}}(z; 0, N) \qquad (5.7.35)$$

so that

$$G_j^*(z; N) \in \{\Gamma(N), 2, 1\} \quad (2 \leq j \leq N). \qquad (5.7.36)$$

The value taken by G_j^* at the cusp $L_i\infty$ is the value taken by $G_j^*|L_i$ at ∞ and so is, by (5.7.28),

$$\delta_{L_i} - \delta_{L_j^{-1}L_i} \quad (1 \leq i \leq N, 2 \leq j \leq N).$$

This is non-zero only for $i = 1$ and $i = j$, being $+1$ and -1 in the two cases, respectively. In particular the sum of $\text{res}(G_j, \zeta, \bar{\Gamma}(N))$ over all points ζ of a proper fundamental region for $\hat{\Gamma}(N)$ is zero as we should expect from theorem 4.4.2.

It is clear that the following analogue of theorem 5.1.3 holds.

Theorem 5.7.2. *Let $f \in H(\bar{\Gamma}(N), 2, 1)$. Then there exists a finite linear combination $F(z)$ of the Poincaré series*

$$G_{L_i-1}(z; m, N) \quad (m<0, 1 \le i \le \hat{\lambda}(N)), \; G_j^*(z; N) \quad (2 \le j \le \hat{\lambda}(N))$$

such that

$$f(z) - F(z) \in \{\bar{\Gamma}(N), 2, 1\}_0.$$

We observe that for $2 \le N \le 5$, $1 + \frac{1}{6}\hat{\mu}(N) = \hat{\lambda}(N) - 1$ and therefore

$$\dim\{\bar{\Gamma}(N), 2, 1\}_0 = 0 \quad \text{for } 1 \le N \le 5. \tag{5.7.37}$$

Further, theorem 5.1.4 extends to the case $k = 2$ in the form:

$$\dim\{\bar{\Gamma}(N), 2, 1\}_0 = \dim\{\bar{\Gamma}(N), 2, 1\} - \hat{\lambda}(N) + 1 \quad (N \ge 1). \tag{5.7.38}$$

We now introduce the inner product (with $\mu = \hat{\mu}(N)$)

$$(G_L(z; m, N), f(z)) = \frac{1}{\mu} \iint_{\mathbb{F}} G_L(z; m, N) \overline{f(z)}\, dx\, dy, \tag{5.7.39}$$

where $\mathbb{F}$ is a fundamental region for $\hat{\Gamma}(N)$, $m \ge 0$ and $f \in \{\bar{\Gamma}(N), 2, 1\}_0$. This is valid even when $m = 0$ although $G_L(z; 0, N)$ is not then a holomorphic function of z. This can be justified, as in §5.2, by splitting $\mathbb{F}$ up into a finite number of transforms of $\mathbb{F}_I$ and transforming the variable. More generally, we can define, for $s \ge 0$ and $m \ge 0$

$$(G_L(z; m, N; s), f(z)) = \frac{1}{\mu} \iint_{\mathbb{F}} G_L(z; m, N; s) \overline{f(z)} y^s\, dx\, dy \tag{5.7.40}$$

as a convergent integral, by appealing to (5.7.25) and (5.7.16), and can show that (5.7.39) is the limit of (5.7.40) as $s \to 0+$.

We then have, in analogy to theorem 5.2.2,

Theorem 5.7.3. *Suppose that $f \in \{\Gamma(N), 2, 1\}_0$, $L = M^{-1} \in \Gamma(1)$ and that*

$$f_M(z) = \sum_{r=1}^{\infty} b_r e^{2\pi i r z/N}.$$

Then for all $s>0$ we have

$$(f(z), G_L(z; m, N; s)) = \begin{cases} \dfrac{b_m N^{2+s}\Gamma(1+s)}{\mu(4\pi m)^{1+s}} & \text{if } m>0, \\ 0 & \text{if } m=0. \end{cases}$$

Proof. For $s>0$ the proof proceeds exactly as in the proof of theorem 5.2.2, the details being simpler because of the even weight and constant MS. In the double integrals the main change is that y^{k-2} is replaced by y^s. The corresponding result for $s=0$ is obtained by letting $s\to 0+$.

Theorem 5.7.3 can be used to show that theorems 5.2.3, 4, 5 and 6 all remain true $k=2$, $\Gamma=\bar{\Gamma}(N)$ and constant MS.

Similar results can be obtained for the series $G_L(z; m; \Gamma, 2, v)$, where $\bar{\Gamma}(N)\subseteq\Gamma=\bar{\Gamma}$ and v is a MS on Γ that is constant on $\Gamma(N)$. This is done most conveniently by applying theorem 5.1.5, with $\Delta=\bar{\Gamma}(N)$, to the results already obtained.

5.8. Further results. The properties of Eisenstein series were first studied by Hecke (1927), who showed that, if f is an entire modular form of weight k ($k\geq 2$), there exists a linear combination F of Eisenstein series such that $f-F$ is a cusp form. He also evaluated the Fourier coefficients of Eisenstein series; see also Petersson (1931).

The foundations of the theory of general Poincaré series were laid by Petersson (1930), who considered series of a more general type than (5.1.9), the numerator being replaced by $F(LTz)$, where $F(t)$ is an arbitrary function meromorphic for $|t|<1$. Petersson's work also applies more generally to automorphic forms on any horocyclic group (*Grenzkreisgruppe*) having a finite number of generators. Formulae for the Fourier coefficients of Poincaré series are given in Petersson (1931, 1932*a*, 1938*b*); see also Selberg (1939). An alternative method, applied to the full modular group, is given by Schwandt (1972). The estimate (4.6.3) is obtained by applying the best known estimates for Kloosterman sums to series of the form (5.5.5). It is to Petersson also (1939, 1940) that the idea of metrizing the space of cusp forms by introducing an inner product is due and this has transformed the whole theory.

The general Riemann–Roch theorem for automorphic forms with arbitrary multiplier systems will be found in Satz 9 on p. 194 of Petersson (1938*a*) and this is applied in Satz 20 (p. 683) to $\{\Gamma(N), k, v\}$ and $\{\Gamma(N), k, v\}_0$, yielding lower bounds for the number of independent forms in these spaces. In theorems 5.2.3 we showed that there exists a half-open interval $I =]-\kappa, N_0 - \kappa]$ with the property that the forms $G_L(z; m)$ with $m \in I$ span $\{\Gamma, k, v\}_0$. As we shall see, we can reduce this interval to $]-\kappa, N-\kappa]$, where $N = \dim\{\Gamma, k, v\}_0$, when $\Gamma = \Gamma(1)$. A similar reduction is not possible in general for subgroups Γ of $\Gamma(1)$ since the 'Weierstrass character' of the points $L^{-1}\infty$ is involved; see Petersson (1940), Sätze 1, 18.

We have used Hecke's limiting process to obtain Poincaré series of weight 2 when the group is a principal congruence group and this can, in certain circumstances, be generalized to other groups and even to weight 1. The magnitude of the Kloosterman sums is critical here. See Hecke (1924, 1925, 1927) and Petersson (1940, 1948, 1956). It may be observed that we could, when considering the functions $G_L(z; m, N; s)$, have assumed that s was a complex variable, in which case these functions are holomorphic functions of s on a neighbourhood of $s = 0$.

An alternative method of dealing with the case $k = 2$ has been developed by Lehner (1959), who uses the fact that the Poincaré series are, in this case, conditionally convergent. Further, Knopp and Lehner (1962) have shown that 'complementary forms' F and G for which FG has weight 2 have a number of interesting properties in relation to each other.

The methods of this chapter do not enable us to construct modular forms of negative weight, except in so far as such forms are reciprocals of forms of positive weight. Petersson (1950) has developed a method of using generalized Poincaré series of arbitrary real weight $k > 2$ and MS v to construct modular forms of negative weight $2-k$ and MS v^{-1} having prescribed poles in their fundamental regions. These forms are complementary in the sense defined by Knopp and Lehner (1962).

It was mentioned in §5.4 that the estimates of the order of magnitude obtained there could be improved. Thus, as a consequence of his deep and difficult work on algebraic curves over a finite field, Weil (1948) proved that

$$|S(u, v; p)| < 2p^{\frac{1}{2}},$$

where, as in (5.4.25), the prime p does not divide uv. More recently Stepanov (1971) has succeeded in obtaining this same estimate by elementary (but complicated) methods. On the other hand, the magnitude of $S(u, v; p^n)$ can be found quite easily for $n \geq 2$; see, for example, Williams (1971).

6: Functions belonging to the full modular group

6.1. Modular forms of even weight with constant multiplier system. In this chapter we shall be concerned with modular forms belonging to the full modular group $\Gamma(1)$ and having various real weights and multiplier systems. We begin by considering the simplest case, which is when the weight is an even integer and the MS is constant; thus $v(T)=1$ for all $T\in\Gamma(1)$. We write

$$M(\Gamma(1),k),\quad H(\Gamma(1),k),\quad \{\Gamma(1),k\},\quad \{\Gamma(1),k\}_0$$

in place of $M(\Gamma(1),k,1)$, etc., and put

$$d_k=\dim\{\Gamma(1),k\},\quad \delta_k=\dim\{\Gamma(1),k\}_0. \tag{6.1.1}$$

Then, by corollary 4.2.1,

$$d_0=1,\quad \delta_0=d_2=\delta_2=0, \tag{6.1.2}$$

the only entire modular functions being constants.

We now consider the case when $k>2$. The Poincaré series discussed in the previous chapter define functions on $H(\Gamma(1),k)$ and we begin by considering the Eisenstein series

$$E_k(z)=\sum_{T\in\mathscr{R}_L}(T:z)^{-k},\quad E_k^*(z)=\sum_{m=-\infty}^{\infty}\sum_{n=-\infty}^{\infty}{}'(mz+n)^{-k}\quad (k>2). \tag{6.1.3}$$

Here $\hat{\Gamma}(1)=\hat{\Gamma}_U\cdot\mathscr{R}_L$ and the dash denotes the omission of the term when $m=n=0$. By (5.5.16), we have $E_k^*(z)=2\zeta(k)E_k(z)$.

The Fourier coefficients of $E_k(z)$ can be obtained from (5.5.3, 4) and (5.4.13). Alternatively, and more simply, we may find the Fourier coefficients of $E_k^*(z)$ from (5.5.19, 23). This gives

$$E_k(z)=1+(-1)^{\frac{1}{2}k}\alpha_k\sum_{n=1}^{\infty}\sigma_{k-1}(n)\,e^{2\pi inz}\quad (z\in\mathbb{Z},k>2) \tag{6.1.4}$$

where $\sigma_{k-1}(n)$ is defined by (5.5.20) and

$$\alpha_k=\frac{(2\pi)^k}{\Gamma(k)\zeta(k)}=\frac{2k}{|B_k|}. \tag{6.1.5}$$

We have used here the fact that, when k is an even positive integer, $\zeta(k)$ can be expressed in terms of the Bernoulli numbers B_k which occur as coefficients in the expansion

$$\frac{t}{e^t-1}=\sum_{n=0}^{\infty}\frac{B_n t^n}{n!} \quad (|t|<2\pi),$$

by means of the formula

$$\zeta(k)=\frac{(2\pi)^k|B_k|}{2(k!)}.$$

The first few values of α_k are given in table 9.

Table 9

k	2	4	6	8	10	12	14	16
α_k	24	240	504	480	264	$\frac{65520}{691}$	24	$\frac{16320}{3617}$

k	18	20	22	24	26
α_k	$\frac{28728}{43867}$	$\frac{13200}{174611}$	$\frac{552}{77683}$	$\frac{131040}{236364091}$	$\frac{24}{657931}$

We have included the value of α_2, since (5.7.30), (5.4.13) and (5.7.13) yield, in exactly the same way,

$$E_2(z):=G_L(z;0,1)=-\frac{3}{\pi y}+1-\alpha_2\sum_{r=1}^{\infty}\sigma_1(r)\,e^{2\pi i r z}. \tag{6.1.6}$$

We recall that $E_2(z)$ is not a holomorphic function of z, although it satisfies the functional equation

$$E_2(Tz)=(T:z)^2E_2(z). \tag{6.1.7}$$

It is clear from corollary 4.2.1 that for $k=4, 6, 8, 10$ and 14 the vector space $\{\Gamma(1), k\}$ consists of multiples of the Eisenstein series E_k and that there are no non-zero cusp forms. From this we deduce that

$$E_8=E_4^2, \quad E_{10}=E_4E_6 \tag{6.1.8}$$

and

$$E_{14}=E_4E_{10}=E_4^2E_6=E_6E_8. \tag{6.1.9}$$

For $E_8-E_4^2$ vanishes at every cusp and so must equal zero; similarly in the other cases.

We now consider the missing case $k=12$. By corollary 4.2.1(ii) we know that there exists at most one non-zero cusp form of weight 12 (apart from constant factors) and such a cusp form certainly exists. For we may take $g=E_4$ in theorem 4.3.1(iii) and we then conclude that

$$h=4E_4E_4''-5(E_4')^2\in\{\Gamma(1),12\}_0 \tag{6.1.10}$$

and is not the zero form. We shall show that this cusp form is a constant multiple of the function Δ defined explicitly as follows:

$$\Delta(z)=e^{2\pi iz}\left\{\prod_{m=1}^{\infty}(1-e^{2\pi imz})\right\}^{24}\quad(z\in\mathbb{H}). \tag{6.1.11}$$

The convergence of the infinite product is uniform on compact subsets of $\mathbb{H}$, and $\Delta(z)$ is therefore holomorphic on $\mathbb{H}$ and has no zeros on $\mathbb{H}$. Also the series

$$2\pi iz+24\sum_{m=1}^{\infty}\log(1-e^{2\pi imz})=2\pi iz-24\sum_{m=1}^{\infty}\sum_{n=1}^{\infty}n^{-1}e^{2\pi imnz}$$

converges to one of the values of $\log\Delta(z)$. Thus on differentiation we obtain

$$\frac{1}{2\pi i}\frac{\Delta'(z)}{\Delta(z)}=1-24\sum_{m=1}^{\infty}\sum_{n=1}^{\infty}m\,e^{2\pi imnz}=1-24\sum_{r=1}^{\infty}\sigma_1(r)\,e^{2\pi irz};$$

$$\frac{1}{2\pi i}\frac{\Delta'(z)}{\Delta(z)}=E_2(z)+\frac{3}{\pi y}=:E_2^{+}(z). \tag{6.1.12}$$

All these processes are justified because of the uniform absolute convergence on compact subsets of $\mathbb{H}$.

It follows from (6.1.12) that

$$\Delta(z)=\Delta(i)\exp\left\{2\pi i\int_i^z E_2^{+}(t)\,dt\right\}.$$

Hence, for all $T\in\Gamma(1)$,

$$\Delta(Tz)=\Delta(Ti)\exp\left\{2\pi i\int_{Ti}^{Tz}E_2^{+}(t)\,dt\right\}.$$

We now argue as in §4.4. We put $t = Tw$, where $v = \operatorname{Im} w$, in the integral and observe that, by (6.1.7, 12)

$$E_2^+(Tw) = (T\colon w)^2 E_2^+(w) + \frac{3}{\pi v}\{|T\colon w|^2 - (T\colon w)^2\}$$

$$= (T\colon w)^2 E_2^+(w) - \frac{6ic}{\pi}(T\colon w).$$

Hence

$$\Delta(Tz) = \Delta(Ti)\exp\left\{2\pi i\int_i^z E_2^+(w)\,dw + 12\int_i^z \frac{c\,dw}{cw+d}\right\}$$

$$= \frac{\Delta(Ti)}{(T\colon i)^{12}\Delta(i)}(T\colon z)^{12}\Delta(z).$$

In particular,

$$\Delta(Uz) = \Delta(z), \quad \Delta(Vz) = (V\colon z)^{12}\Delta(z),$$

and therefore, since $\Gamma(1)$ is generated by U and V, we deduce that

$$\Delta(Tz) = (T\colon z)^{12}\Delta(z).$$

Thus $\Delta \in \{\Gamma(1), 12\}_0$. The function Δ is called the (*modular*) *discriminant*. We also put

$$\Delta(z) = \sum_{n=1}^{\infty} \tau(n)\, e^{2\pi i n z} \tag{6.1.13}$$

so that, by (6.1.11), $\tau(1) = 1$. The Fourier coefficient $\tau(n)$ is called *Ramanujan's function* $\tau(n)$.

By using the fact that all cusp forms of weight 12 are constant multiples of Δ we obtain the following relations

$$\Delta = \frac{1}{1728}(E_4^3 - E_6^2), \tag{6.1.14}$$

$$= \frac{691}{432000}(E_4E_8 - E_{12}), \tag{6.1.15}$$

$$= \frac{691}{209088}(E_{12} - E_6^2). \tag{6.1.16}$$

For in each case the power series of the expression on the right begins with $e^{2\pi i z}$, as is easily checked by means of (6.1.4). Finally, by (6.1.10),

$$\Delta = \frac{5(E_4')^2 - 4E_4E_4''}{3840\pi^2}. \tag{6.1.17}$$

We now state some simple results about zeros of Δ, E_4 and E_6.

Theorem 6.1.1. (i) *The discriminant Δ has no zeros in* $\mathbb{H}$. (ii) *E_4 has a simple zero at each point of $\mathbb{E}_3$ and no other zeros; i.e.* $\operatorname{ord}(E_4, \rho, \Gamma(1)) = \frac{1}{3}$. (iii) *$E_6$ has a simple zero at each point of $\mathbb{E}_2$ and no other zeros; i.e.* $\operatorname{ord}(E_6, i, \Gamma(1)) = \frac{1}{2}$.

Proof. By theorem 4.1.4, the total orders of the functions Δ, E_4 and E_6 modulo $\Gamma(1)$ are 1, $\frac{1}{3}$ and $\frac{1}{2}$, respectively. Since $\operatorname{ord}(\Delta, \infty, \Gamma(1)) = 1$, Δ cannot have a zero at any point of $\mathbb{H}$, and this is, in any case, obvious from (6.1.11). By corollary 4.1.2 and (4.1.20),

$$\operatorname{ord}(E_4, \rho, \Gamma(1)) \geq \tfrac{1}{3}, \quad \operatorname{ord}(E_6, i, \Gamma(1)) \geq \tfrac{1}{2},$$

and it follows that E_4 and E_6 can only vanish on $\mathbb{E}_3$ and $\mathbb{E}_2$, respectively, and that equality must hold in these inequalities.

In the following theorem we show that the upper bounds for d_k and δ_k given in corollary 4.2.1 are actually attained. For this purpose it is convenient to make the convention that

$$E_0(z) = 1 \quad (z \in \mathbb{H}'). \tag{6.1.18}$$

Theorem 6.1.2. *For even $k \geq 2$,*

$$d_k = \left[\frac{k}{12}\right] + 1 \quad (k \not\equiv 2 \pmod{12}), \quad d_k = \left[\frac{k}{12}\right] \quad (k \equiv 2 \pmod{12}),$$

and $\delta_k = d_k - 1$ for $k \geq 4$. Further, the d_k forms

$$E_{k-12r}\Delta^r \quad (0 \leq r \leq k/12, r \neq (k-2)/12) \tag{6.1.19}$$

form a basis for $\{\Gamma(1), k\}$. If $k \geq 12$, $k \neq 14$, the $d_k - 1$ forms listed in (6.1.19) for $r > 0$, form a basis for $\{\Gamma(1), k\}_0$.

Proof. Since the number of forms in (6.1.19) is $[k/12] + 1$ when $k \not\equiv 2 \pmod{12}$ and is $[k/12]$ when $k \equiv 2 \pmod{12}$, it follows from corollary 4.2.1 that we need only show that these forms are linearly independent. This follows from the fact that the first term in the Fourier expansion of $E_{k-12r}(z)\Delta^r(z)$ $(r \neq (k-2)/12)$ is $e^{2\pi i r z}$.

Theorem 6.1.3. *Every member of* $\{\Gamma(1), k\}$ $(k \geq 2)$ *can be expressed as a polynomial in* E_4 *and* E_6.

Proof. The theorem is true for $2 \leq k \leq 14$ by (6.1.8, 9, 14, 15). We assume its truth for $2 \leq k < r$, where $r \geq 16$, and prove it true for $k = r$. By theorem 6.1.2, it is enough to show that E_r is expressible as a polynomial in E_4 and E_6. Choose non-negative integers p, q such that $4p + 6q = r$; this is always possible with $q = 0$ or 1. Then

$$E_r - E_4^p E_6^q = \Delta F_{r-12},$$

where $F_{r-12} \in \{\Gamma(1), r-12\}$, and so is expressible as a polynomial in E_4 and E_6. The theorem follows.

We now define Klein's *absolute invariant* J by

$$J(z) := \frac{E_4^3(z)}{1728\,\Delta(z)} = 1 + \frac{E_6^2(z)}{1728\,\Delta(z)} \quad (z \in \mathbb{H}). \qquad (6.1.20)$$

It is also convenient to put

$$j(z) := 1728\,J(z). \qquad (6.1.21)$$

Since $\Delta(z) \neq 0$ for $z \in \mathbb{H}$, it follows that J is a modular function belonging to $\Gamma(1)$ and that it is holomorphic on $\mathbb{H}$. Further, $\operatorname{ord}(J, \infty, \Gamma(1)) = -1$, so that J has a simple pole at ∞. The Fourier expansion of $j(z)$ has the form

$$j(z) = e^{-2\pi i z} + 744 + 196884\, e^{2\pi i z} + \cdots. \qquad (6.1.22)$$

The coefficients of higher powers become very large; since $j(z) = E_4^3(z)/\Delta(z)$, it is clear that the coefficients are all positive integers.

Further, it follows from theorem 6.1.1 that $J(z)$ has zeros of order 3 at points of $\mathbb{E}_3$ and $J(z) - 1$ has zeros of order 2 at points of $\mathbb{E}_2$. These are the only zeros of $J(z)$ and $J(z) - 1$, respectively.

For every $w_0 \in \mathbb{C}$, $J(z) - w_0$ is a modular function having a simple pole at ∞. It follows from theorem 4.1.4 that the sum of the orders of the zeros of $J(z) - w_0$ in a fundamental region is 1. Because of the above results on the form of $J(z)$ for $z \in \mathbb{E}$, it follows that for each $w_0 \in \mathbb{C}$ there exists exactly one point z_0 in $\mathbb{F}_I$ such that $J(z_0) = w_0$. Accordingly, the transformation $w = J(z)$ maps $\mathbb{F}_I$ onto $\mathbb{C}$ and is a one-to-one mapping, which is conformal at all points of $\mathbb{F}_I$ except ∞, ρ, i, since ρ and i are the only points of $\mathbb{F}_I$ at which J' vanishes. The images of ∞, ρ and i are the points ∞, 0 and 1.

It follows from (6.1.21, 22) that, for large $y = \text{Im } z$ and $-\frac{1}{2} < x = \text{Re } z < 0$, $\text{Im } J(z)$ has the same sign as

$$\text{Im } e^{-2\pi i z} = -e^{2\pi y} \sin 2\pi x,$$

and so is positive. Further, from (6.1.4),

$$\overline{E_r(z)} = E_r(-\bar{z}) \quad \text{for } r > 2,$$

so that

$$\overline{J(z)} = J(-\bar{z}). \tag{6.1.23}$$

Also $E_r(z)$ and $J(z)$ are real for $x = 0, \pm\frac{1}{2}$. It follows that $\text{Im } J(z) > 0$ for all z in the interior of the left-hand triangle $\mathbb{F}_I^{(1)}$. For otherwise, by continuity, $J(z)$ would be real for some z in the interior of $\mathbb{F}_I^{(1)}$ and would therefore, by (6.1.23), take the same value at $-\bar{z}$, which is an incongruent point in the interior of $\mathbb{F}_I^{(2)}$. Similarly, $\text{Im } J(z) < 0$ for all z in the interior of $\mathbb{F}_I^{(2)}$. Further, since $J(Vz) = J(z)$, it follows that $J(z)$ is also real for $|z| = 1$, and it follows by continuity that (i) $J(z) \leq 0$ for $z \in l_U$ (see (2.4.6, 7)), (ii) $0 \leq J(z) \leq 1$ for $z \in l_V$ and (iii) $J(z) \geq 1$ for $z = iy$ where $y \geq 1$. The transformation $w = J(z)$ therefore maps $\bar{\mathbb{F}}_I^{(1)}$ onto $\bar{\mathbb{H}}$. These mapping properties could have been used, by Riemann's mapping theorem, to define $J(z)$ uniquely.

From theorem 4.3.4 we deduce that any modular form $f \in M(\Gamma(1), k)$, where k is any even integer, can be expressed in the form

$$f = (J')^{\frac{1}{2}k} P(J)/Q(J),$$

where P and Q are polynomials. In particular, if $k = 0$ and f has poles of total order q in $\mathbb{F}_I$, then the degrees of P and Q do not exceed q.

For example, if $f = \Delta$, then $g = \Delta(J')^{-6}$ has a zero of order 7 at ∞ and poles of order 12 and 6 at ρ and i. Hence $\text{ord}(g, \zeta, \Gamma(1)) = 7, -4$ and -3 for $\zeta = \infty, \rho, i$, respectively. Also $\text{ord}(J^{-4}(J-1)^{-3}, \zeta, \Gamma(1))$ takes the same values, and it follows that $gJ^4(J-1)^3$ is a modular function with no poles or zeros and so is a constant. From this we find that

$$\Delta = \frac{-1}{(48\pi^2)^3} \frac{(J')^6}{J^4(J-1)^3}. \tag{6.1.24}$$

It can be proved similarly that

$$E_4 = -\frac{(J')^2}{4\pi^2 J(J-1)} \tag{6.1.25}$$

and

$$E_6 = \frac{(J')^3}{(2\pi i)^3 J^2(J-1)}. \tag{6.1.26}$$

Since J is a rational function of E_4 and E_6, and so is J', by (6.1.25, 26), we obtain the following theorem:

Theorem 6.1.4. *If $f \in M(\Gamma(1), k)$, then f is a rational function of E_4 and E_6.*

6.2. Poincaré series. For even $k > 2$ and $m \in \mathbb{Z}$ we write

$$G_k(z, m) = G_I(z; m; \Gamma(1), k, 1). \tag{6.2.1}$$

If $m \leq 0$, it follows from (5.3.29) that $G_k(z, m)$ has the Fourier expansion

$$G_k(z, m) = e^{2\pi imz} + \sum_{n=1}^{\infty} a_n e^{2\pi inz}, \tag{6.2.2}$$

where the coefficients a_n depend upon m and k as well as on n. We have, of course, $E_k(z) = G_k(z, 0)$.

Theorem 5.1.3 shows that, by subtracting a suitable linear combination of forms $G_k(z, m)$ with $m \leq 0$ from a given modular form, we obtain a cusp form, and we shall therefore assume in what follows that $m > 0$. It is convenient to write

$$g_k(z, m) = m^{k-1} G_k(z, m), \tag{6.2.3}$$

$$= \sum_{r=1}^{\infty} c_k(r, m) e^{2\pi irz}. \tag{6.2.4}$$

Theorem 6.2.1. *The functions $g_k(z, m)$ $(m > 0)$ vanish identically if $k = 4, 6, 8, 10, 14$. If $k = 12$ or $k \geq 16$ then $\delta_k > 0$ and the δ_k functions $g_k(z, m)$ $(m = 1, 2, \ldots, \delta_k)$ form a basis for $\{\Gamma(1), k\}_0$. Also*

$$c_k(r, m) = c_k(m, r) \quad (r, m \in \mathbb{Z}^+) \tag{6.2.5}$$

and the coefficients $c_k(r, m)$ are real.

Proof. Since $N=N_0=\delta_k$, theorem 4.2.3 shows that the δ_k functions $g_k(z, m)$ $(m=1, 2, \ldots, \delta_k)$ form a basis, and theorem 5.2.5 (with $L=S=I$, $n_1=n_2=1$, $\kappa_1=\kappa_2=0$) shows that

$$c_k(r, m)=c_k(m, r) \quad (r, m \in \mathbb{Z}^+).$$

It remains to prove that $c_k(r, m)$ is real. Since

$$\overline{g_k(-\bar{z}, m)}=\sum_{r=1}^{\infty} \overline{c_k(r, m)}\, e^{2\pi irz},$$

we need only prove that $\overline{g_k(-\bar{z}, m)}=g_k(z, m)$.

Let $\Gamma(1)=\Gamma_U \cdot \mathcal{R}$. Write

$$T^*=\begin{bmatrix} a & -b \\ -c & d \end{bmatrix}.$$

Then as T runs through $\mathcal{R}$, T^* runs through $\mathcal{R}^*$, where $\Gamma(1)=\Gamma_U \cdot \mathcal{R}^*$. Since $\overline{-T(-\bar{z})}=T^*(z)$ and $\overline{T:(-\bar{z})}=T^*:z$, we have

$$m^{1-k}\,\overline{g_k(-\bar{z}, m)}=\sum_{T\in\mathcal{R}} \frac{e^{2\pi imT^*(z)}}{(T^*:z)^k}=m^{1-k}g_k(z, m),$$

as required.

Now write, for any $k>2$,

$$\omega_k=\frac{\Gamma(k-1)}{(4\pi)^{k-1}}, \tag{6.2.6}$$

and suppose that $f\in\{\Gamma(1), k\}$, where

$$f(z)=\sum_{n=0}^{\infty} b_n e^{2\pi inz}.$$

It then follows from theorem 5.2.2 and (6.2.3) that

$$(f(z), g_k(z, m))=\omega_k b_m \quad (m>0). \tag{6.2.7}$$

6.3 The case $\delta_k=1$. It follows from theorem 6.1.2 that $\delta_k=1$ if and only if

$$k=12, 16, 18, 20, 22 \quad \text{or} \quad 26, \tag{6.3.1}$$

and we restrict our attention to these values of k in the present section. Write

$$F_k(z)=E_{k-12}(z)\Delta(z)=\sum_{n=1}^{\infty} \gamma_k(n)\, e^{2\pi inz}, \tag{6.3.2}$$

so that $F_{12}=\Delta$ and $\gamma_{12}(n)=\tau(n)$. Then every cusp form in $\{\Gamma(1), k\}_0$ is a constant multiple of F_k and so, in particular,

$$g_k(z, m)=c_k(1, m)F_k \quad (m \in \mathbb{Z}^+) \tag{6.3.3}$$

and

$$c_k:=c_k(1, 1)\neq 0. \tag{6.3.4}$$

It follows from (6.2.5) and (6.3.2, 3) that

$$c_k(r, m)=c_k(1, m)\gamma_k(r)=c_k(m, 1)\gamma_k(r)=c_k(1, 1)\gamma_k(m)\gamma_k(r),$$

so that

$$c_k(r, m)=c_k(m, r)=c_k\gamma_k(m)\gamma_k(r). \tag{6.3.5}$$

In particular,

$$c_k(1, m)=c_k\gamma_k(m), \tag{6.3.6}$$

$$g_k(z, m)=c_k\gamma_k(m)F_k(z). \tag{6.3.7}$$

We now obtain a formula for c_k. Write

$$e(0)=1, \quad e(m)=\alpha_4\sigma_3(m) \quad (m\geq 1),$$

so that

$$E_4(z)=\sum_{n=0}^{\infty} e(m)\, e^{2\pi imz}.$$

Then

$$\begin{aligned}\sum_{m=0}^{\infty} e_m G_k(z, m) &= \sum_{m=0}^{\infty} e_m \sum_{T\in\mathcal{R}} \frac{e^{2\pi imTz}}{(T:z)^k}\\ &= \sum_{T\in\mathcal{R}} \frac{E_4(Tz)}{(T:z)^k}.\end{aligned} \tag{6.3.8}$$

The inversion of the order of summation is justified since, by theorem 4.5.1,

$$\begin{aligned}\sum_{T\in\mathcal{R}}\sum_{m=0}^{\infty} |e_m|\left|\frac{e^{2\pi imTz}}{(T:z)^k}\right| &= \sum_{T\in\mathcal{R}} \frac{E_4(i\operatorname{Im}Tz)}{|T:z|^k}\\ &\leq C\sum_{T\in\mathcal{R}}\left\{\frac{1}{|T:z|^k}+\frac{y^{-4}}{|T:z|^{k-8}}\right\},\end{aligned}$$

which is convergent. Hence, by (6.3.7, 8),

$$E_k(z)+c_k\alpha_4 \sum_{n=1}^{\infty} \sigma_3(m)m^{1-k}\gamma_k(m)F_k(z)=E_4(z) \sum_{T\in\mathcal{R}} (T:z)^{4-k}$$
$$=E_4(z)E_{k-4}(z).$$

Now, since $E_4E_{k-4}-E_k$ is a cusp form,

$$E_4E_{k-4}-E_k=\{\alpha_4+(-1)^{\frac{1}{2}k}\alpha_{k-4}-(-1)^{\frac{1}{2}k}\alpha_k\}F_k$$

and so

$$c_k \sum_{m=1}^{\infty} \frac{\sigma_3(m)\gamma_k(m)}{m^{k-1}}=1+(-1)^{\frac{1}{2}k}\frac{\alpha_{k-4}-\alpha_k}{\alpha_4}=:\beta_k, \quad (6.3.9)$$

say.

We shall prove in §10.2 that $\gamma_k(n)$ is a multiplicative function, and a consequence is that it can be shown that

$$\sum_{m=1}^{\infty} \frac{\sigma_3(m)\gamma_k(m)}{m^{k-1}}=\frac{\phi_k(k-1)\phi_k(k-4)}{\zeta(k-4)}, \quad (6.3.10)$$

where

$$\phi_k(s)=\sum_{n=1}^{\infty} \frac{\gamma_k(m)}{m^s}. \quad (6.3.11)$$

This series is absolutely convergent for $\operatorname{Re} s>\frac{1}{2}(k+1)$, by theorem 4.5.2(iv). Hence we obtain the explicit formula

$$c_k=\frac{\beta_k^{-1}\zeta(k-4)}{\phi_k(k-1)\phi_k(k-4)}, \quad (6.3.12)$$

where β_k is defined by (6.3.9).

We can use (6.3.9) to evaluate (f,f) for any $f\in\{\Gamma(1),k\}_0$; it is sufficient to find (F_k,F_k). We have, by (6.2.7) and (6.3.3),

$$c_k(F_k,F_k)=(F_k(z),g_k(z,1))=\omega_k,$$

so that

$$(F_k,F_k)=\frac{\omega_k}{c_k}=\frac{\omega_k\beta_k\phi_k(k-1)\phi_k(k-4)}{\zeta(k-4)}. \quad (6.3.13)$$

In particular,

$$(\Delta,\Delta)=\frac{10!\,14511\phi_{12}(8)\phi_{12}(11)}{2^{24}\pi^{19}}=10^{-6}\times 1.035\ldots, \quad (6.3.14)$$

since

$$\phi_{12}(8) = 0.9307\ldots, \quad \phi_{12}(11) = 0.989\,433\ldots,$$

as can be calculated from Lehmer's (1942) or Watson's (1949) tables of $\tau(n)$. A more accurate value of $10^{-6} \times 1.035\,290\,48\ldots$ has been obtained by Lehmer (1942) by direct evaluation of the double integral.

We note in conclusion that, by (6.3.7), $\gamma_k(m) = 0$ if and only if $g_k(z, m)$ vanishes identically. It is conjectured that $\tau(m) \neq 0$ for all $m \in \mathbb{Z}^+$. This would be proved if we could show that

$$\sum_{T \in \mathfrak{R}} \frac{e^{2\pi i m T z}}{(T:z)^{12}}$$

does not vanish identically for any positive integer m. The difficulty in proving this is illustrated by the fact that, by theorem 6.2.1, the corresponding series with 12 replaced by 14 does vanish identically for every positive integer m.

6.4. Modular forms of any real weight. Put $t = e^{2\pi i z}$ and write

$$D(t) = \sum_{n=1}^{\infty} \tau(n) t^{n-1} = e^{-2\pi i z} \Delta(z).$$

Then $D(t)$ is a holomorphic function of t for $|t| < 1$ with no zeros in this circle and $D(0) = 1$. Let κ be any real number. It follows that there exists a unique function F_κ with the following three properties: (i) $F_\kappa(t)$ is holomorphic for $|t| < 1$, (ii) $F_\kappa(0) = 1$, and (iii) $F_\kappa(t)$ is a value of $\{D(t)\}^\kappa$ for all t with $|t| < 1$. It follows that $e^{2\pi i \kappa z} F_\kappa(e^{2\pi i z})$ is a holomorphic function for $z \in \mathbb{H}$ and is a value of $\{\Delta(z)\}^\kappa$ for all $z \in \mathbb{H}$ we can therefore denote this function by $\Delta^\kappa(z)$, without ambiguity. Accordingly

$$\Delta^\kappa(z) = e^{2\pi i \kappa z} \sum_{n=0}^{\infty} \delta_\kappa(n)\, e^{2\pi i n z} = e^{2\pi i \kappa z} \left\{ \prod_{m=1}^{\infty} (1 - e^{2\pi i m z} \right\}^{24\kappa}, \tag{6.4.1}$$

so that

$$\delta_\kappa(0) = 1, \quad \delta_\kappa(1) = -24\kappa.$$

We now investigate the behaviour of Δ^κ under transformations belonging to $\Gamma(1)$. For each $T \in \Gamma(1)$,

$$\Delta^\kappa(Tz) \quad \text{is a value of} \quad \{(T:z)^{12} \Delta(z)\}^\kappa,$$

so that we may write

$$\Delta^\kappa(Tz) = \nu_T(z)\Delta^\kappa(z), \tag{6.4.2}$$

where $\nu_T(z)$ is holomorphic for $z \in \mathbb{H}$ and

$$|\nu_T(z)| = |T : z|^{12\kappa}.$$

Further, for any S and T in $\Gamma(1)$,

$$\nu_{ST}(z)\Delta^\kappa(z) = \Delta^\kappa(STz) = \nu_S(Tz)\Delta^\kappa(Tz) = \nu_S(Tz)\nu_T(z)\Delta^\kappa(z),$$

so that ν_T is an AF of weight 12κ for $\Gamma(1)$.

If v is the corresponding MS, then, by (6.4.1),

$$\Delta^\kappa(Uz) = e^{2\pi i\kappa}\,\Delta^\kappa(z),$$

so that

$$v(U) = e^{2\pi i\kappa}. \tag{6.4.3}$$

The fractional part of κ is therefore the cusp parameter at ∞; this is true even when $\kappa < 0$. It follows from (3.4.2, 7) that

$$v(P) = e^{-4\pi i\kappa}, \quad v(V) = e^{-6\pi i\kappa}. \tag{6.4.4}$$

Now let k be any real number and put $\kappa = k/12$. It follows that there exists a MS v of weight k for $\Gamma(1)$ having the properties

$$v(U) = e^{\pi ik/6}, \quad v(V) = e^{-\pi ik/2}, \quad v(P) = e^{-\pi ik/3}. \tag{6.4.5}$$

Further, for any $T \in \Gamma(1)$,

$$v(T) = \frac{\Delta^{k/12}(Tz)}{(T:z)^k \Delta^{k/12}(z)}. \tag{6.4.6}$$

It follows from §3.4 that there are exactly six different multiplier systems for $\Gamma(1)$ of each weight k, their values differing from (6.4.6) by sixth roots of unity. These MS can be constructed explicitly by putting

$$\mathcal{R} := \{0, 4, 6, 8, 10, 14\} \tag{6.4.7}$$

and observing that, for each $r \in \mathcal{R}$,

$$f_r = E_r\,\Delta^{(k-r)/12} \in H(\Gamma(1), k, v^{(r)}) \tag{6.4.8}$$

where, for $T \in \Gamma(1)$,

$$v^{(r)}(T) = \frac{\Delta^{(k-r)/12}(Tz)}{(T:z)^{k-r}\Delta^{(k-r)/12}(z)}. \tag{6.4.9}$$

For $r \neq 0$, the arguments required are similar to those used for the case $r=0$ and, in particular, $v^{(0)}(T)=v(T)$. Further,

$$v^{(r)}(U)=e^{\pi i(k-r)/6}, \tag{6.4.10}$$

so that, if κ_r is the associated cusp parameter,

$$\kappa_r=\left\{\frac{k-r}{12}\right\} \quad \text{(fractional part)} \tag{6.4.11}$$

and we have

$$v^{(r)}(U)=e^{2\pi i\kappa_r}=v^{(0)}(U)\,e^{-\pi ir/6}. \tag{6.4.12}$$

The values taken by $e^{-\pi ir/6}$ are, by (6.4.7), different sixth roots of unity, as required. We obtain the following table from theorem 4.1.2 and the definition of $\varepsilon(\zeta)$; see before corollary 4.1.2. The order of f_r at i and ρ (mod $\Gamma(1)$) is obtained from theorem 6.1.1 and (6.1.8, 9). Table 10 shows that

$$\varepsilon(i)+\varepsilon(\rho)=r/12$$

Table 10

r	0	4	6	8	10	14
$\operatorname{ord}(f_r, i, \Gamma(1))$	0	0	$\frac{1}{2}$	0	$\frac{1}{2}$	$\frac{1}{2}$
s_I	0	0	1	0	1	1
$\varepsilon(i)$	0	0	$\frac{1}{2}$	0	$\frac{1}{2}$	$\frac{1}{2}$
$\operatorname{ord}(f_r, \rho, \Gamma(1))$	0	$\frac{1}{3}$	0	$\frac{2}{3}$	$\frac{1}{3}$	$\frac{2}{3}$
m_I	0	2	0	1	2	1
$\varepsilon(\rho)$	0	$\frac{1}{3}$	0	$\frac{2}{3}$	$\frac{1}{3}$	$\frac{2}{3}$

and therefore, by (4.2.1–3), $\xi=\kappa_r+r/12$. It follows from theorem 4.2.1 and (6.4.11) that

$$d_k^{(r)}:=\dim\{\Gamma(1), k, v^{(r)}\}\le \max\left\{0, 1+\left[\frac{k-r}{12}\right]\right\}. \tag{6.4.13}$$

The same upper bound holds for

$$\delta_k^{(r)}:=\dim\{\Gamma(1), k, v^{(r)}\}_0 \tag{6.4.14}$$

except when $k\equiv r \pmod{12}$, $1+[(k-r)/12]$ then being replaced by $[(k-r)/12]$. That these bounds are exact is shown by the fact that, if $k \geq r$, the forms

$$E_{r+12s}\Delta^{-s+(k-r)/12} \quad \left(0 \leq s \leq \frac{k-r}{12}\right) \tag{6.4.15}$$

are linearly independent forms belonging to $\{\Gamma(1), k, v^{(r)}\}$. Accordingly, we have proved the following theorem, which extends theorem 6.1.2.

Theorem 6.4.1. *Let $k \in \mathbb{R}$ and let $v^{(r)}$ be a MS of weight k for $\Gamma(1)$, where $r \in \mathfrak{R}$. Put*

$$d_k^{(r)} = \dim\{\Gamma(1), k, v^{(r)}\}, \quad \delta_k^{(r)} = \dim\{\Gamma(1), k, v^{(r)}\}_0. \tag{6.4.16}$$

Then $d_k^{(r)} = 0$ if $k < r$. If $k \geq r$,

$$d_k^{(r)} = 1 + \left[\frac{k-r}{12}\right],$$

and the $d_k^{(r)}$ forms (6.4.11) *form a basis. If $k \not\equiv r$* (mod 12), *these forms are all cusp forms; if $k \equiv r$* (mod 12), *an Eisenstein series exists and $\delta_k^{(r)} = d_k^{(r)} - 1$.*

The MS $v^{(r)}$ is constant only when $k \equiv r$ (mod 12).

We conclude this section by remarking that a MS that is not constant on $\Gamma(1)$ may be constant on some subgroup of $\Gamma(1)$. We illustrate this by considering Δ^κ for fractional κ. We therefore have $r = 0 = m_L = s_L$ for all $L \in \Gamma(1)$ and so, by (3.1.13) and (3.3.9, 10), for $k = 12\kappa$,

$$v(-I) = e^{-12\pi i\kappa}, \quad v(LP^sL^{-1}) = e^{-4\pi i\kappa s} \quad (-2 \leq s \leq 3)$$

and

$$v(LV^sL^{-1}) = e^{-6\pi i\kappa s} \quad (-1 \leq s \leq 2).$$

We deduce that v is constant on Γ^2 or Γ^3 if $\kappa = \frac{1}{2}$ or $\frac{1}{3}$, respectively, so that

$$\Delta^{1/2} \in \{\Gamma^2, 6, 1\}_0, \quad \Delta^{1/3} \in \{\Gamma^3, 4, 1\}_0.$$

It follows that

$$\Delta^{1/6} \in \{\Gamma'(1), 2, 1\}_0.$$

If we take $\kappa = \frac{1}{4}$, we have $k = 3$ and

$$v(-I) = -1, \quad v(LP^sL^{-1}) = (-1)^s.$$

Then $v(T) = 1$ for all $T \in \Gamma^4$ and $v(T) = -1$ for all $T \in \Gamma^2 - \Gamma^4$. We may therefore write

$$\Delta^{1/4} \in \{\Gamma^2, 3, \pm 1\}_0.$$

It follows also that

$$\Delta^{1/12} \in \{\bar{\Gamma}'(1), 1, \pm 1\}_0.$$

Here the MS v is such that $v(T) = 1$ for all $T \in \Gamma'(1)$ and $v(T) = -1$ for all $T \in \bar{\Gamma}'(1) - \Gamma'(1)$.

6.5. Modular equations. In this section we apply the theory developed in §4.3 to $\Gamma(1)$ and to the function J.

Let $n \in \mathbb{Z}^+$. Then, since J has valence 1 with respect to $\Gamma(1)$, we deduce from theorems 4.3.7, 8 that there exists an irreducible polynomial $\Phi_n(\tau, t)$ of degrees not exceeding $\psi(n)$ in each of τ and t such that

$$\Phi_n(J, J_T) = 0 \quad \text{for all } T \in \boldsymbol{T}_n. \tag{6.5.1}$$

Here $\Omega_n^* = \Gamma(1) \cdot \boldsymbol{T}_n$ and $|\boldsymbol{T}_n| = \psi(n)$.

We prove the following theorem:

Theorem 6.5.1. *The irreducible polynomial $\Phi_n(\tau, t)$ $(n > 1)$ has degree $\psi(n)$ in each of its two variables. Further*

$$\Phi_n(\tau, t) = \Phi_n(t, \tau) \tag{6.5.2}$$

and, by multiplication by a suitable non-zero constant, we may express $\Phi_n(J(z), t)$ in the form

$$\Phi_n(J(z), t) = \prod_{T \in \boldsymbol{T}_n} \{t - J_T(z)\}. \tag{6.5.3}$$

Proof. We show, first of all, that the $\psi(n)$ functions $J_T(T \in \boldsymbol{T}_n)$ are distinct. The set of points z in $\mathbb{H}$ for which $Tz \in \mathbb{E}$ for $T \in \boldsymbol{T}_n$ is enumerable and we can therefore choose $z \in \mathbb{H}$ such that $Tz \notin \mathbb{E}(T \in \boldsymbol{T}_n)$. Then, if, for T_1 and T_2 in T_n,

$$J(T_2 z) = J(T_1 z),$$

we have $T_2 z = ST_1 z$ for some $S \in \Gamma(1)$ and hence $T_2 = ST_1$. This can only hold if $T_2 = T_1$. It follows from this that the degree of $\Phi_n(\tau, t)$ in t is precisely $\psi(n)$.

Now take any

$$T = \begin{bmatrix} a & b \\ c & d \end{bmatrix} \in \boldsymbol{T}_n$$

and put

$$T^* = \begin{bmatrix} d & -b \\ -c & a \end{bmatrix}, \tag{6.5.4}$$

so that $T^*T = nI$ and therefore $T^*Tz = z$. Since $T^* \in \Omega_n^*$ we have $T^* = SL$ for some $L \in \boldsymbol{T}_n$ and $S \in \Gamma(1)$. Hence

$$\begin{aligned}\Phi_n(J_T(z), J(z)) &= \Phi_n(J(Tz), J(T^*Tz)) = \Phi_n(J(Tz), J(SLTz)) \\ &= \Phi_n(J(Tz), J(LTz)) = \Phi_n(J(Tz), J_L(Tz)) \\ &= 0,\end{aligned}$$

by (6.5.1). Thus the polynomial equation

$$\Phi_n(\tau, J(z)) = 0$$

has at least $\psi(n)$ different roots, namely $\tau = J_T(z)$ $(T \in \boldsymbol{T}_n)$ and, since the degree of $\Phi_n(\tau, t)$ is not more than $\psi(n)$ in τ, it follows that $\Phi_n(\tau, t)$ has exact degree $\psi(n)$ in each variable.

Moreover, since $\Phi_n(t, J(z)) = 0$ whenever $\Phi_n(J(z), t) = 0$, it follows that

$$\Phi_n(t, J(z)) = f(J(z))\Phi_n(J(z), t),$$

i.e.

$$\Phi_n(t, \tau) = f(\tau)\Phi_n(\tau, t),$$

where $f(\tau)$ is a rational function of τ. Applying this result twice we deduce that

$$f(\tau)f(t) = 1,$$

from which it follows that $f(\tau) = \pm 1$. We cannot have $f(\tau) = -1$ since then we should have $\Phi_n(t, t) = 0$. From this it would follow that $\Phi_n(\tau, t)$ had the factor $t - \tau$, which is false since Φ_n is irreducible of degree $\psi(n) > 1$. Hence (6.5.2) holds.

Now put

$$G(z) = \prod_{T \in \boldsymbol{T}_n} J_T(z).$$

Each $S \in \Gamma(1)$ defines a map $\pi_S: \boldsymbol{T}_n \to \boldsymbol{T}_n$ defined by

$$\pi_S(T) = T_1, \tag{6.5.5}$$

where

$$TS = S_1 T_1 \quad (T_1 \in \boldsymbol{T}_n, \quad S \in \Gamma(1)). \tag{6.5.6}$$

Since $\pi_S(T) = \pi_S(T')$ implies that $T = T'$, π_S is a permutation and it follows that $G(Sz) = G(z)$. Hence

$$G \in M(\Gamma(1), 0, 1).$$

Let $q = V(G, \Gamma(1))$. It was shown at the end of §4.3 that the transformation groups Γ_T are conjugate to $\Gamma_0(n)$ when $\Gamma = \Gamma(1)$. We therefore have, by theorem 4.3.8,

$$\begin{aligned}\hat{\mu}(n)q = V(G, \Gamma(n)) &= \sum_{T \in \boldsymbol{T}_n} V(J_T, \Gamma(n)) \\ &= [\Gamma_0(n):\Gamma(n)] \sum_{T \in \boldsymbol{T}_n} V(J_T, \Gamma_T) \\ &= \psi(n)[\Gamma_0(n):\Gamma(n)][\Gamma(1):\Gamma_0(n)]\end{aligned}$$

so that $q = \psi(n)$. Since G has no poles in $\mathbb{H}$, we deduce from theorem 4.3.3 that G is a polynomial of degree $\psi(n)$ in J.

Now we can clearly express $\Phi_n(J(z), t)$ in the form

$$\Phi_n(J(z), t) = g(J(z)) \prod_{T \in \boldsymbol{T}_n} \{t - J(Tz)\}$$

where $g(\tau)$ is a polynomial in τ of degree at most $\psi(n)$. The 'constant term' on the right-hand side is $\pm g(J(z))G(z)$ and it follows that $g(\tau)$ is a constant which may be taken to be 1.

We now apply theorem 4.3.9 to the discriminant function Δ. For each $T \in \Omega_n^*$ ($n \in \mathbb{Z}^+$) let

$$D_T(z) := \frac{\Delta_T(z)}{\Delta(z)} = \frac{n^6 \Delta(Tz)}{(T:z)^{12} \Delta(z)} \quad (z \in \mathbb{H}). \tag{6.5.7}$$

Theorem 6.5.2. *For each $T \in \Omega_n^*$, $D_T \in H(\Gamma_T, 0, 1)$ where*

$$\Gamma_T = T^{-1}\Gamma(1)T \cap \Gamma(1)$$

and is a subgroup of $\Gamma(1)$ conjugate to $\Gamma_0(n)$.

Let $\Omega_n^* = \Gamma(1) \cdot \mathbf{T}_n$. Then, for each $S \in \Gamma(1)$,

$$D_T(Sz) = D_{T_1}(z),$$

where T and $T_1 = \pi_S(T)$ belong to $\mathbf{T}_n$; see (6.5.5, 6).

Further, $D_T(z)$ can be expressed as a rational function of $J(z)$ and $J_T(z)$.

Proof. We have, in the notation of (6.5.6), by (4.3.19),

$$\begin{aligned} D_T(Sz) &= \frac{n^6\,\Delta(TSz)}{(T\!:Sz)^6\,\Delta(Sz)} = \frac{n^6\,\Delta(S_1T_1z)}{(TS\!:z)^6\,\Delta(z)} \\ &= \frac{n^6(S_1\!:T_1z)^6\,\Delta(T_1z)}{(S_1T_1\!:z)^6\,\Delta(z)} \\ &= \frac{n^6\,\Delta(T_1z)}{(T_1\!:z)^6\,\Delta(z)} = D_{T_1}(z). \end{aligned}$$

Clearly $D_T(z)$ is holomorphic on $\mathbb{H}$.

Finally consider the expression

$$\Phi_n(J(z), t) \sum_{T \in \mathbf{T}_n} \frac{D_T(z)}{t - J_T(z)}. \tag{6.5.8}$$

It is clear from (6.5.3, 10) that this is a polynomial of degree $\psi(n)-1$ in t and that it remains unaltered when z is replaced by Sz. Its coefficients are therefore rational functions of $J(z)$ and we therefore denote (6.5.8) by

$$\Psi_n(J(z), t).$$

We then have

$$\Psi_n(J(z), J_T(z)) = D_T(z) \frac{\partial}{\partial t} \Phi_n(J(z), t),$$

the derivative being taken at the point $t = J_T(z)$. Since this derivative is not zero for $z \in \mathbb{H}$ and is a rational function of $J(z)$ and $J_T(z)$, the last part of the theorem follows.

6.6. Further remarks. The discriminant function Δ defined by (6.1.11), is so called since it is, apart from a factor, the discriminant of the cubic appearing on the right-hand side of the equation

$$\left(\frac{du}{dt}\right)^2 = 4u^3 - g_2u - g_3$$

satisfied by the Weierstrass elliptic function $u=\wp(t)$. The argument z in $\Delta(z)$ is the ratio z_1/z_2 of a pair of primitive periods z_1, z_2 of $\wp(t)$, chosen so that this ratio lies in the half-plane $\mathbb{H}$. In fact,

$$\Delta(z)=\left(\frac{z_2}{2\pi}\right)^{12}(g_2^3-27g_3^2),$$

where the 'invariants' g_2 and g_3 are expressed in terms of Eisenstein series by the formulae

$$z_2^4g_2=60E_4^*(z),\quad z_2^6g_3=140E_6^*(z).$$

The function $\Delta^{1/24}$ (see §6.4 for fractional powers of Δ) is usually denoted by η and is known as Dedekind's eta function, since it was Dedekind who first completely solved the problem of specifying its multiplier system; see Ayoub (1963) chapter 3, Knopp (1970) and Rademacher and Grosswald (1972) for detailed information and further references.

The arithmetical function $\tau(n)$, occurring as the nth Fourier coefficient of $\Delta(z)$, was so named by Ramanujan (1916), who was the first of many mathematicians to make a study of its properties. This is one of several similar functions discussed in chapters 9 and 10.

From (6.3.2, 3, 4) with $k=12$ it follows that

$$g_{12}(z,m)=c_{12}\tau(m)\Delta(z),$$

so that $\tau(m)=0$ if and only if the Poincaré series $G_{12}(z,m)$ vanishes identically. This fact has not, however, been of any assistance in determining whether integers m exist for which $\tau(m)=0$. By other methods, using the multiplicative properties of $\tau(n)$ proved in §10.2, Lehmer (1947, 1959) has shown that $\tau(n)\neq 0$ for $n\leq 113\ 740\ 230\ 287\ 998$.

In theorem 6.1.1 information is given on the location of the zeros of the Eisenstein series E_4 and E_6. The orders of the zeros of E_k for $k\geq 8$ at elliptic fixed points can be determined; see Atkin (1969). Moreover it can be shown that all the zeros of E_k lie on the unit circle $\{z\in\mathbb{H}: |z|=1\}$ or its transforms by mappings belonging to $\hat{\Gamma}(1)$; see Rankin and Swinnerton-Dyer (1970).

The Fourier coefficients of the invariant function j, defined by (6.1.21), have interesting properties. See, for example, Lehner (1969), chapters 5 and 6.

The Fourier coefficients $c_k(m, n)$ of the Poincaré series $g_k(z, m)$, defined by (6.2.3), have been studied systematically by Petersson (1939). The method of evaluating $c_k(1, 1)$ and of finding inner products explicitly, which is described in §6.3, is due to Rankin (1952).

The six multiplier systems on $\Gamma(1)$ described in §6.4 were first discussed by Petersson (1938*b*); see also van Lint (1957).

§6.5 is based on Fueter (1924), chapter 1.

7: Groups of level 2 and sums of squares

7.1. Hermite and theta functions. For $\mu, \nu \in \mathbb{R}$ and $z \in \mathbb{H}$ put

$$\Theta_{\mu,\nu}(z) = \sum_{n=-\infty}^{\infty} \exp\{\pi i(n+\tfrac{1}{2}\mu)^2 z + \pi i n\nu\}. \tag{7.1.1}$$

The series is clearly uniformly convergent for $y = \operatorname{Im} z \geq \eta > 0$, and so defines $\Theta_{\mu,\nu}(z)$ as a holomorphic function of z on $\mathbb{H}$.

Theorem 7.1.1. *If $z \in \mathbb{H}$,*

$$\Theta_{\mu,\nu}(z) = \frac{e^{-\frac{1}{2}\pi i\mu\nu}}{(-iz)^{\frac{1}{2}}}\Theta_{\nu,-\mu}\left(-\frac{1}{z}\right).$$

Proof. We note that, for $z \in \mathbb{H}$, $|\operatorname{ph}(-iz)| < \frac{1}{2}\pi$. If we prove the result stated for $z = iy$, where $y > 0$, its truth for all $z \in \mathbb{H}$ will follow by analytic continuation; for the difference between the two sides is holomorphic on $\mathbb{H}$. Accordingly, we have to prove that, for $y > 0$,

$$\sum_{n=-\infty}^{\infty} \exp\{-\pi y(n+\tfrac{1}{2}\mu)^2 + i\pi n\nu\}$$
$$= y^{-\frac{1}{2}} e^{-\frac{1}{2}\pi i\mu\nu} \sum_{n=-\infty}^{\infty} \exp\{-\pi(n+\tfrac{1}{2}\nu)^2 y^{-1} - i\pi\mu n\}. \tag{7.1.2}$$

Write

$$J = \int_{-\infty}^{\infty} e^{-\pi u^2}\, du.$$

Actually $J = 1$, but this will emerge in the course of the proof. We use the fact that

$$\int_{-\infty+\beta i}^{\infty+\beta i} e^{-\pi y u^2}\, du = J y^{-\frac{1}{2}} = \int_{-\infty}^{\infty} e^{-\pi y t^2}\, dt, \tag{7.1.3}$$

the integral on the left being taken along the line $\operatorname{Im} u = \beta$. This holds, by Cauchy's theorem, since it is easily shown that

$$\left|\int_X^{X+\beta i} e^{-\pi y u^2}\, du\right| \to 0 \quad \text{as } |X| \to \infty.$$

To prove (7.1.2) define

$$F(z)=\frac{\exp\{-\pi ys^2-\pi\mu ys+i\pi\nu s\}}{e^{2\pi is}-1}\quad(s=\sigma+it)$$

for s inside and on the rectangle R_N with vertices at the points $\pm(N+\frac{1}{2})\pm i$, where N is a positive integer. Then F is holomorphic with the exception of simple poles at integral points $s=n$ $(-N\leq n\leq N)$ of residue

$$\frac{1}{2\pi i}\exp\{-\pi n^2y-\pi\mu ny+i\pi\nu n\}.$$

Hence

$$\sum_{n=-N}^{N}\exp\{-\pi yn^2-\pi y\mu n+i\pi\nu n\}=\oint_{R_N}F(s)\,ds.$$

On the vertical sides

$$s=\pm(N+\tfrac{1}{2})+it\quad(-1\leq t\leq 1),$$

so that

$$e^{2\pi is}=e^{\pm 2\pi i(N+\frac{1}{2})-2\pi t}=-e^{-2\pi t},$$

and therefore

$$|F(s)|=\frac{\exp\{-\pi y[(N+\frac{1}{2})^2-t^2]\mp\pi\mu y(N+\frac{1}{2})-\pi\nu t\}}{1+e^{-2\pi t}}$$

$$\leq\exp\{-\pi N^2y+\pi|\mu|(N+\tfrac{1}{2})y+\pi|\nu|\}.$$

The length of each vertical side is 2, so that the integrals over them tend to zero as $N\to\infty$.

On the horizontal sides $t=\pm1$ and the integrals over them from $-\infty\pm i$ to $\infty\pm i$ are absolutely convergent since

$$|F(s)|\leq\begin{cases}\dfrac{\exp\{-\pi y\sigma^2+\pi y-\pi\mu y\sigma+\pi|\nu|\}}{1-e^{-2\pi}} & (t=1),\\[2ex] \dfrac{\exp\{-\pi y\sigma^2+\pi y-\pi\mu y\sigma+\pi|\nu|\}}{e^{2\pi}-1} & (t=-1).\end{cases}$$

Hence we have

$$\exp(\tfrac{1}{4}\pi\mu^2y)\Theta_{\mu,\nu}(iy)=\int_{-\infty-i}^{\infty-i}F(s)\,ds-\int_{-\infty+i}^{\infty+i}F(s)\,ds.$$

In the first integral $t=-1$ and $|e^{2\pi is}|=e^{2\pi}>1$, so that

$$\frac{e^{i\pi\nu s}}{e^{2\pi is}-1}=\sum_{n=1}^{\infty} e^{-2\pi i(n-\frac{1}{2}\nu)s}=\sum_{n=-1}^{-\infty} e^{2\pi i(n+\frac{1}{2}\nu)s}.$$

Thus

$$F(s)=\sum_{n=-1}^{-\infty}\exp\{-\pi ys^2-\pi y\mu s+2\pi i(n+\tfrac{1}{2}\nu)s\}.$$

For $|n|>\frac{1}{2}\nu$ the nth term has absolute value not greater than

$$\exp\{-\pi y\sigma^2+\pi y-\pi y\mu\sigma-2\pi|n+\tfrac{1}{2}\nu|\},$$

so that the series may clearly be integrated term by term from $s=-\infty-i$ so $s=\infty-i$. In a similar way we have, when $t=1$, $|e^{2\pi is}|=e^{-2\pi}<1$, so that

$$\frac{e^{i\pi\nu s}}{e^{2\pi is}-1}=-\sum_{n=0}^{\infty} e^{2\pi i(n+\frac{1}{2}\nu)s}$$

and the corresponding series for $F(s)$ may again be integrated term by term. We obtain, by (7.1.3),

$$\begin{aligned}\exp(\tfrac{1}{4}\pi\mu^2y)\Theta_{\mu,\nu}(iy)&=\sum_{n=-1}^{-\infty}\exp\left\{\pi y\left(\frac{\mu}{2}-\frac{n+\frac{1}{2}\nu}{y}i\right)^2\right\}\\&\quad\times\int_{-\infty-i}^{\infty-i}\exp\left\{-\pi y\left(s+\frac{\mu}{2}-\frac{n+\frac{1}{2}\nu}{y}i\right)^2\right\}ds\\&\quad+\sum_{n=0}^{\infty}\exp\left\{\pi y\left(\frac{\mu}{2}-\frac{n+\frac{1}{2}\nu}{y}i\right)^2\right\}\\&\quad\times\int_{-\infty+i}^{\infty+i}\exp\left\{-\pi y\left(s+\frac{\mu}{2}-\frac{n+\frac{1}{2}\nu}{y}i\right)^2\right\}ds\\&=Jy^{-\frac{1}{2}}\sum_{n=-\infty}^{\infty}\exp\left\{\pi y\left(\frac{\mu}{2}-\frac{n+\frac{1}{2}\nu}{y}i\right)^2\right\}.\end{aligned}$$

From this (7.1.2) follows; for we can prove that $J=1$ by taking $\mu=\nu=0$. We get

$$\sum_{n=-\infty}^{\infty} e^{-\pi n^2y}=Jy^{-\frac{1}{2}}\sum_{n=-\infty}^{\infty} e^{-\pi n^2/y}=J^2\sum_{n=-\infty}^{\infty} e^{-\pi n^2y},$$

and since $J>0$, we must have $J=1$. This completes the proof of theorem 7.1.1.

We note that the following functional equations are satisfied:

$$\Theta_{\mu,\nu+2}(z)=\Theta_{\mu,\nu}(z), \quad \Theta_{\mu+2,\nu}(z)=e^{-i\pi\nu}\Theta_{\mu,\nu}(z). \tag{7.1.4}$$

Also the series for $\Theta_{\mu,\nu}(z)$ and $\Theta_{\nu,-\mu}(-1/z)$ are uniformly convergent in μ and ν when μ and ν lie in bounded closed intervals. Accordingly, by differentiating with respect to μ, we obtain

$$i\pi z \sum_{n=-\infty}^{\infty} (n+\tfrac{1}{2}\mu)\exp\{i\pi z(n+\tfrac{1}{2}\mu)^2+i\pi n\nu\}=-\tfrac{1}{2}\pi i\nu\Theta_{\mu,\nu}(z)$$
$$-i\pi\frac{e^{-\frac{1}{2}i\pi\mu\nu}}{(-iz)^{\frac{1}{2}}}\sum_{n=-\infty}^{\infty} n\exp\{-i\pi(n+\tfrac{1}{2}\nu)^2/z-i\pi\mu n\}.$$

Take $\mu=\nu=1$ and note that $\Theta_{1,1}(z)$ vanishes identically. We obtain

$$\sum_{n=-\infty}^{\infty}(2n+1)(-1)^n e^{i\pi(n+\frac{1}{2})^2 z}=\frac{1}{(-iz)^{3/2}}\sum_{n=-\infty}^{\infty}2n(-1)^n e^{-i\pi(n+\frac{1}{2})^2/z}$$
$$=\frac{1}{(-iz)^{3/2}}\sum_{n=-\infty}^{\infty}(2n+1)(-1)^n e^{-i\pi(n+\frac{1}{2})^2/z}. \tag{7.1.5}$$

We now write, as is customary,

$$q=e^{\pi iz} \quad (z\in\mathbb{H}) \tag{7.1.6}$$

and put

$$\vartheta_1'(0|z)=\vartheta_1'(z)=\pi\sum_{n=-\infty}^{\infty}(-1)^n(2n+1)q^{(n+\frac{1}{2})^2}, \tag{7.1.7}$$

$$\vartheta_2(0|z)=\vartheta_2(z):=\sum_{n=-\infty}^{\infty}q^{(n+\frac{1}{2})^2}=\Theta_{1,0}(z), \tag{7.1.8}$$

$$\vartheta_3(0|z)=\vartheta_3(z):=\sum_{n=-\infty}^{\infty}q^{n^2}=\Theta_{0,0}(z), \tag{7.1.9}$$

$$\vartheta_4(0|z)=\vartheta_4(z)=\sum_{n=-\infty}^{\infty}(-1)^n q^{n^2}=\Theta_{0,1}(z). \tag{7.1.10}$$

Some remarks on these and other notations will be found in §7.5. Then we deduce

Theorem 7.1.2. *We have, for all $z\in\mathbb{H}$,*

(i) $\vartheta_1'(z+1)=e^{\frac{1}{4}\pi i}\vartheta_1'(z)$, $\vartheta_2(z+1)=e^{\frac{1}{4}\pi i}\vartheta_2(z)$,
$\vartheta_3(z+1)=\vartheta_4(z)$, $\vartheta_4(z+1)=\vartheta_3(z)$,

(ii) $\vartheta_1'(z+2)=e^{\frac{1}{4}\pi i}\vartheta_1'(z),\quad \vartheta_2(z+2)=e^{\frac{1}{4}\pi i}\vartheta_2(z),$
$\vartheta_3(z+2)=\vartheta_3(z),\quad \vartheta_4(z+2)=\vartheta_4(z),$

(iii) $\vartheta_1'(-1/z)=(-iz)^{3/2}\vartheta_1'(z),\quad \vartheta_2(-1/z)=(-iz)^{\frac{1}{2}}\vartheta_4(z),$
$\vartheta_3(-1/z)=(-iz)^{\frac{1}{2}}\vartheta_3(z),\quad \vartheta_4(-1/z)=(-iz)^{\frac{1}{2}}\vartheta_2(z).$

Proof. The results for ϑ_2, ϑ_3 and ϑ_4 follow from theorem 7.1.1 and the definitions (7.1.8, 9, 10). To obtain the results for ϑ_1', use (7.1.7, 1, 5).

Before stating the next theorem we recall that the groups $\Gamma_U(2)$, $\Gamma_V(2)$, $\Gamma_W(2)$ are conjugate subgroups of index 3 in $\Gamma(1)$ and that, as follows from (1.5.3),

$$\Gamma_U(2)=P\Gamma_V(2)P^{-1},\quad \Gamma_W(2)=P^2\Gamma_V(2)P^{-2}.$$

Since $\Gamma_V(2)$ is generated by U^2 and V (see table 3), it follows that $\Gamma_U(2)$ is generated by

$$PU^2P^{-1}=W^{-2}\quad\text{and}\quad PVP^{-1}=-W^{-2}U,\tag{7.1.11}$$

and that $\Gamma_W(2)$ is generated by

$$P^2U^2P^{-2}=-U^{-2}W^2\quad\text{and}\quad P^2VP^{-2}=U^{-2}W.\tag{7.1.12}$$

Theorem 7.1.3. *The functions ϑ_2, ϑ_3 and ϑ_4 are entire modular forms of weight $\frac{1}{2}$ for the groups $\Gamma_U(2)$, $\Gamma_V(2)$ and $\Gamma_W(2)$, respectively. Their associated multiplier systems are u, v and w respectively and are defined by*

$$u(-I)=v(-I)=w(-I)=-i,\tag{7.1.13}$$

$$v(U^2)=u(PU^2P^{-1})=w(P^2U^2P^{-2})=1,\tag{7.1.14}$$

$$v(V)=u(PVP^{-1})=w(P^2VP^{-2})=e^{-\frac{1}{4}\pi i}.\tag{7.1.15}$$

Further,

$$\vartheta_3|P=e^{-\frac{1}{4}\pi i}\vartheta_4,\quad \vartheta_3|P^2=e^{-\frac{1}{4}\pi i}\vartheta_2.\tag{7.1.16}$$

Finally, the three functions have no zeros in $\mathbb{H}$.

Proof. By theorem 7.1.2 (ii) and (iii),

$$\vartheta_3(U^2z)=\vartheta_3(z),\quad \vartheta_3(Vz)=e^{-\frac{1}{4}\pi i}(V:z)^{\frac{1}{2}}\vartheta_3(z),$$

so that ϑ_3 is an unrestricted modular form for $\Gamma_V(2)$ of dimension $-\frac{1}{2}$ with a MS v defined by (3.1.13, 14) and

$$v(U^2)=1,\quad v(V)=e^{-\frac{1}{4}\pi i}.$$

Further,

$$\begin{aligned}\vartheta_3(Pz) &= \vartheta_3(VUz) = v(V)(V:Uz)^{\frac{1}{2}}\vartheta_3(Uz)\\ &= v(V)(z+1)^{\frac{1}{2}}\vartheta_4(z) = v(V)(P:z)^{\frac{1}{2}}\vartheta_4(z),\end{aligned}$$

and

$$\begin{aligned}\vartheta_3(P^2z) &= \vartheta_3(U^{-1}Vz) = \vartheta_4(Vz) = v(V)(V:z)^{\frac{1}{2}}\vartheta_2(z)\\ &= v(V)(P^2:z)^{\frac{1}{2}}\vartheta_2(z),\end{aligned}$$

from which (7.1.16) follows. The series expansions (7.1.8, 9, 10) now show that ϑ_3 is a modular form, since ∞, $P\infty$ and $P^2\infty$ are the three cusps of the fundamental region $\mathbb{F}_3$. It now follows from (7.1.16) and theorem 4.1.1(i) that ϑ_2 and ϑ_4 are entire modular forms of weight $\frac{1}{2}$ belonging to $\Gamma_U(2)$ and $\Gamma_W(2)$, respectively. Their MS are u and w where, by (3.1.20),

$$u(PTP^{-1}) = v(T)\sigma(T,P)/\sigma(P,PTP^{-1})$$

and

$$w(P^2TP^{-2}) = v(T)\sigma(T,P^2)/\sigma(P^2,P^2TP^{-2})$$

for $T\in\Gamma_V(2)$.

From these definitions the remaining parts of (7.1.13–15) follow, or they can be deduced directly from theorem 7.1.2. For example, taking $T=U^2$ we get, by (7.1.11),

$$u(PU^2P^{-1}) = v(U^2)\sigma(U^2,P)/\sigma(P,W^{-2})$$

and

$$\sigma(U^2,P) = \sigma(P,W^{-2}) = 1,$$

by (3.2.1, 17).

Finally, the total order of each of the three functions is, by theorem 4.1.4, $\frac{1}{8}$. Since $U\in\Gamma_U(2)$, $n(\infty,\Gamma_U(2))=1$ and the associated cusp parameter κ is, by theorem 7.1.2(i), $\frac{1}{8}$. Hence ϑ_2 has no zero in $\mathbb{H}$ and the same therefore holds for ϑ_3 and ϑ_4.

Because of this fact, we can define ϑ_2^s, ϑ_3^s, ϑ_4^s as modular forms for any real s, exactly as was done for Δ^2 in §6.4. We make these definitions unique by assuming that ϑ_3^s and ϑ_4^s take the value 1 at ∞ while the Laurent series for ϑ_2^s begins with the term $e^{\frac{1}{4}\pi isz}$. Each of the forms ϑ_2^s, ϑ_3^s, ϑ_4^s has weight $\frac{1}{2}s$ and we denote their associated

MS by u_s, v_s and w_s, respectively. Then

$$u_s(-I) = v_s(-I) = w_s(-I) = e^{-\frac{1}{2}\pi is}. \tag{7.1.17}$$

Also

$$u_s(U) = e^{\frac{1}{4}\pi is}, \quad v_s(U^2) = w_s(U^2) = 1. \tag{7.1.18}$$

Further,

$$v_s(V) = \frac{\vartheta_3^s(Vz)}{(V:z)^{\frac{1}{2}s}\vartheta_3^s(z)}$$

so that, by putting $z = i$, we obtain

$$v_s(V) = e^{-\frac{1}{4}\pi is}$$

and we have, similarly,

$$u_s(PVP^{-1}) = w_s(P^2VP^{-2}) = v_s(V) = e^{-\frac{1}{4}\pi is}, \tag{7.1.19}$$

and

$$v_s(U^2) = u_s(PU^2P^{-1}) = w_s(P^2U^2P^{-2}) = 1. \tag{7.1.20}$$

When $s \geq 0$ the forms ϑ_2^s, ϑ_3^s and ϑ_4^s are entire functions, and we now calculate the dimensions of the spaces

$$\{\Gamma_U(2), \tfrac{1}{2}s, u_s\}, \quad \{\Gamma_V(2), \tfrac{1}{2}s, v_s\}, \quad \{\Gamma_W(2), \tfrac{1}{2}s, w_s\},$$

and of the corresponding spaces of cusp forms. It is enough to do this for $\{\Gamma_V(2), \frac{1}{2}s, v_s\}$ and $\{\Gamma_V(2), \frac{1}{2}s, v_s\}_0$.

By table 3 (p. 63), the fundamental region $\mathbb{F}_3$ contains exactly one elliptic fixed point of $\Gamma_V(2)$, namely i. By (3.3.10) with $L = I$, $r = 1$, and (7.1.19), we see that $s_I = 0$ so that $\varepsilon(i) = 0$ and therefore $\kappa_2 = \kappa_3 = 0$ in (4.2.2).

There are two incongruent cusps, namely ∞ and $P^2\infty$, of widths 2 and 1 respectively. By (3.3.3),

$$e^{2\pi i\kappa_L} = v_s(LU^{n_L}L^{-1}).$$

Thus $\kappa_I = 0$ and, since $P^2UP^{-2} = -U^{-2}V$ and, by (3.3.10),

$$v_s(-U^{-2}V) = v_s(-V) = v_s(V^{-1}) = e^{\frac{1}{4}\pi is},$$

we have $\kappa_{P^2} = \{\frac{1}{8}s\}$ (fractional part). This can also be obtained directly from (7.1.16) and the expansion of ϑ_2^s. It follows from (4.2.1, 3) that

$$1 - \xi + k\mu/12 = 1 - \{\tfrac{1}{8}s\} + \tfrac{1}{8}s = 1 + [\tfrac{1}{8}s].$$

This is therefore (when $s \geq 0$) an upper bound for the dimension of $\{\Gamma_V(2), \frac{1}{2}s, v_s\}$, by theorem 4.2.1. That it is exact is shown by the following theorem.

Theorem 7.1.4. *We have, for* $s \geq 0$,

$$\dim\{\Gamma_U(2), \tfrac{1}{2}s, u_s\} = \dim\{\Gamma_V(2), \tfrac{1}{2}s, v_s\} = \dim\{\Gamma_W(2), \tfrac{1}{2}s, w_s\}$$
$$= 1 + [\tfrac{1}{8}s]. \tag{7.1.21}$$

The following $1+[\frac{1}{8}s]$ *forms constitute a basis for* $\{\Gamma_V(2), \frac{1}{2}s, v_s\}$:

$$\vartheta_3^{s-8r}(\vartheta_2\vartheta_4)^{4r} \quad (0 \leq r \leq \tfrac{1}{8}s). \tag{7.1.22}$$

The forms in (7.1.22) are cusp forms when $0 < r < \frac{1}{8}s$ *and span* $\{\Gamma_V(2), \frac{1}{2}s, v_s\}_0$. *Similar results hold for* $\Gamma_U(2)$ *and* $\Gamma_W(2)$, *the suffixes* 3,2,4 *in (7.1.22) being replaced by* 2,4,3 *and* 4,3,2 *respectively.*

Proof. From the series expansions (7.1.8, 9, 10) it is clear that $f = [\vartheta_2\vartheta_4/\vartheta_3^2]^4$ is not a constant and so the $1+[\frac{1}{8}s]$ functions in (7.1.22) are linearly independent functions. Now f is a modular function belonging to $\Gamma_V(2)$ and it is clear from theorem 7.1.2 that

$$f(U^2z) = f(z), \quad f(Vz) = f(z),$$

and so $f(Tz) = f(z)$ for all $T \in \Gamma_V(2)$. It follows that the forms in (7.1.22) are all functions in $\{\Gamma_V(2), \frac{1}{2}s, v_s\}$. Hence $\dim\{\Gamma_V(2), \frac{1}{2}s, v_s\} = 1 + [\frac{1}{8}s]$. When $r = 0$ the form is ϑ_3^s and takes the value 1 at ∞ and the value 0 at all other incongruent cusps. When $s \equiv 0 \pmod 8$ and $r = \frac{1}{8}s$, the form is $(\vartheta_2\vartheta_4)^{\frac{1}{2}s}$, which vanishes at ∞ and is non-zero at $P^2\infty$. This shows that $\{\Gamma_V(2), \frac{1}{2}s, v_s\}_0$ is spanned by the forms with $0 < r < \frac{1}{8}s$.

The corresponding results for the groups $\Gamma_U(2)$ and $\Gamma_W(2)$ are proved similarly.

Theorem 7.1.5. *We have* $\vartheta_1' = \pi\vartheta_2\vartheta_3\vartheta_4$ *and*

$$\Delta(z) = \{\tfrac{1}{2}\vartheta_2(z)\vartheta_3(z)\vartheta_4(z)\}^8. \tag{7.1.23}$$

More generally, for any $s \in \mathbb{R}$,

$$\Delta^s = 2^{-8s}\vartheta_2^{8s}\vartheta_3^{8s}\vartheta_4^{8s}. \tag{7.1.24}$$

Also

$$\vartheta_3^4 = \vartheta_2^4 + \vartheta_4^4. \tag{7.1.25}$$

It follows from theorem 7.1.2 that, if $g(z)$ is the right-hand side of (7.1.23), then

$$g(Uz)=g(z) \quad \text{and} \quad g(Vz)=(V:z)^{12}g(z).$$

Also g vanishes at every cusp and since its Fourier expansion begins with the term q^2 and $\dim\{\Gamma(1), 12\}_0=1$, it follows that $g(z)=\Delta(z)$. From this (7.1.24) follows since both sides have expansions beginning with q^{2s}. A similar argument shows that $(\vartheta_1')^8 \in \{\Gamma(1), 12\}_0$ and from this and (7.1.23) the first part of the theorem follows.

When $s=4$ the space $\{\Gamma_V(2), 2, v_4\}$ has dimension 1 and, by theorem 7.1.2, both ϑ_3^4 and $\vartheta_2^4+\vartheta_4^4$ belong to this space. As each takes the value 1 at ∞, (7.1.25) follows.

We know from (7.1.19) that $v_s(V)=e^{-\frac{1}{4}\pi is}$. Also it follows from (3.3.10) that there are only two possible values of $v(V)$ for a MS of weight $\frac{1}{2}s$ belonging to $\Gamma_V(2)$, namely $v_s(V)$ and $-v_s(V)$. Write v_s^* for the MS defined by the equations

$$v_s^*(-I)=e^{-\frac{1}{2}\pi is}, \quad v_s^*(U^2)=1, \quad v_s^*(V)=-e^{-\frac{1}{4}\pi is} \tag{7.1.26}$$

and (3.1.14), and let $\{\Gamma_V(2), \frac{1}{2}s, v_s^*\}$ be the vector space of entire forms of weight $\frac{1}{2}s$ having this MS. We prove

Theorem 7.1.6. *We have, for* $s\geq 0$,

$$\dim\{\Gamma_V(2), \tfrac{1}{2}s, v_s^*\}=\left[\frac{s+4}{8}\right]. \tag{7.1.27}$$

The following forms constitute a basis for $\{\Gamma_V(2), \frac{1}{2}s, v_s^*\}$:

$$\vartheta_3^{s-4-8r}(\vartheta_2\vartheta_4)^{4r}(\vartheta_4^4-\vartheta_2^4) \quad (0\leq r\leq (s-4)/8). \tag{7.1.28}$$

The forms in (7.1.27) are cusp forms for $0<r<(s-4)/8$ *and span* $\{\Gamma_V(2), \frac{1}{2}s, v_s^*\}_0$.

Proof. This is proved in the same way as theorem 7.1.4. We need only remark that $s_I=1$, so that $\kappa_2=\frac{1}{2}, \kappa_3=0$, while $\kappa_I=0$ and $\kappa_{P^2}=\{(s+4)/8\}$, since

$$v_s^*(P^2UP^{-2})=v_s^*(-U^{-2}V)=v_s^*(V^{-1})=-e^{\frac{1}{4}\pi is}.$$

Hence

$$1-\xi+k\mu/12=\left[\frac{s+4}{8}\right]$$

and it is easily checked that the forms in (7.1.28) belong to the group $\Gamma_V(2)$ with this MS, and that the final statement is true.

We note that

$$v_s(U^2)=v_s^*(U^2)=v_s(W^2)=v_s^*(W^2)=1,$$

so that both multipliers v_s and v_s^* agree on the subgroup $\Gamma^*(2)$.

Theorem 7.1.7. *For $s\geq 0$*

$$\dim\{\Gamma(2), \tfrac{1}{2}s, v_s\}=1+[\tfrac{1}{4}s].$$

Further, $\{\Gamma(2), \frac{1}{2}s, v_s\}=:\mathcal{V}_s$ is the direct sum of the spaces

$$\mathcal{B}_s=\{\Gamma_V(2), \tfrac{1}{2}s, v_s\} \quad \textit{and} \quad \mathcal{B}_s^*=\{\Gamma_V(2), \tfrac{1}{2}s, v_s^*\}.$$

These spaces are orthogonal; i.e., if $f\in\mathcal{B}_s$ and $g\in\mathcal{B}_s^*$, at least one being a cusp form, then $(f, g)=0$.

Proof. For $\mathcal{V}_s$ we have $\kappa_2=\kappa_3=0$, since there are no elliptic fixed points, and $\kappa_I=\kappa_P=0$ while $\kappa_{P^2}=\{\frac{1}{4}s\}$, since

$$v_s(P^2U^2P^{-2})=v_s(-U^{-2}W^2)=e^{\frac{1}{2}\pi is}.$$

Hence $1-\xi+k\mu/12=1+[\frac{1}{4}s]$. On the other hand it is clear that $\mathcal{B}_s\cap\mathcal{B}_s^*$ consists of the zero form only and that

$$\mathcal{B}_s\oplus\mathcal{B}_s^*\subseteq\mathcal{V}_s.$$

By theorems 7.1.4, 6, the dimension of $\mathcal{B}_s\oplus\mathcal{B}_s^*$ is

$$1+[\tfrac{1}{8}s]+\left[\frac{s+4}{8}\right]=1+[\tfrac{1}{4}s]$$

and it follows that $\dim\mathcal{V}_s=1+[\frac{1}{4}s]$.

Finally, let $\mathbb{F}_3$ be the fundamental region for $\hat{\Gamma}_V(2)$ given by (2.4.15). Since $UP^r=VP^{r+1}$, $U\mathbb{F}_3=V\mathbb{F}_3$ and accordingly, by theorem 2.3.5,

$$\mathbb{F}^*=\mathbb{F}_3\cup V\mathbb{F}_3$$

is a fundamental region for $\hat{\Gamma}(2)$. If $f\in\mathcal{B}_s$, $g\in\mathcal{B}_s^*$ and at least one is a cusp form, then their scalar product as forms of $\mathscr{V}_s$ is

$$(f,g)=\frac{1}{6}\iint_{F_3} f(z)\overline{g(z)}y^{\frac{1}{2}s-1}\,dx\,dy+\frac{1}{6}\iint_{VF_3} f(z)\overline{g(z)}y^{\frac{1}{2}s-1}\,dx\,dy.$$

Since

$$f(Vz)\overline{g(Vz)}=v_s(V)\overline{v_s^*(V)}f(z)\overline{g(z)}|V:z|^2$$

and $v_s(V)\overline{v_s^*(V)}=-1$, it follows that $(f,g)=0$.

Theorem 7.1.8. *For all $z\in\mathbb{H}$,*

$$2\vartheta_2(2z)\vartheta_3(2z)=\vartheta_2^2(z),\qquad 2\vartheta_2^2(2z)=\vartheta_3^2(z)-\vartheta_4^2(z),$$
$$2\vartheta_3^2(2z)=\vartheta_3^2(z)+\vartheta_4^2(z),\qquad \vartheta_4^2(2z)=\vartheta_3(z)\vartheta_4(z).$$

Proof. To prove the first relation it is enough to show that

$$f(z)=2\vartheta_2(2z)\vartheta_3(2z)\in\{\Gamma_U(2),1,u_2\},$$

since the Fourier series of each side begins with $4q^{\frac{1}{2}}$ and since the dimension of this space is 1 by theorem 7.1.4. This is the case since, by theorem 7.1.2,

$$F(z+1)=iF(z),\quad \vartheta_2^2(z+1)=i\vartheta_2^2(z)$$

and

$$F(W^2z)=2\vartheta_2\left(\frac{2z}{2z+1}\right)\vartheta_3\left(\frac{2z}{2z+1}\right)=2e^{\frac{1}{4}\pi i}\vartheta_2\left(\frac{-1}{2z+1}\right)\vartheta_4\left(\frac{-1}{2z+1}\right)$$
$$=2\,e^{\frac{1}{4}\pi i}\{-i(2z+1)\}\vartheta_4(2z+1)\vartheta_2(2z+1)=(2z+1)F(z),$$

while, by (7.1.11, 14),

$$\vartheta_2^2(W^2z)=\vartheta_2^2(PU^{-2}P^{-1}z)=(2z+1)\vartheta_2^2(z).$$

The remaining results can be proved similarly.

Comparison of (7.1.23) with (6.1.11) suggests that the functions ϑ_3, ϑ_3 and ϑ_4 may possess infinite product expansions. We prove this in our final theorem.

Theorem 7.1.9. *If $z\in\mathbb{H}$ and $q=e^{\pi iz}$, then*

$$\vartheta_1'(z)=2\pi q^{\frac{1}{4}}q_0,\quad \vartheta_2(z)=2q^{\frac{1}{4}}q_0q_1^2,\quad \vartheta_3(z)=q_0q_2^2,\quad \vartheta_4(z)=q_0q_3^2,$$
$$(7.1.29)$$

where

$$\left.\begin{aligned} q_0 &= \prod_{n=1}^{\infty} (1-q^{2n}), \quad q_1 = \prod_{n=1}^{\infty} (1+q^{2n}), \\ q_2 &= \prod_{n=1}^{\infty} (1+q^{2n-1}), \quad q_3 = \prod_{n=1}^{\infty} (1-q^{2n-1}). \end{aligned}\right\} \tag{7.1.30}$$

The products are absolutely convergent for $z \in \mathbb{H}$ and satisfy the identity

$$q_1 q_2 q_3 = 1. \tag{7.1.31}$$

Proof. We show that

$$\vartheta_3(z) = \frac{\Delta^{5/24}(z)}{\Delta^{1/12}(\frac{1}{2}z)\Delta^{1/12}(2z)}. \tag{7.1.32}$$

It can be checked that these both belong to $\{\Gamma_V(2), \frac{1}{2}, v_1\}$, but it is slightly easier to consider

$$F(z) = \frac{\Delta^{5/6}(z)}{\Delta^{1/3}(\frac{1}{2}z)\,\Delta^{1/3}(2z)}$$

and check from §6.4 that

$$F(U^2 z) = F(z) \quad \text{and} \quad F(Vz) = -(V:z)^2 F(z),$$

so that $F \in \{\Gamma_V(2), 2, v_4\}$. By theorem 7.1.4 this space has dimension 1 and, by considering the constant term, we deduce that $\vartheta_3^4(z) = F(z)$, from which (7.1.32) follows. From this and (6.1.11) we deduce that

$$\vartheta_3(z) = \frac{q^{5/12} q_0^5}{q^{1/12}(q_0 q_3)^2 \cdot q^{1/3}(q_0 q_1)^2} = \frac{q_0}{(q_1 q_3)^2} = q_0 q_2^2,$$

since (7.1.31) is obvious. Since $\vartheta_4(z) = \vartheta_3(z+1)$, the corresponding formula for $\vartheta_4(z)$ follows and the results for ϑ_1' and ϑ_2 are then obtained from the first part of theorem 7.1.5.

7.2. Modular functions of level 2. In this section we give applications of some of the general results obtained in §4.3. We define a modular function λ on $\Gamma(2)$ by the formula

$$\lambda(z) = \vartheta_2^4(z)/\vartheta_3^4(z). \tag{7.2.1}$$

From theorem 7.1.2 and (7.1.25) we deduce that

$$\left.\begin{aligned} &\lambda|U=\frac{1}{\lambda-1}, \quad \lambda|V=1-\lambda, \quad \lambda|W=\frac{1}{\lambda}, \\ &\lambda|P=\frac{1}{1-\lambda}, \quad \lambda|P^2=1-\frac{1}{\lambda}. \end{aligned}\right\} \tag{7.2.2}$$

From these we deduce that $\lambda|U^2=\lambda|W^2=\lambda$. Also λ has no zeros or poles in $\mathbb{H}$, since the same holds for ϑ_2 and ϑ_3, and λ takes the values 0, 1 and ∞ at the three incongruent cusps ∞, $P\infty=0$ and $P^2\infty=-1$ of the fundamental region $\mathbb{F}=\mathbb{F}_3\cup U\mathbb{F}_3$ of $\hat{\Gamma}(2)$; see table 3. It follows that $\lambda\in M(\Gamma(2),0,1)$ and that λ has valence 1.

The formulae (7.2.2) give us a particular case of theorem 4.3.5. They, together with $\lambda|I=\lambda$, display a group of bilinear mappings isomorphic to the factor group $\Gamma(1)/\Gamma(2)$.

From theorem 4.3.3 we deduce that every function f in $M(\Gamma(2),0,1)$ is a rational function of λ. In particular, the function J, defined by (6.1.20), being a member of $M(\Gamma(1),0,1)$, is also a member of $M(\Gamma(2),0,1)$ and so is a rational function of λ. We can easily find the form of this rational function.

We observe that the points of $\mathbb{E}_3$ are not fixed points of $\Gamma(2)$ and that $\mathbb{F}\cap\mathbb{E}_3$ consists of the two points ρ, $-\rho^2=U\rho$; these are fixed points for the mappings P and UPU^{-1}. From (7.2.2) we deduce that

$$\lambda^2(z)-\lambda(z)+1=0 \quad \text{for } z=\rho \text{ and } z=-\rho^2.$$

Hence

$$\{\lambda(z)-\lambda(\rho)\}\{\lambda(z)-\lambda(-\rho^2)\}=\lambda^2(z)-\lambda(z)+1. \tag{7.2.3}$$

Further, J has poles at each cusp of $\mathbb{F}$ and λ takes the values 0, 1 and ∞ at these cusps. Also, each zero is a triple zero and each pole is a double pole in the appropriate local uniformizing variable. From this we deduce that

$$J=A\frac{(\lambda^2-\lambda+1)^3}{\lambda^2(\lambda-1)^2}$$

for some constant A. We now use the fact that $J(i)=1$ and that $\lambda(i)=\frac{1}{2}$; for $\lambda(i)=\lambda(Vi)=1-\lambda(i)$, by (7.2.2). This determines the constant A and shows that

$$J=\frac{4}{27}\frac{(\lambda^2-\lambda+1)^3}{\lambda^2(\lambda-1)^2}. \tag{7.2.4}$$

By considering the form of the Fourier series for $\lambda(z)$ one can show, in the same way as we did for $J(z)$ in §6.1, that $\lambda(z)$ takes real values on the boundaries of $\mathbb{F}_3$ and $U\mathbb{F}_3$ and that its imaginary part is positive on the interior of $U\mathbb{F}_3$ and is negative on the interior of $\mathbb{F}_3$. It follows from (7.2.3) that $\lambda(\rho)=-\rho$ and $\lambda(-\rho^2)=-\rho^2$.

Similar methods can be used to obtain functions of valence 1 for the subgroups between $\Gamma(2)$ and $\Gamma(1)$. Thus

$$f_1=(\lambda-1)\lambda^{-2},\quad f_2=\lambda(1-\lambda),\quad f_3=-\lambda(1-\lambda)^{-2} \tag{7.2.5}$$

are modular functions of valence 1 for $\Gamma_U(2)$, $\Gamma_V(2)$ and $\Gamma_W(2)$, respectively, and $f_2=f_1|P, f_3=f_2|P$. From (7.2.4) we deduce that

$$J=\frac{4}{27}\frac{(1-f_\nu)^3}{f_\nu^2}\quad(\nu=1,2,3). \tag{7.2.6}$$

Now consider the three functions J_1, J_2, J_3 defined by

$$J_1(z)=J(2z),\quad J_2(z)=J(\tfrac{1}{2}z+\tfrac{1}{2}),\quad J_3(z)=J(\tfrac{1}{2}z). \tag{7.2.7}$$

It can be deduced from theorem 4.3.7 that these three functions belong to the groups $\Gamma_U(2)$, $\Gamma_V(2)$ and $\Gamma_W(2)$, respectively. Alternatively, we may observe that U and V_2 – see (1.3.10) – generate $\Gamma_U(2)$ and

$$J_1(Uz)=J(2z+2)=J(2z)=J_1(z),$$

$$J_1(V_2z)=J(2V_2z)=J(V_12z)=J(2z)=J_1(z),$$

while $J_2=J_1|P$ and $J_3=J_2|P$.

It is easily verified that J_1 has a pole of order 2 at ∞ and a pole of order 1 at 0. Also, since ρ is not a fixed point for $\Gamma_U(2)$, J_1 has a zero of order 3 at $\frac{1}{2}\rho$ and no other zeros in its fundamental region. Put $\alpha=f_1(\frac{1}{2}\rho)$. Then

$$J_1(z)=A_1\frac{(f_1(z)-\alpha)^3}{f_1(z)} \tag{7.2.8}$$

for some constant A_1. We can express A_1 in terms of α by observing that, since $J(z)-1$ has a double zero at i, $J_1'(\frac{1}{2}i)=0$. We find that

$$J_1'=A_1f_1'f_1^{-2}(f_1-\alpha)^2(2f_1+\alpha)$$

and, since $f_1'(\frac{1}{2}i) \neq 0$, it follows that $f_1(\frac{1}{2}i) = -\frac{1}{2}\alpha$. Putting $z = \frac{1}{2}i$ in (7.2.8) we have

$$1 = \frac{27}{4} A_1 \alpha^2$$

and so

$$J_1 = \frac{4}{27} \frac{(f_1 - \alpha)^3}{\alpha^2 f_1}.$$

To evaluate α we apply theorem 7.1.8 to obtain the equation

$$\lambda(2z) = \left\{ \frac{\vartheta_3^4(z) - \vartheta_4^4(z)}{\vartheta_2^4(z)} \right\}^4.$$

If we put $\sqrt{\lambda} = \vartheta_2^2/\vartheta_3^2$, $\sqrt{(1-\lambda)} = \vartheta_4^2/\vartheta_3^2$ this becomes

$$\lambda(2z) = \left\{ \frac{1 - \sqrt{(1-\lambda)}}{\sqrt{\lambda}} \right\}^4 = \left\{ \frac{\sqrt{\lambda}}{1 + \sqrt{(1-\lambda)}} \right\}^4,$$

where λ denotes $\lambda(z)$. Hence

$$\lambda(2z) + \frac{1}{\lambda(2z)} = \lambda^{-2}[\{1 - \sqrt{(1-\lambda)}\}^2 + \{1 + \sqrt{(1-\lambda)}\}^2]$$

$$= 2 + 16\lambda^{-2}(1-\lambda).$$

Write $\lambda_0 = \lambda(\frac{1}{2}i)$. Since $\frac{1}{2}i$ lies on the imaginary axis between 0 and i, λ_0 is a real number between 1 and 0 and satisfies the quadratic equation

$$\lambda_0^2 + 32\lambda_0 - 32 = 0.$$

Accordingly, $\lambda_0 = 4(3\sqrt{2} - 4)$ and so

$$\frac{1}{\lambda_0} = \frac{1}{8}(3\sqrt{2} + 4).$$

From this we deduce that $f_1(\frac{1}{2}i) = -\frac{1}{32}$ and so $\alpha = \frac{1}{16}$.

Accordingly,

$$J_1 = \frac{(16f_1 - 1)^3}{2^2 \cdot 3^6 f_1}$$

and therefore

$$J_\nu = \frac{(16f_\nu - 1)^3}{2^2 \cdot 3^6 f_\nu} \quad (\nu = 1, 2, 3). \tag{7.2.9}$$

We can use equations (7.2.6, 9) to obtain a modular equation connecting J and J_ν. It is convenient to put

$$\tau = j(z) = 1728\,J(z), \quad t = 1728\,J_\nu(z). \tag{7.2.10}$$

We then have

$$t+\tau = 16f_\nu^{-2}\{16(1-f_\nu)^3 + f_\nu(16f_\nu - 1)^3\}$$

and

$$(t\tau)^{1/3} = 16f_\nu^{-1}(1-f_\nu)(16f_\nu - 1),$$

from which we obtain the following equation in irrational form, on eliminating f_ν:

$$t+\tau-(t\tau)^{2/3}+3^2\,.\,5\,.\,11(t\tau)^{1/3}-2^4\,.\,3^3\,.\,5^3=0. \tag{7.2.11}$$

The rational form of this equation is

$$\begin{aligned}\Phi_2(\tau, t) = \tau^3 + t^3 - \tau^2 t^2 + 1488\tau t(\tau+t) - 162\,000(\tau^2+t^2)\\ +40\,773\,375\tau t + 8748\,.\,10^6(\tau+t) - 157\,464\,.\,10^9 = 0.\end{aligned} \tag{7.2.12}$$

Observe that $\Phi_2(\tau, t)$ is symmetric in τ and t and has degree $\psi(2)=3$ in each variable, as expected from theorem 6.5.1.

The functions f_1, f_2 and f_3 are closely connected with the functions D_T defined in (6.5.7), in the case when $n=2$. For it follows from (7.1.23) and theorem 7.1.8 that

$$\frac{\Delta(2z)}{\Delta(z)} = \frac{\{\frac{1}{2}\vartheta_2(z)\}^8}{\vartheta_3^4(z)\vartheta_4^4(z)} = \frac{2^{-8}\lambda^2(z)}{1-\lambda(z)} = -\frac{1}{2^8 f_1(z)}. \tag{7.2.13}$$

Theorem 6.5.2 states that $\Delta(2z)/\Delta(z)$ can be expressed as a rational function of $J(z)$ and $J(2z)$, and this may be done by using equations (7.2.6, 9) to express f_1 in terms of J and J_1.

In general, if F is any modular function belonging to $M(\Gamma_U(2), 0, 1)$, then since $V(J, \Gamma_U(2)) = 3$, we deduce from theorem 4.3.2 that there exists an irreducible polynomial $\Phi(x_1, x_2)$ of degree at most 3 in x_2 such that

$$\Phi(J, F) = 0$$

and it is clear that this equation is satisfied also by $F|P$ and $F|P^2$.

In the same way we can show that the function

$$g=\left(\frac{\lambda+\rho}{\lambda+\rho^2}\right)^3 \tag{7.2.14}$$

belongs to $\hat{\Gamma}^2$ and has valence 1. Further, it follows from (7.2.4) that

$$J=\frac{-4g}{(1-g)^2} \tag{7.2.15}$$

and this equation is satisfied not only by g but also by

$$g|V=1/g. \tag{7.2.16}$$

We can also apply theorems 4.3.4, 5 to express modular forms of even weight in terms of modular forms and their derivatives. From theorem 4.3.1 we know that $\lambda'\in M(\Gamma(2), 2, 1)$ and we can find out how λ' behaves under transformations belong to $\Gamma(1)$ by applying theorem 4.3.5 to the relations (7.2.2). We note, in particular, that

$$\lambda'|V=-\lambda',$$

so that $\lambda'\in M(\Gamma_V(2), 2, v_4)$, in the notation of (7.1.17–19).

Also, by using theorem 4.3.5 we can show that

$$i\pi\vartheta_2^4=\frac{\lambda'}{1-\lambda},\quad i\pi\vartheta_3^4=\frac{\lambda'}{\lambda(1-\lambda)},\quad i\pi\vartheta_4^4=\frac{\lambda'}{\lambda}, \tag{7.2.17}$$

so that, by (7.1.23),

$$\Delta^{\frac{1}{2}}=\frac{1}{16}\frac{(\lambda'/i\pi)^3}{\lambda^2(1-\lambda^2)}. \tag{7.2.18}$$

We conclúde by obtaining the modular equation connecting $\lambda(z)$ and $\lambda(3z)$, since this can be obtained quite easily from the properties of theta functions that we have established.

If we put

$$\vartheta(z)=\vartheta_3(z)\vartheta_3(3z)-\vartheta_4(z)\vartheta_4(3z)-\vartheta_2(z)\vartheta_2(3z),$$

we find from theorem 7.1.2 that

$$\vartheta(Uz)=-\vartheta(z),\quad \vartheta(W^3z)=-(W^3:z)\vartheta(z).$$

Now the group $\Gamma_U(3)=\Gamma_0(3)$ defined by (1.6.2) has as a fundamental region

$$\mathbb{F}_I\cup\mathbb{F}_V\cup\mathbb{F}_{VU}\cup\mathbb{F}_{VU^{-1}};$$

see fig. 4 (p. 53). The two pairs of sides of this region are transformed into each other by the mappings U and W^3. It follows that

$$\Gamma_0(3) = \langle U, W^3 \rangle$$

and that

$$\vartheta \in \{\Gamma_0(3)m\ 1, v\},$$

where the MS v is defined on $\Gamma_0(3)$ by

$$v(U) = v(W^3) = -1.$$

It follows that $\kappa_I = \frac{1}{2}$ and therefore, by (4.2.1), $\xi \geq \frac{1}{2}$. Hence

$$1 - \xi + k\mu/12 = \tfrac{4}{3} - \xi < 1$$

and so, by theorem 4.2.1,

$$\dim\{\Gamma_0(3), 1, v\} = 0.$$

Hence $\vartheta(z) = 0$ for all $z \in \mathbb{H}$. Thus

$$\frac{\vartheta_2(z)}{\vartheta_3(z)} \cdot \frac{\vartheta_2(3z)}{\vartheta_3(3z)} + \frac{\vartheta_4(z)}{\vartheta_3(z)} \cdot \frac{\vartheta_4(3z)}{\vartheta_3(3z)} = 1,$$

from which we deduce the irrational modular equation

$$\lambda^{\frac{1}{4}}\lambda_3^{\frac{1}{4}} + (1-\lambda)^{\frac{1}{4}}(1-\lambda_3)^{\frac{1}{4}} = 1,$$

where

$$\lambda_3(z) = \lambda(3z).$$

7.3. Eisenstein series belonging to $\Gamma(2)$. Our object is to consider more closely the Fourier coefficients of Eisenstein series belonging to $\Gamma(2)$ and having integral weight k. When k is even we shall suppose, as in §5.5, that the multiplier system is constant on $\Gamma(2)$; when k is odd we shall suppose, as in §5.6, that it is constant on $\Gamma(4)$.

We require certain ancillary number-theoretic functions, which we now introduce. We denote by χ_0 and χ the two Dirichlet characters modulo 4. Their values can be expressed concisely by the formulae

$$\chi_0(n) = \sin^2 \tfrac{1}{2}\pi n, \quad \chi(n) = \sin \tfrac{1}{2}\pi n \quad (n \in \mathbb{Z}). \qquad (7.3.1)$$

With them are associated the Dirichlet L-series

$$L(s,\chi_0)=\sum_{n=1}^{\infty}\chi_0(n)n^{-s},\quad L(s,\chi)=\sum_{n=1}^{\infty}\chi(n)n^{-s}\quad (s>1). \tag{7.3.2}$$

We note that

$$L(k,\chi_0)=(1-2^{-k})\zeta(k)\quad (k\geq 2), \tag{7.3.3}$$

so that the values of $L(k,\chi_0)$ can be expressed for even k in terms of the Bernoulli numbers; see (6.1.5). For odd k the values of $L(k,\chi)$ can be expressed in terms of the Euler numbers E_n, which appear as coefficients in the expansion

$$\frac{2e^t}{e^{2t}+1}=\sum_{n=0}^{\infty}\frac{E_n t^n}{n!}\quad (|t|<\pi). \tag{7.3.4}$$

The clash of notation between these coefficients and Eisenstein series belonging to the full modular group should not cause confusion. For odd k we shall write

$$l=\tfrac{1}{2}(k-1). \tag{7.3.5}$$

Then we have

$$L(k,\chi)=L(2l+1,\chi)=\tfrac{1}{2}(\tfrac{1}{2}\pi)^k|E_{k-1}|/\Gamma(k)\quad (k\geq 1) \tag{7.3.6}$$

The first few values of $|E_{k-1}|$ are given in table 11.

Table 11

k	1	3	5	7	9	11	13
$\lvert E_{k-1}\rvert$	1	1	5	61	1385	50521	27027765

We also need certain divisor functions in addition to the functions $\sigma_{k-1}(n)$ and $\sigma_{k-1}(n;C,D;2)$ given in (5.5.20, 26). We put

$$\sigma^*_{k-1}(n)=\sum_{d\mid n}\chi_0(n/d)d^{k-1},\quad \sigma^\dagger_{k-1}(n)=\sum_{d\mid n}(-1)^d d^{k-1}, \tag{7.3.7}$$

$$\sigma'_{k-1}(n)=\sum_{d\mid n}\chi(d)d^{k-1},\qquad \sigma''_{k-1}(n)=\sum_{d\mid n}\chi(n/d)d^{k-1}, \tag{7.3.8}$$

and

$$\sigma'''_{k-1}(n) = \sum_{d|n} (-1)^d \chi(n/d) d^{k-1}. \tag{7.3.9}$$

Observe that $\sigma''_{k-1}(n) = n^{k-1}\sigma'_{1-k}(n)$, and that $\sigma^{\dagger}_{k-1}(n) = -\sigma_{k-1}(n)$ when n is odd. In (7.3.7–9) the divisors summed over are positive.

We first take the case of even $k > 2$ and use the notation of §5.5. We write

$$E_k(z; L) := E_k(z; C, D; 2) = G_L(z; 0; \Gamma(2), k, 1) \tag{7.3.10}$$

$$= \delta_L + \sum_{n=1}^{\infty} \alpha_k(n, L)\, e^{\pi i n z}, \tag{7.3.11}$$

where, by (5.3.24) and (5.5.2),

$$\delta_L = \begin{cases} 1 & \text{if } C \equiv 0, D \equiv 1 \pmod 2, \\ 0 & \text{otherwise.} \end{cases} \tag{7.3.12}$$

Note that, by (5.1.27, 28),

$$E_k(z; \pm U^q L) = E_k(z; L) \quad (q \in \mathbb{Z},\ k \text{ even}) \tag{7.3.13}$$

for any choice of the ambiguous sign. Also, by (5.5.25, 26) and (6.1.5),

$$\alpha_k(n, L) = \alpha_k(n, \pm U^q L) = \frac{(-1)^{\frac{1}{2}k}\alpha_k}{2^k - 1} \sum_{\substack{d|n \\ (n/d) \equiv C \pmod 2}} (-1)^{dD} d^{k-1}, \tag{7.3.14}$$

where α_k is defined by (6.1.5).

Since a fundamental region of $\hat{\Gamma}(2)$ has three cusps congruent to the points $P^{-r}\infty$ $(r = 0, 1, 2)$, there are only three different Eisenstein series for each even $k > 2$ and we can restrict our attention to the cases when $L = I$, P and P^2. From (7.3.14) we deduce

Theorem 7.3.1. *Let k be an even integer greater than 2. Then the coefficients $\alpha_k(n, P^r)$ $(n \geq 1, r = 0, 1, 2)$ of the Eisenstein series $E_k(z; P^r)$ are given by*

$$\alpha_k(n, I) = \frac{(-1)^{\frac{1}{2}k}\alpha_k}{2^k - 1}\sigma^{\dagger}_{k-1}(\tfrac{1}{2}n), \tag{7.3.15}$$

$$\alpha_k(2^{\mu}m, P) = \frac{(-1)^{\frac{1}{2}k}\alpha_k}{2^k - 1}\sigma_{k-1}(m) \quad \text{or} \quad \frac{(-1)^{\frac{1}{2}k+1}}{2^k - 1}\alpha_k\sigma_{k-1}(m) \ (m \text{ odd}), \tag{7.3.16}$$

according as $\mu>0$ or $\mu=0$, and

$$\alpha_k(n, P^2) = \frac{(-1)^{\frac{1}{2}k}\alpha_k}{2^k-1}\sigma^*_{k-1}(n). \tag{7.3.17}$$

Moreover,

$$\alpha_k(n, I)+\alpha_k(n, P)+\alpha_k(n, P^2) = (-1)^{\frac{1}{2}k}\alpha_k\sigma_{k-1}(\tfrac{1}{2}n). \tag{7.3.18}$$

In (7.3.15, 18) $\sigma^\dagger_{k-1}(\frac{1}{2}n)$ and $\sigma_{k-1}(\frac{1}{2}n)$ are zero when n is odd.

The last formula may be verified directly from (7.3.15–17) or, alternatively, from the fact that

$$E_k(z) = E_k(z; I)+E_k(z; P)+E_k(z; P^2). \tag{7.3.19}$$

Similar results can be deduced in the case when $k=2$. For it has already been observed in §5.7 that the Fourier coefficients of the non-analytic function $G_L(z; 0, N)$ defined by (5.7.32, 13) agree with (5.5.4) when k is put equal to 2. We can then use (5.5.17) with $k=2$ to obtain corresponding formulae for the Fourier coefficients of $E_2(z; C, D; 2)$ and so deduce that

$$E_2(z; L) := G_L(z; 0, 2)$$

$$= -\frac{1}{\pi y}+\delta_L+\sum_{n=1}^{\infty}\alpha_2(n, L)\,e^{\pi i n z}, \tag{7.3.20}$$

where the coefficients $\alpha_2(n, L)$ are given by (7.3.15–17) with $k=2$. In this way we can find the Fourier coefficients of the analytic functions

$$E_2(z; I)-E_2(z; P), \quad E_2(z; I)-E_2(z; P^2)$$

which span $\{\Gamma(2), 2, 1\}$; see (5.7.37).

We now consider the case when k is odd ($k>2$). This is more complicated because there is no constant MS on $\Gamma(2)$. We make use of the results of §5.6, from which we see that Eisenstein series only exist in the spaces $\{\Gamma(2), k, v^{(s)}\}$ $(s=0, 1, 2)$, namely the functions $E_k(z; P^s, v^{(s)})$ and $E_k(z; P^{s-1}, v^{(s)})$ defined in (5.6.14, 15). Accordingly we have the three pairs of Eisenstein series:

$$\left.\begin{aligned} E_k(z; I, v^{(0)}) &= E_k(z; 0, 1; 4)+E_k(z; 2, 1; 4),\\ E_k(z; P^2, v^{(0)}) &= E_k(z; 1, 0; 4)-E_k(z; -1, -2; 4),\end{aligned}\right\} \tag{7.3.21}$$

$$\left.\begin{aligned} E_k(z; P, v^{(1)}) &= E_k(z; 1, 1; 4) + E_k(z; 1, -1; 4), \\ E_k(z; I, v^{(1)}) &= E_k(z; 0, 1; 4) - E_k(z; 2, 1; 4), \end{aligned}\right\} \quad (7.3.22)$$

$$\left.\begin{aligned} E_k(z; P^2, v^{(2)}) &= E_k(z; 1, 0; 4) + E_k(z; -1, -2; 4), \\ E_k(z; P, v^{(2)}) &= E_k(z; 1, 1; 4) - E_k(z; 1, -1; 4). \end{aligned}\right\} \quad (7.3.23)$$

We therefore examine the Fourier coefficients $a(r; C, D; 4)$ of the functions on the right-hand sides. By (5.5.21), for $h = \pm 1$, we have

$$\sigma_{k-1}(r; hC, hC; 4) = \chi(h)\sigma_{k-1}(r; C, D; 4)$$

and therefore, by (5.5.24),

$$\begin{aligned} a(r; C, D; 4) &= \frac{(\pi/2i)^k}{\Gamma(k)} \sum_{m=1}^{\infty} \chi(m)\mu(m)m^{-k}\sigma_{k-1}(r; C, D; 4) \\ &= \frac{(\pi/2i)^k}{\Gamma(k)L(k, \chi)}\sigma_{k-1}(r; C, D; 4) \\ &= \frac{2}{i^k|E_{k-1}|}\sigma_{k-1}(r; C, D; 4). \end{aligned} \quad (7.3.24)$$

We now examine the sum $\sigma_{k-1}(r; C, D; 4)$. We have, by (5.5.21),

$$\sigma_{k-1}(r; C, D; 4) = \sum_{\substack{d|r \\ (r/d) \equiv C \pmod 4}} d^{k-1} i^{dD} - \sum_{\substack{d|r \\ (r/d) \equiv -C \pmod 4}} d^{k-1} i^{-dD}, \quad (7.3.25)$$

where in each summation d is positive.

We consider various cases. When C is even D must be odd, so that the sum (7.3.25) vanishes except when $r = 2n$ and $n \equiv \frac{1}{2}C \pmod 2$. For $C \equiv -C \pmod 4$ and

$$i^{dD} - i^{-dD} = 2i \sin(\tfrac{1}{2}\pi\, dD) = 2i\chi(dD),$$

which vanishes when d is even. Accordingly, $a(r; C, D; 4) = 0$ except when $r = 2n$ and $n \equiv \frac{1}{2}C \pmod 2$ and then

$$a(r; C, D; 4) = \frac{2\chi(D)}{i^k|E_{k-1}|}\sigma'_{k-1}(n). \quad (7.3.26)$$

Similarly, if D is even then C is odd and $i^{dD} = i^{-dD} = (-1)^{\frac{1}{2}Dd}$, so that

$$\sigma_{k-1}(r; C, D; 4) = \sum_{d|r} (-1)^{\frac{1}{2}Dd}\chi(Cr/d)d^{k-1}$$

and therefore

$$a(r;C,D;4)=\frac{2\chi(C)}{i^k|E_{k-1}|}\sum_{d|r}(-1)^{\frac{1}{2}Dd}\chi(r/d)d^{k-1}$$

$$=\begin{cases}\dfrac{2\chi(C)}{i^k|E_{k-1}|}\sigma'''_{k-1}(r) & (D\equiv 2 \pmod 4)\\[2ex] \dfrac{2\chi(C)}{i^k|E_{k-1}|}\sigma''_{k-1}(r) & (D\equiv 0 \pmod 4).\end{cases} \tag{7.3.27}$$

Finally, suppose that both C and D are odd. Then, by (7.3.25),

$$\sigma_{k-1}(r;C,D;4)=i\sum_{\substack{d|r\\ d\text{ odd}}}\chi(dD)\chi_0(Cr/d)d^{k-1}$$

$$+\sum_{\substack{d|r\\ d\text{ even}}}(-1)^{\frac{1}{2}d}\chi(Cr/d)d^{k-1},$$

so that

$$\sigma_{k-1}(r;C,D;4)=\begin{cases}i\chi(D)\sigma'_{k-1}(r) & (r\text{ odd})\\ \chi(C)2^{k-1}\sigma'''_{k-1}(\tfrac{1}{2}r) & (r\text{ even}).\end{cases}$$

Hence

$$a(r;C,D;4)=\begin{cases}\dfrac{2\chi(D)}{i^{k-1}|E_{k-1}|}\sigma'_{k-1}(r) & (r\text{ odd})\\[2ex] \dfrac{2^k(\chi(C)}{i^k|E_{k-1}|}\sigma'''_{k-1}(\tfrac{1}{2}r) & (r\text{ even}).\end{cases} \tag{7.3.28}$$

We therefore have, by (7.3.5), (7.3.21–23) and (7.3.26–28),

$$E_k(z;I,v^{(0)})=1+\frac{4(-1)^l}{|E_{k-1}|}\sum_{n=1}^{\infty}\sigma'_{k-1}(n)\,e^{\pi inz}, \tag{7.3.29}$$

$$E_k(z;P^2,v^{(0)})=-\frac{i(-1)^l2^{k+1}}{|E_{k-1}|}\sum_{n=1}^{\infty}\sigma''_{k-1}(n)\,e^{\pi inz}, \tag{7.3.30}$$

$$E_k(z;P,v^{(1)})=-\frac{i(-1)^l2^{k+1}}{|E_{k-1}|}\sum_{n=1}^{\infty}\sigma'''_{k-1}(n)\,e^{\pi inz}, \tag{7.3.31}$$

$$E_k(z;I,v^{(1)})=1+\frac{4(-1)^l}{|E_{k-1}|}\sum_{n=1}^{\infty}(-1)^n\sigma'_{k-1}(n)\,e^{\pi inz}, \tag{7.3.32}$$

$$E_k(z;P^2,v^{(2)}) = -\frac{4i(-1)^l}{|E_{k-1}|}\sum_{n=1}^{\infty}\sigma''_{k-1}(2n-1)\,e^{\pi i(n-\frac{1}{2})z}, \tag{7.3.33}$$

$$E_k(z;P,v^{(2)}) = \frac{4(-1)^l}{|E_{k-1}|}\sum_{n=1}^{\infty}\sigma'_{k-1}(2n-1)\,e^{\pi i(n-\frac{1}{2})z}. \tag{7.3.34}$$

7.4. Functions belonging to $\Gamma_V(2)$; sums of squares. In this section we study Poincaré series belonging to the spaces $\{\Gamma_V(2), \frac{1}{2}s, v_s\}$ and $\{\Gamma_V(2), \frac{1}{2}s, v_s^*\}$ for arbitrary real s. We are particularly interested in Eisenstein series belonging to these spaces and we give explicit formulae for their Fourier coefficients, in the case of the former space, when s is an even integer greater than 2. This is because the Fourier coefficients of the function ϑ_3^s, which belongs to $\{\Gamma_V(2), \frac{1}{2}s, v_s\}$, enumerate the number of representations of a number as a sum of s squares.

We shall, as usual, take

$$k = \tfrac{1}{2}s \quad (k \in \mathbb{R}).$$

We recall the definitions of the multiplier systems v_s and v_s^* given in §7.1:

$$v_s(-I) = v_s^*(-I) = e^{-\frac{1}{2}\pi i s}, \quad v_s(U^2) = v_s^*(U^2) = 1 \tag{7.4.1}$$

$$v_s(V) = e^{-\frac{1}{4}\pi i s}, \quad v_s^*(V) = -e^{-\frac{1}{4}\pi i s} \tag{7.4.2}$$

and we deduce, since $W^2 = -U^{-2}VU^{-2}V$, that

$$v_s(W^2) = v_s^*(W^2) = 1. \tag{7.4.3}$$

From (7.4.1, 3) we see that, when k is an even integer, v_s and v_s^* are constant on $\Gamma(2)$, while, when k is odd, v_s and v_s^* coincide on $\Gamma^*(2)$ with the MS $v^{(0)}$ introduced in §5.6. More generally, for all real s, the two multiplier systems v_s and v_s^* agree on $\Gamma(2)$.

We now apply theorem 5.1.5 to obtain Poincaré series for $\Gamma_V(2)$ in terms of Poincaré series for $\Gamma(2)$. We take any real $s>4$ and apply the theorem with $\Gamma = \Gamma_V(2)$, $\Delta = \Gamma(2)$ and (i) $\zeta = \infty$, (ii) $\zeta = -1 = P^{-1}\infty$. For these are incongruent cusps modulo $\Gamma_V(2)$.

In the first case we take $L = I$ and find that $n' = n = 2$, $\kappa' = \kappa = 0$. These hold for both multiplier systems, by (7.3.1). We take $L_1 = I$, $L_2 = V$ and deduce that, for any $m \in \mathbb{Z}$,

$$G_I(z;m;\Gamma_V(2),\tfrac{1}{2}s,v_s) = G_I(z;m;\Gamma(2),\tfrac{1}{2}s,v_s) + e^{\frac{1}{4}\pi i s}G_V(z;m;\Gamma(2),\tfrac{1}{2}s,v_s) \tag{7.4.4}$$

and

$$G_I(z;m;\Gamma_V(2),\tfrac{1}{2}s,v_s^*)=G_I(z;m;\Gamma(2),\tfrac{1}{2}s,v_s^*)$$
$$-e^{\frac{1}{4}\pi is}G_V(z;m;\Gamma(2),\tfrac{1}{2}s,v_s^*). \quad (7.4.5)$$

We now take $L=P$. For both MS we have $n'=2$, $n=1$ so that we can take $L=I$, since $l=1$. We have

$$v_s(P^{-1}UP)=e^{\frac{1}{4}\pi is},\quad v_s^*(P^{-1}UP)=-e^{\frac{1}{4}\pi is}$$

and

$$v_s(P^{-1}U^2P)=v_s^*(P^{-1}U^2P)=e^{\frac{1}{2}\pi is}.$$

Accordingly, we have

$$\kappa=\left\{\frac{s}{8}\right\}=:\kappa_s,\qquad \kappa'=\left\{\frac{s}{4}\right\}\quad \text{for } v_s,$$

$$\kappa=\left\{\frac{s}{8}+\frac{1}{2}\right\}=:\kappa_s^*,\quad \kappa'=\left\{\frac{s}{4}\right\}\quad \text{for } v_s^*,$$

and so

$$G_P(z;m;\Gamma_V(2),\tfrac{1}{2}s,v_s)=G_P(z;2m+[2\kappa_s];\Gamma(2),\tfrac{1}{2}s,v_s) \quad (7.4.6)$$

and

$$G_P(z;m;\Gamma_V(2),\tfrac{1}{2}s,v_s^*)=G_P(z;2m+[2\kappa_s^*];\Gamma(2),\tfrac{1}{2}s,v_s^*). \quad (7.4.7)$$

If we take $m=0$ in these formulae we see that both

$$E_k(z;I,v_{2k}):=G_I(z;0;\Gamma_V(2),\tfrac{1}{2}s,v_s)\quad (k>2) \quad (7.4.8)$$

and

$$E_k(z;I,v_{2k}^*):=G_I(z;0;\Gamma_V(2),\tfrac{1}{2}s,v_s^*)\quad (k>2) \quad (7.4.9)$$

are Eisenstein series. However, (7.4.6) gives an Eisenstein series if and only if $s=8q$ for some $q\in\mathbb{Z}^+$, while (7.4.7) gives an Eisenstein series if and only if $s=8q+4$ for some $q\in\mathbb{Z}^+$.

We can, in fact, also apply (7.4.4) when $m=0$ and $k=2$, i.e. $s=4$. The non-analytic terms on the right-hand side then cancel out and we have

$$E_2(z;I,v_4)=E_2(z;I)-E_2(z;V)$$
$$=E_2(z;I)-E_2(z;P^2) \quad (7.4.10)$$

in the notation of (7.3.20).

Write, for $z \in \mathbb{H}$ and real $k \geq 2$,

$$E_k(z; I, v_{2k}) = 1 + \sum_{n=1}^{\infty} \beta_k(n)\, e^{\pi i n z} \quad (z \in \mathbb{H}). \tag{7.4.11}$$

Then, when k is an even integer, we have, by (7.3.10) and (7.4.4),

$$E_k(z; I, v_{2k}) = E_k(z; I) + (-1)^k E_k(z; P^2), \tag{7.4.12}$$

since, by (7.3.13), $E_k(z; V) = E_k(z; P^2)$. When k is an odd integer, we have, by (5.6.14, 15), (7.3.21) and (7.4.4),

$$\begin{aligned} E_k(z; I, v_s) &= G_I(z; 0; \Gamma(2), k, v^{(0)}) + i^k G_{P^2}(z; 0; \Gamma(2), k, v^{(0)}) \\ &= E_k(z; I, v^{(0)}) + i^k E_k(z; P^2, v^{(0)}). \end{aligned} \tag{7.4.13}$$

Hence we have, by (7.3.15, 17, 20),

$$\begin{aligned} \beta_k(n) &= \alpha_k(n, I) + (-1)^{\frac{1}{2}k} \alpha_k(n, P^2) \\ &= \frac{\alpha_k}{2^k - 1} \{\sigma^*_{k-1}(n) + (-1)^{\frac{1}{2}k} \sigma^\dagger_{k-1}(\tfrac{1}{2}n)\} \quad (k \text{ even}, k \geq 2). \end{aligned} \tag{7.4.14}$$

By (7.3.29, 30),

$$\beta_k(n) = \frac{4}{|E_{k-1}|} \{2^{k-1} \sigma''_{k-1}(n) + (-1)^{\frac{1}{2}(k-1)} \sigma'_{k-1}(n)\} \quad (k \text{ odd}, k > 2). \tag{7.4.15}$$

From these formulae we obtain the values of $\beta_k(1)$ and $\beta_k(2)$ listed in table 12.

Table 12. Values of $\beta_k(1)$ and $\beta_k(2)$

k	2	3	4	5	6	7
$\beta_k(1)$	8	12	16	$\frac{68}{5}$	8	$\frac{252}{61}$
$\beta_k(2)$	24	60	112	$\frac{1028}{5}$	264	$\frac{16380}{61}$

k	8	9	10	11	12
$\beta_k(1)$	$\frac{32}{17}$	$\frac{1028}{1385}$	$\frac{8}{31}$	$\frac{4092}{50521}$	$\frac{16}{691}$
$\beta_k(2)$	$\frac{4064}{17}$	$\frac{262148}{1385}$	$\frac{4104}{31}$	$\frac{4194300}{50521}$	$\frac{32752}{691}$

We now resume our study of the function ϑ_3^s, where $s \in \mathbb{R}$. We write

$$\vartheta_3^s(z) = 1 + \sum_{n=1}^{\infty} r_s(n) q^n \quad (q = e^{\pi i z}) \tag{7.4.16}$$

for $|q| < 1$. Since $\vartheta_3^s(z) = (1 + 2q + 2q^4 + \cdots)^s$, it follows that

$$r_s(1) = 2s = 4k, \quad r_s(2) = 2s(s-1) = 4k(2k-1), \tag{7.4.17}$$

where, as usual, $k = \frac{1}{2}s$. For positive integral s, ϑ_3^s is the product of s absolutely convergent power series and it is clear that $r_s(n)$ is just the number of solutions of the equation

$$n = n_1^2 + n_2^2 + \cdots + n_s^2$$

in integers $n_1, n_2, \ldots, n_s$.

For any real $k \geq 2$, both $\vartheta_3^{2k}(z)$ and $E_k(z; I, v_{2k})$, as defined by (7.4.8), belong to $\{\Gamma_V(2), k, v_{2k}\}$ and they both take the value 1 at ∞ and 0 at -1. Hence we have

$$\vartheta_3^s(z) = E_k(z; I, v_{2k}) + R_k(z) \quad (k = \tfrac{1}{2}s) \tag{7.4.18}$$

where $R_k \in \{\Gamma_V(2), k, v_{2k}\}_0$. We write

$$R_k(z) = \sum_{n=1}^{\infty} \rho_k(n)\, e^{2\pi i n z}, \tag{7.4.19}$$

so that

$$r_{2k}(n) = \beta_k(n) + \rho_k(n) \quad (s \geq 4, n \geq 1). \tag{7.4.20}$$

From the formulae for $\beta_k(n)$ it is clear that, for large n, $\beta_k(n) = O(n^{k-1})$ when $k > 2$, while, by theorem 4.5.2 (i) $\rho_k(n) = O(n^{\frac{1}{2}k})$. Thus the first term on the right of (7.4.20) forms a good approximation to the magnitude of $r_{2k}(n)$. For small values of k we can improve on this statement.

For, by theorem 7.1.4, the dimension of $\{\Gamma_V(2), k, v_{2k}\}_0$ is zero for $2 \leq k \leq 4$ and we deduce that

$$\vartheta^{2k}(z) = E_k(z; I, v_{2k}) \quad (2 \leq k \leq 4) \tag{7.4.21}$$

and

$$r_{2k}(n) = \beta_k(n) \quad (2 \leq k \leq 4). \tag{7.4.22}$$

In particular, we have, by (7.4.14, 15), for $n \geq 1$,

$$r_4(n) = 8\{\sigma_1^*(n) - \sigma_1^\dagger(\tfrac{1}{2}n)\}, \tag{7.4.23}$$

$$r_6(n) = 4\{4\sigma_2''(n) - \sigma_2'(n)\}, \tag{7.4.24}$$

and

$$r_8(n) = 16\{\sigma_1^*(n) + \sigma_1^\dagger(\tfrac{1}{2}n)\}. \tag{7.4.25}$$

When $5 \le k \le 8$, $\dim\{\Gamma_V(2), k, v_{2k}\}_0 = 1$ and every cusp form is a multiple of

$$\psi_k = \tfrac{1}{16}\vartheta_3^{2k-8}(\vartheta_2\vartheta_4)^4. \tag{7.4.26}$$

In his papers on sums of squares, Glaisher (1907*a*,*b*,*c*) named the coefficients of q^n in ψ_5, ψ_6, ψ_7 and ψ_8

$$\chi_4(n), \quad \Omega(n), \quad W(n) \quad \text{and} \quad (-1)^{n-1}\Theta(n), \tag{7.4.27}$$

respectively. Clearly

$$R_k(z) = \{r_{2k}(1) - \beta_k(1)\}\psi_k(z) \quad (k = 5, 6, 7, 8) \tag{7.4.28}$$

so that

$$r_{10}(n) = \tfrac{4}{5}\{16\sigma_4''(n) + \sigma_4'(n) + 8\chi_4(n)\}, \tag{7.4.29}$$

$$r_{12}(n) = 8\{\sigma_5^*(n) - \sigma_s^\dagger(\tfrac{1}{2}n) + 2\Omega(n)\}, \tag{7.4.30}$$

$$r_{14}(n) = \tfrac{4}{61}\{\sigma_6''(n) - \sigma_6'(n) + 364\,W(n)\}, \tag{7.4.31}$$

and

$$r_{16}(n) = \tfrac{32}{17}\{\sigma_7^*(n) + \sigma_7^\dagger(\tfrac{1}{2}n) + 16(-1)^{n-1}\Theta(n)\}. \tag{7.4.32}$$

Observe that, by (7.1.23),

$$\psi_6(z) = \Delta^{\frac{1}{2}}(z). \tag{7.4.33}$$

It also follows from (7.1.32) that

$$\psi_8(z) = \left\{\frac{\Delta^4(z)}{\Delta(\frac{1}{2}z)\,\Delta(2z)}\right\}^{1/3}. \tag{7.4.34}$$

For values of $k > 8$ the dimension of $\{\Gamma_V(2), k, v_{2k}\}_0$ exceeds 1 and it is necessary to look at the coefficients of powers of q in order to express $R_k(z)$ as a linear combination of known cusp forms. For example, if $k = 9, 10, 11, 12$, the functions ψ_k and $\phi_k = 2^{-8}\vartheta_3^{2k-16}(\vartheta_3\vartheta_4)^8$ span $\{\Gamma_V(2), k, v_{2k}\}_0$ and we find that

$$R_k(z) = E_k(z; I, v_{2k}) + \{4k - \beta_k(1)\}\psi_k(z)$$
$$+\{4k(23-2k) + 4(k-6)\beta_k(1) - \beta_k(2)\}\phi_k(z). \tag{7.4.35}$$

The values of the coefficients of ψ_k and ϕ_k may be calculated from this and table 12. However, the functions ψ_k and ϕ_k, and more generally the functions listed in (7.1.22), need not always be the most convenient to choose to span the space of cusp forms, and this is the case for $k=9$, 10 and 11. On the other hand, it is easily shown from (7.1.23) and theorem 7.1.8 that

$$\psi_{12}(z)=-\Delta(\tfrac{1}{2}z+\tfrac{1}{2}),\quad \phi_{12}(z)=\Delta(z) \tag{7.4.36}$$

and these are convenient forms to choose to span $\{\Gamma_V(2),-12,v_{24}\}_0$. We find that

$$\begin{aligned} r_{24}(n)=\tfrac{16}{691}\{\sigma^*_{11}(n)+\sigma^\dagger_{11}(\tfrac{1}{2}n)\}\\ +\tfrac{128}{691}\{259(-1)^{n-1}\tau(n)-512\tau(\tfrac{1}{2}n)\}. \end{aligned} \tag{7.4.37}$$

We note in conclusion that formulae (7.4.25, 32 and 37) become slightly simpler when it is observed that

$$\sigma^*_{k-1}(n)+\sigma^\dagger_{k-1}(\tfrac{1}{2}n)=\begin{cases}\sigma_{k-1}(n) & (n \text{ odd}),\\ \sum\limits_{\substack{d\mid n\\ d \text{ even}}} d^{k-1}-\sum\limits_{\substack{d\mid n\\ d \text{ odd}}} d^{k-1} & (n \text{ odd}).\end{cases}$$

7.5. Further remarks. Various different notations for theta functions are commonly used. The one adopted here is based on that employed by Tannery and Molk (1893–1902) in their comprehensive treatise on elliptic functions. From the point of view of that theory each theta function is, in their notation, a function $\vartheta_\alpha(\nu|\tau)$ of two complex variables ν and τ and $\alpha=1$, 2, 3 or 4. In elliptic function theory the variable τ (which we have called z) is the ratio of two primitive periods, is normally regarded as constant, and therefore is often suppressed, while interest attaches to the dependence of the functions on the first variable ν; thus, for example, in (7.1.7) the prime denotes differentiation with respect to this variable.

From the point of view of the theory of modular forms, and for the purposes for which we require theta functions, it is the second variable τ (or z) that is of importance and we take $\nu=0$. Thus what we denote by $\vartheta_\alpha(z)$ is, in Tannery and Molk's notation, $\vartheta_\alpha(0|z)$.

We have used the 'Hermite functions' $\Theta_{\mu,\nu}(z)$ (again with the variable ν put equal to zero) to provide a proof of the functional equations of the theta functions. A variety of other methods can be used, such as the Poisson summation formula.

In Tannery and Molk will be found explicit formulae for the multiplier systems $u(T)$, $v(T)$, $w(T)$ for arbitrary T in the various groups. The transformation formulae given in theorem 7.1.2 are, however, sufficient for our purpose.

The multiplier systems u_s, v_s and w_s are associated with the functions ϑ_2^s, ϑ_3^s and ϑ_4^s, while the multiplier system v_s^* is associated with $\vartheta_3^{s-4}(\vartheta_4^4-\vartheta_2^4)$. Conjugate multiplier systems u_s^* and w_s^* can be defined similarly, as in (7.1.19, 20), namely by taking

$$u_s^*(PTP^{-1}) = v_s^*(T) = w_s^*(P^2TP^{-2}) \quad (T \in \Gamma(1). \qquad (7.5.1)$$

These hold for all $s \in \mathbb{R}$. If s is an integer divisible by 4, so that $k = \frac{1}{2}s$ is even, the multiplier systems v_s, v_s^* are constant on $\Gamma(2)$. If $s \equiv 2 \pmod 4$, so that k is odd, then v_s and v_s^* are constant on $\Gamma^*(2)$ and agree on that group with the MS $v^{(0)}$ defined in table 8 (p. 182). Similar results hold for u_s, u_s^* and w_s, w_s^*.

The modular function λ, introduced in §7.2, plays an important role in one of the proofs of Picard's theorem; see, for example, Copson (1935), chapter 15. There is a vast literature on the subject of modular and transformation equations, particularly in older works on elliptic functions. See Weber (1909) and Fueter (1924, 1927).

The problem of finding the number of representations of a positive integer n as a sum of s squares reduces essentially to that of specifying the cusp form $R_k(z)$ ($k = \frac{1}{2}s$) in (7.4.18). For an interesting brief account of work on this subject see the notes at the end of chapter 9 of Hardy's *Ramanujan* (1940).

The formulae (7.4.23–25) are due to Jacobi, (7.4.29) was found by Liouville, (7.4.30–32) were derived by Glaisher (1970*a*,*b*,*c*), and (7.4.37) was obtained by Ramanujan (1916). For other references to work on the representation of integers by sums of squares and other positive definite quadratic forms of various kinds see Dickson, vol. 2 (1934) and LeVeque (1974), §E24.

The formula (7.4.18) holds for any real $s \geq 4$. It was shown by Rankin (1965) that for $s > 8$ the cusp form $R_k(z)$ is never the zero form. A kind of generalization of this has been proved by Kogan (1969).

8: The structure of the space of modular forms of level N and integral weight

8.1. Forms of fixed character or divisor. Our object in this chapter is to break up the space of modular forms of integral weight belonging to the principal congruence group into smaller subspaces corresponding to groups lying between $\Gamma(N)$ and $\Gamma(1)$ and to introduce various families of linear operators which are useful for this purpose. Throughout the chapter we shall assume that N is a fixed positive integer, and that k is an integer satisfying the condition

$$k \text{ is even when } N=1 \text{ or } 2. \tag{8.1.1}$$

The multiplier system v is taken to be constant on $\Gamma(N)$, so that

$$v(S)=\begin{cases}1 & \text{for } S\in\Gamma(N),\\ (-1)^k & \text{for } S\in\bar{\Gamma}(N)-\Gamma(N).\end{cases} \tag{8.1.2}$$

We write

$$M(N, k) := M(\bar{\Gamma}(N), k, v), \tag{8.1.3}$$

and, in a similar way, we denote by

$$H(N, k), \quad \{N, k\} \quad \text{and} \quad \{N, k\}_0 \tag{8.1.4}$$

the subspaces of holomorphic forms, entire forms and cusp forms. Observe that the condition (8.1.1) may be omitted if we make the convention that, for odd k, $M(N, k)$ consists solely of the zero form when $N=1$ or 2.

We shall define a number of linear operators that act on the space $M(N, k)$. These will be expressed as linear combinations of the general stroke operator introduced in (4.3.21). We shall consider various subspaces M of $M(N, k)$ that are invariant under the operators O introduced. Thus $f|O\in M$, whenever $f\in M$; i.e. $M|O\subseteq M$.

We examine the effect on $M(N, k)$ of transformations T belonging to the group $\Gamma_0^0(N)$ defined by (1.4.21). For $f\in M(N, k)$, $f|T$ is determined solely by the fourth entry d of T modulo N, since $b \equiv c \equiv 0 \pmod N$, and therefore $ad \equiv 1 \pmod N$. Accordingly, it is convenient to denote by R_d any element of $\Gamma(1)$ satisfying the congruence

$$R_d \equiv \begin{bmatrix} a & 0 \\ 0 & d \end{bmatrix} \pmod N, \tag{8.1.5}$$

where d is prime to N and a is chosen to make $ad \equiv 1 \pmod N$.

The factor group $\Gamma_0^0(N)/\Gamma(N)$ is clearly abelian, being isomorphic to the multiplicative group of residue classes prime to N and being generated by the cosets $\Gamma(N)R_d$, for $(d, N) = 1$. Let χ be any character modulo N. Then, for any $T \in \Gamma_0^0(N)$, we define

$$\chi(T) = \chi(d). \tag{8.1.6}$$

Then it is clear that χ is a MS of weight k on $\Gamma_0^0(N)$ if and only if $\chi(-I) = (-1)^k$, i.e., if and only if

$$\chi(-1) = (-1)^k. \tag{8.1.7}$$

χ is clearly an extension of the MS v. We write

$$M(N, k, \chi) := M(\Gamma_0^0(N), k, \chi) \tag{8.1.8}$$

and, as in (8.1.4), use the notations

$$H(N, k, \chi), \quad \{N, k, \chi\}, \quad \{N, k, \chi\}_0 \tag{8.1.9}$$

for the corresponding subspaces of holomorphic, entire and cusp forms. These clearly are subspaces of the corresponding spaces in (8.1.3, 4). We can dispense with the condition (8.1.7) by agreeing that these spaces reduce to the zero form when the condition is violated.

For each character χ modulo N we now define a linear operator R^χ on $M(N, k)$ as follows:

$$f|R^\chi = \frac{1}{\phi(N)} \sum_{\substack{m=1 \\ (m,N)=1}}^{N} \bar{\chi}(m)(f|R_m), \tag{8.1.10}$$

where $\bar{\chi}$ is the conjugate character. Thus $f|R^\chi = f_\chi$ in the notation used in the proof of theorem 4.2.2, where in this case $\Delta = \bar{\Gamma}(N)$ and $\Gamma = \Gamma_0^0(N)$. Accordingly, when $f \in M(N, K)$, $f_\chi \in M(N, k, \chi)$; it is also clear that, when f belongs to one of the three spaces in (8.1.4), f_χ will belong to the corresponding space in (8.1.9). We call f_χ a *form of character* χ.

Theorem 8.1.1. *Let M be a subspace of $M(N, k)$ that is invariant under the operators R^χ and write*

$$M^\chi := M|R^\chi,$$

where χ is any character modulo N. Then M can be expressed as a direct sum

$$M = \bigoplus_\chi M^\chi.$$

In particular, $M(N, k)$ is the direct sum of the spaces

$$M(N, k, \chi) = M(N, k)|R^\chi, \tag{8.1.11}$$

and similar results hold for the spaces listed in (8.1.9).

The operator R^χ is self-adjoint on $\{N, k\}_0$ and, if $M \subseteq \{N, k\}_0$, the subspaces M^χ are mutually orthogonal.

Proof. The statement that R^χ is self-adjoint on $\{N, k\}_0$ means that, for any f and g in $\{N, k\}_0$,

$$(f, g|R^\chi) = (f|R^\chi, g). \tag{8.1.12}$$

This follows immediately from (8.1.10), (5.2.8) and theorem 5.2.1. The remaining parts of the theorem follow from theorems 4.2.2 and 5.2.7.

We now decompose $M(N, k)$ into subspaces of a different kind by examining the effect on $M(N, k)$ of the translation operators U^q ($q \in \mathbb{Z}$). Let $f \in M(N, k)$, so that, for $y = \operatorname{Im} z > y_0$, say, $f(z)$ has a Fourier expansion

$$f(z) = \sum_{r=-\infty}^{\infty} a_r e^{2\pi i r z/N}, \tag{8.1.13}$$

where the series contains only a finite number of terms with negative r for which $a_r \neq 0$.

Because the conjugate MS v^U is identical with v, $f|U$ also belongs to $M(N, k)$. By taking a linear combination of forms $f|U^q$ ($0 < q \leq N$), we can obtain a form with Fourier coefficients b_r, where b_r is non-zero only when r belongs to certain residue classes modulo N. More precisely, take any positive divisor t of N and write

$$N = tt_1. \tag{8.1.14}$$

Consider the linear operator D_t defined by

$$D_t := \frac{1}{N} \sum_{\substack{\nu=1 \\ (\nu,N)=t}}^{N} \sum_{l=1}^{N} e^{-2\pi i\nu l/N} U^l \tag{8.1.15}$$

$$= \frac{1}{N} \sum_{l=1}^{N} \Big(\sum_{d|(l,t_1)} d\mu(t_1/d) \Big) U^l, \tag{8.1.16}$$

by (5.4.13). Then $f|D_t \in M(N, k)$ when $f \in M(N, k)$. Also, by (8.1.13), for $y > y_0$,

$$f(z)|D_t = \frac{1}{N} \sum_{\substack{\nu=1 \\ (\nu,N)=t}}^{N} \sum_{l=1}^{N} e^{-2\pi i\nu l/N} f(z+l)$$

$$= \frac{1}{N} \sum_{\substack{\nu=1 \\ (\nu,N)=t}}^{N} \sum_{r=-\infty}^{\infty} a_r \sum_{l=1}^{N} \exp\{2\pi i(rz + rl - \nu l)/N\}$$

$$= \sum_{\substack{\nu=1 \\ (\nu,N)=t}}^{N} \sum_{\substack{r=-\infty \\ r\equiv\nu \,(\mathrm{mod}\, N)}}^{\infty} a_r e^{2\pi i r z/N}$$

$$= \sum_{\substack{r=-\infty \\ (r,N)=t}}^{\infty} a_r e^{2\pi i r z/N}. \tag{8.1.17}$$

Moreover,

$$f(z) = \sum_{t|N} \{f(z)|D_t\}. \tag{8.1.18}$$

Any member of $M(N, k)$ having the property that its Fourier coefficients a_r are zero except possibly when $(r, N) = t$ is called a *form of divisor t*. Such a form has a Fourier expansion of the form

$$\sum_{r=-\infty}^{\infty} A_r e^{2\pi i r z/t_1},$$

where A_r is non-zero only when $(r, t_1) = 1$. The parameter t_1 is, in some respects, more basic than t and we call t_1 the *codivisor* of t.

Thus the linear operator D_t maps a modular form f into one of divisor t and the forms of divisor t constitute a subspace $M_t(N, k)$ of $M(N, k)$. It is clear also that $H(N, k)$, $\{N, k\}$ and $\{N, k\}_0$ are mapped by D_t into themselves.

Since it is evident from their Fourier expansions that the forms in $M_t(N, k)$ are invariant under the operator U^{t_1}, they belong to the group

$$\Gamma_t(N) := \bar{\Gamma}(N)\langle U^{t_1}\rangle. \tag{8.1.19}$$

More precisely, if $f \in M_t(N, k)$, then

$$f \in M(\Gamma_t(N), k, v) =: M^{(t)}(N, k), \tag{8.1.20}$$

where the MS v has been extended to $\Gamma_N(N)$ (which includes all the $\Gamma_t(N)$ for $t|N$) by taking $v(U)=1$. Observe that $f \in M^{(t)}(N, k)$ whenever $f \in M_\tau(N, k)$, where $t|\tau$ and $\tau|N$.

If $f \in M^{(t)}(N, k)$ and (8.1.13) holds, then in the definition of $f|D_t$ we may replace U^l in (8.1.15) by U^{ln} for any fixed integer n prime to t_1. That this is so is easily verified on carrying out the analysis leading to (8.1.17), on observing that a_r is non-zero only when t divides r.

From these results we easily deduce

Theorem 8.1.2. *Let M be a subspace of $M(N, k)$ that is invariant under the operators D_t and write*

$$M_t := M|D_t, \quad M^{(t)} = M \cap M^{(t)}(N, k), \tag{8.1.21}$$

for any positive divisor t of N. Then

$$M = \bigoplus_{t|N} M_t, \tag{8.1.22}$$

and

$$M^{(t)} = \bigoplus_{t|\tau|N} M_\tau, \tag{8.1.23}$$

where τ runs through all divisors of N that are divisible by t.

We may note that, if a space M is invariant under U, then certainly $M|D_t \subseteq M$.

Theorem 8.1.3. *Let f and g be members of $\{N, k_0\}$ with divisors t and τ, where $t \neq \tau$. Then f and g are orthogonal.*

Proof. Put $N = tt_1 = \tau\tau_1$. We may suppose that, for some prime p dividing N, we have

$$t_1 = p^\alpha t_0, \quad \tau_1 = p^\beta \tau_0, \quad N = p^\gamma N_0,$$

where $p \nmid t_0\tau_0N_0$, and that $\alpha \neq \beta$. Without loss of generality suppose that $\alpha > \beta \geq 0$ and write

$$m = N_0p^\beta, \quad n = p^{\alpha-\beta}.$$

Then $\tau_1|m$ and therefore $g|U^{-lm} = g$ for all $l \in \mathbb{Z}$. Now, if $(r, N) = t$, then $r = ts$, where $(s, t_1) = 1$. Hence $p \nmid s$ and

$$\sum_{l=1}^{n} e^{2\pi irlm/N} = \sum_{l=1}^{n} e^{2\pi isl/n} = 0.$$

It follows from (8.1.17) that

$$\sum_{l=1}^{n} f|U^{lm} = 0.$$

Accordingly, by the properties of inner products,

$$\begin{aligned} n(f, g) &= \sum_{l=1}^{n} (f, g|U^{-lm}) \\ &= \sum_{l=1}^{n} (f|U^{lm}, g) = 0. \end{aligned}$$

8.2. The interaction of the operators R^χ and D_t. We now examine the combined effect of these operators on $M(N, k)$. For this purpose we recall the groups $\Gamma_0(m, n)$ introduced in (1.4.25) and note that, if $T \in \Gamma_0(N, t_1)$, where $tt_1 = N$, then $TU^{rt_1} \in \Gamma_0^0(N)$ for some $r \in \mathbb{Z}$, so that

$$\Gamma_0(N, t_1) = \Gamma_0^0(N)\langle U^{t_1}\rangle. \tag{8.2.1}$$

This is the smallest subgroup of $\Gamma(1)$ containing both of the groups $\Gamma_0^0(N)$ and $\Gamma_t(N)$ considered in the previous section, and it lies between $\Gamma_0^0(N)$ (for $t_1 = N$) and $\Gamma_0(N)$ (for $t_1 = 1$). The MS χ on $\Gamma_0^0(N)$ can be extended to $\Gamma_0(N, t_1)$ by taking

$$\chi(T) = \chi(d) \quad \text{for} \quad T \in \Gamma_0(N, t_1), \tag{8.2.2}$$

and we write

$$M^{(t)}(N, k, \chi) = M(\Gamma_0(N, t_1), k, \chi); \tag{8.2.3}$$

this agrees with the notation defined in (8.1.21); we also write, similarly,

$$H^{(t)}(N, k, \chi), \quad \{N, k, \chi\}^{(t)} \quad \text{and} \quad \{N, k, \chi\}_0^{(t)} \tag{8.2.4}$$

for the subspaces of holomorphic, entire and cusp forms.

The relationship between the various groups and spaces is shown in figs. 9 and 10. Subgroups are normal except when indicated by a dotted line. In fig. 9 the larger groups are above, while in fig. 10 the larger spaces are below. Indices of groups are shown in fig. 9 for the case when $N>2$.

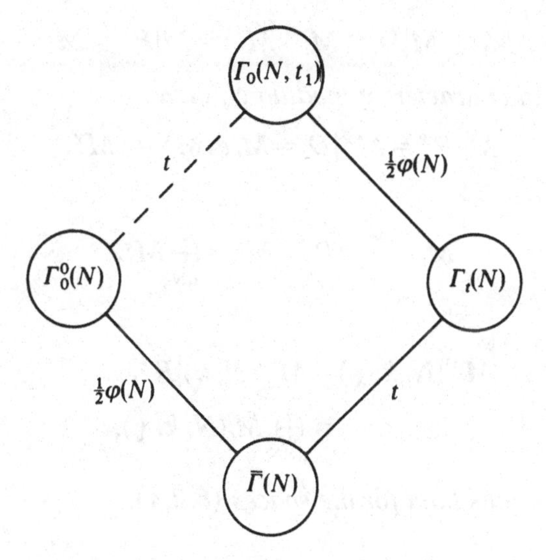

Fig. 9. Congruence groups of level N.

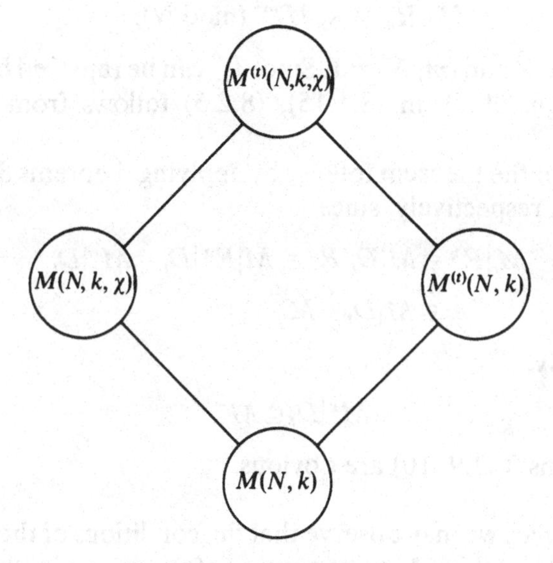

Fig. 10. Corresponding spaces.

Theorem 8.2.1. *The operators D_t and R^χ commute; i.e., for any $f \in M(N, k)$,*

$$(f|D_t)|R^\chi = (f|R^\chi)|D^t \tag{8.2.5}$$

for all $t|N$ and characters χ modulo N. Further, if M is a subspace of $M(N, k)$ for which

$$M_t := M|D_t \subseteq M, \quad M^\chi := M|R^\chi \subseteq M \tag{8.2.6}$$

for all $t|N$ and characters χ modulo N, then

$$M_t|R^\chi = M^\chi|D_t = M_t \cap M^\chi =: M_t^\chi \tag{8.2.7}$$

and

$$M_t = \bigoplus_\chi M_t^\chi, \quad M^\chi = \bigoplus_{t|N} M_t^\chi. \tag{8.2.8}$$

Finally,

$$M^{(t)}(N, k, \chi) = M^{(t)}(N, k)|R^\chi, \tag{8.2.9}$$

$$= \bigoplus_{t|\tau|N} M_t(N, k, \chi), \tag{8.2.10}$$

and similar results hold for the spaces (8.2.4).

Proof. It is easily verified that

$$U^l R_m \equiv R_m U^{lm^2} \pmod N, \tag{8.2.11}$$

whenever $l \in \mathbb{Z}$ and $(m, N) = 1$. Since U^l can be replaced by U^{lm^2} in the definition of D_t in (8.1.15), (8.2.5) follows from this and (8.1.10).

The rest of the theorem follows by applying theorems 8.1.1, 2 to M_t and M^χ, respectively, since

$$M_t|R^\chi = M|D_t|R^\chi = M|R^\chi|D_t = M^\chi|D_t$$
$$\subseteq M|D_t = M_t,$$

and similarly,

$$M^\chi|D_t \subseteq M^\chi.$$

The relations (8.2.9, 10) are obvious.

In particular, we may observe that the conditions of the theorem are satisfied by taking M to be any one of the spaces in (8.1.3, 4) or by

$$M=\{N, k, \varepsilon\}_0,$$

where ε is a fixed character modulo N. For, if f belongs to this space, then $f|R^\chi=0$ or f, according as $\chi\neq\varepsilon$ or $\chi=\varepsilon$, so that $M^\chi\subseteq M$. Further, if $f\in M$, then $f|D_t$ is a cusp form and

$$(f|D_t)|R^\varepsilon=(f|R^\varepsilon)|D_t=f|D_t.$$

It follows that $f|D_t\in M^\varepsilon=M$.

The subspaces of the spaces (8.2.3, 4) or (8.1.8, 9) consisting of forms of divisor t we denote by

$$M(N, k, \chi, t),\quad H(N, k, \chi, t),\quad \{N, k, \chi, t\}\quad\text{and}\quad \{N, k, \chi, t\}_0, \tag{8.2.12}$$

respectively.

Theorem 8.2.2. *Let $f\in M^{(t)}(N, k, \chi)$ and let ε be a character modulo t_1. Then $f^{(\varepsilon)}\in M(N, k, \varepsilon^2\chi, t)$ where, for $y>y_0$,*

$$f^{(\varepsilon)}(z)=\sum_{r=-\infty}^{\infty} a_{tr}\varepsilon(r)\, e^{2\pi i r z/t_1}, \tag{8.2.13}$$

$f(z)$ being defined by (8.1.13). If f is a cusp form, so is $f^{(\varepsilon)}$.

Proof. Consider the operator

$$D_t^{(\varepsilon)}:=\frac{1}{t_1}\sum_{l=1}^{t_1}\sum_{s=1}^{t_1} e^{-2\pi i l s/t_1}\varepsilon(s)U^l. \tag{8.2.14}$$

If $f\in M^{(t)}(N, k)$, clearly so does $f|D_t^{(\varepsilon)}$, and

$$f(z)|D_t^{(\varepsilon)}=\frac{1}{t_1}\sum_{r=-\infty}^{\infty} a_{tr}\sum_{l=1}^{t_1}\sum_{s=1}^{t_1} e^{2\pi i(rz+lr-ls)/t_1}\varepsilon(s).$$

On summing over l, it follows that

$$f|D_t^{(\varepsilon)}=f^{(\varepsilon)}. \tag{8.2.15}$$

It is clear from (8.2.13) that $f^{(\varepsilon)}$ has divisor t; also $f^{(\varepsilon)}$ is a cusp form whenever f is.

Finally, if $(m, N)=1$ and $f|R_m=\chi(m)f$, then, by (8.2.11),

$$f^{(\varepsilon)}|R_m=\frac{\chi(m)}{t_1}\sum_{l=1}^{t_1}\sum_{s=1}^{t_1} e^{-2\pi i l s/t_1}\varepsilon(s)f|U^{lm^2}.$$

If we write $s = m^2 s'$, it follows easily that

$$f^{(\varepsilon)}|R_m = \chi(m)\varepsilon(m^2)f^{(\varepsilon)},$$

and this completes the proof of the theorem.

Observe that $D_t^{(\varepsilon)} = D_t$ when ε is the principal character. If f has divisor t and ε is a real character, then both f and $f^{(\varepsilon)}$ belong to $M(N, k, \chi, t)$.

8.3. The operator J_n. We recall that in (4.3.8) we introduced the matrix

$$J = \begin{bmatrix} 1 & 0 \\ 0 & n \end{bmatrix} \tag{8.3.1}$$

for any positive integer n. We can, however, omit the restriction that n is positive and define J_n by (8.3.1) for any non-zero integer n. Observe that, when $n = -1$, $J_n = J$, in the notation of (1.2.23).

For any $S \in \Gamma(1)$ we have

$$J_n^{-1} S J_n = \begin{bmatrix} \alpha & \beta n \\ \gamma/n & \delta \end{bmatrix}, \tag{8.3.2}$$

as in (4.3.17). We now assume that $(n, N) = 1$ and introduce an automorphism θ_n of the modulary group $G(N) = \Gamma(1)/\Gamma(N)$ as follows:

For any $S \in \Gamma(1)$ we can write, by theorem 1.4.4,

$$S = LS_0,$$

where $L \in \Gamma(N)$ and $S_0 \in \Gamma(n)$. We now put

$$\theta_n(S) = S_n := J_n^{-1} S_0 J_n, \tag{8.3.3}$$

so that, by (8.3.2), $S_n \in \Gamma(1)$. The matrix S_n is not, of course, uniquely determined, but it is clearly uniquely defined modulo N and

$$S_n \equiv \begin{bmatrix} \alpha & \beta n \\ \gamma n' & \delta \end{bmatrix} \pmod{N}, \tag{8.3.4}$$

where n' is any integer such that

$$nn' \equiv 1 \pmod{N}.$$

We define θ_n on $G(N)$ by

$$\theta_n(\Gamma(N)S) = \Gamma(N)S_n. \tag{8.3.5}$$

Then it is clear that θ_n is an automorphism of $G(N)$. Moreover,

$$SJ_n \equiv J_nS_n \pmod N. \tag{8.3.6}$$

We now use the automorphism θ_n to study arbitrary congruence groups of level N.

Theorem 8.3.1. *Let n be any integer prime to N and let Γ be any group satisfying*

$$\Gamma(N) \subseteq \Gamma \subseteq \Gamma(1). \tag{8.3.7}$$

Write

$$\Gamma_n := \Gamma(N)\{\Gamma(1) \cap J_n^{-1}\Gamma J_n\}, \tag{8.3.8}$$

so that

$$\Gamma(N) \subseteq \Gamma_n \subseteq \Gamma(1).$$

Then $\Gamma_n/\Gamma(N)$ is the image of $\Gamma/\Gamma(N)$ under the isomorphism θ_n defined by (8.3.5).

Further, let u be a MS of weight k on Γ whose restriction to $\Gamma(N)$ is the MS v, and define u_n on Γ_n by

$$u_n(S_n) = u(S) \quad \text{for all } S_n \in \Gamma_n, \tag{8.3.9}$$

where S and S_n are related as in (8.3.4). Then u_n is a MS of the weight k on Γ_n and is constant on $\Gamma(N)$.

Proof. It is clear that

$$\theta_n(\Gamma/\Gamma(N)) = \Gamma_n^*/\Gamma(N),$$

where Γ_n^* is some subgroup of $\Gamma(1)$ containing $\Gamma(N)$. We have to show that $\Gamma_n^* = \Gamma_n$, where Γ_n is defined by (8.3.8).

If $S \in \Gamma_n$, then $S = T_1S_0$, where $T_1 \in \Gamma(N)$ and $S_0 \in \Gamma(1) \cap J_n^{-1}\Gamma J_n$. Accordingly, $J_nS_0J_n^{-1} \in \Gamma$ and

$$\Gamma(N)S = \Gamma(N)S_0 = \theta_n(\Gamma(N)J_nS_0J_n^{-1}),$$

so that $\Gamma_n \subseteq \Gamma_n^*$.

Conversely, if $S^* \in \Gamma_n^*$, then $S^* = T\theta_n(S)$ for some $T \in \Gamma(N)$ and $S \in \Gamma$. Put $S = LS_0$, as above, where $L \in \Gamma(N)$ and $S_0 \in \Gamma_0(n) \cap \Gamma$. It follows that

$$\theta_n(S) = J_n^{-1} S_0 J_n \in \Gamma(1) \cap J_n^{-1} \Gamma J_n$$

and so $S^* \in \Gamma_n$.

Finally, it is clear that u_n is a MS of weight k for Γ_n that is constant on $\Gamma(N)$.

In certain circumstances the group Γ_n with MS u_n will coincide with the group Γ with MS u. For our later study of this we introduce the notations:

$$\mathbb{Z}_N = \{n \in \mathbb{Z} : (n, N) = 1\}, \quad \mathbb{Z}_N^+ = \{n \in \mathbb{Z}_N : n > 0\}, \quad (8.3.10)$$

$$\mathbb{Z}_\Gamma = \mathbb{Z}_\Gamma(u) = \{n \in \mathbb{Z}_N : \Gamma_n = \Gamma,\ u_n = u\}, \quad (8.3.11)$$

$$\mathbb{Z}_\Gamma^+ = \mathbb{Z}_\Gamma^+(u) = \{n \in \mathbb{Z}_\Gamma : n > 0\}. \quad (8.3.12)$$

In particular, if $\Gamma = \Gamma_0(N, t_1)$, $\Gamma_0^0(N)$, $\Gamma_0(N)$ or $\Gamma^0(N)$ with MS u defined by $u(T) = \chi(d)$, where χ is a character modulo N satisfying (8.1.7), then $\mathbb{Z}_\Gamma = \mathbb{Z}_N$.

We now turn to analytic applications of J_n and, for this purpose, we shall restrict our attention to positive integers n. If $f \in M(N, k)$, we have, by (4.3.21),

$$f(z)|J_n = n^{-\frac{1}{2}k} f(z/n), \quad f(z)|J_n^{-1} = n^{\frac{1}{2}k} f(nz). \quad (8.3.13)$$

We now apply theorem 4.3.9 to the groups $\Gamma_0(l, m)$ defined in (1.4.25), with $L = J_n$ and $L = nJ_n^{-1}$, so that, in both cases, $L \in \Omega_n^*$. We then obtain the following theorem; by $v^{J_n^{-1}}$ we mean the MS $v^{nJ_n^{-1}}$.

Theorem 8.3.2. *Let $f \in M(\Gamma_0(l, m), k, v)$. Then, for any $n \in \mathbb{Z}^+$,*

$$f|J_n \in M(\Gamma_0(l/(l, n), mn), k, v^{J_n})$$

and

$$f|J_n^{-1} \in M(\Gamma_0(ln, m/(m, n)), k, v^{J_n^{-1}}).$$

We now consider the effect of the operator J_n in the particular case when n is a positive divisor t of N.

Theorem 8.3.3. *Let M denote any of the spaces listed in (8.1.3, 4) or (8.1.8, 9) and let $t = u\tau$ where u is any positive divisor of t and t*

divides N. Then, in the notation of (8.1.21),

$$M_t|J_u \subseteq M^{(t)}|J_u \subseteq M^{(\tau)}. \tag{8.3.14}$$

Moreover, the operators J_u and R_m commute on $M^{(t)}(N, k)$.

Proof. Let $f \in M^{(t)}(N, k, \chi)$ and take any $S \in \Gamma_0(N, \tau_1)$, where $N = tt_1 = \tau\tau_1$. Then $J_u S J_u^{-1} \in \Gamma_0(N, t_1)$ and so

$$(f|J_u)|S = (f|J_u S J_u^{-1})|J_u = \chi(J_u S J_u^{-1}) f|J_u = \chi(S) f|J_u.$$

Thus $f|J_u \in M^{\tau)}(N, k, \chi)$. To complete the proof of (8.3.14) in all cases we note that, by theorem 4.3.9(v), holomorphic, entire and cusp forms retain these properties after operation by J_u.

Finally, it is easily verified that

$$J_u S J_u^{-1} S^{-1} = [J_u, S] \equiv U^{\alpha\beta/u} \pmod N,$$

when $S \in \Gamma_0^0(N)$, so that, by (8.1.19),

$$[J_u, S] \in \Gamma_t(N) \quad \text{and} \quad v([J_u, S]) = 1.$$

Accordingly, since $R_m \in \Gamma_0^0(N)$ when $(m, N) = 1$, we have

$$(f|J_u)|R_m = (f|R_m)|J_u,$$

whenever $f \in M^{(t)}(N, k)$, and this completes the proof of the theorem.

We use this theorem to define an operator D_τ^t that generalizes the operator D_τ and maps forms of divisor t into forms of divisor τ, where τ divides t.

If $f \in M^{(t)}(N, k)$ and τ is any positive divisor of t, we define

$$f|D_\tau^t = u^{\frac{1}{2}k}(f|J_u)|D_\tau \quad (t = u\tau), \tag{8.3.15}$$

so that $f|D_\tau^t \in M_\tau(N, k)$, by (8.3.14). Note that

$$D_t^t = D_t, \quad D_1^t = t^{\frac{1}{2}k} J_t D_1.$$

If

$$f(z) = \sum_{r=-\infty}^{\infty} A_r e^{2\pi i r z/t_1}, \tag{8.3.16}$$

then

$$f(z)|D_\tau^t = \sum_{\substack{r=-\infty \\ (r,\tau_1)=1}}^{\infty} A_r e^{2\pi i r z/\tau_1}, \tag{8.3.17}$$

where $\tau\tau_1 = N$, as in the theorem. Moreover, it follows easily that whenever $\tau|\sigma$ and $\sigma|t$,

$$(f|D_\sigma^t)D_\tau^\sigma = f|D_\tau^t. \tag{8.3.18}$$

For the purposes of the next theorem we recall certain facts about characters. If χ is any character modulo N, then its conductor N_χ is the least positive integer m such that there exists a character χ' modulo m with

$$\chi(n) = \chi'(n)\chi_0(n) \quad (n \in \mathbb{Z}),$$

where χ_0 is the principal character modulo N. When $N_\chi = N$, χ is called a *primitive character* modulo N.

Further if, for some divisor m of N, $\chi(l) = 1$ whenever both $(l, N) = 1$ and $l \equiv 1 \pmod m$, then $N_\chi | m$; see Ayoub (1963), lemma 4.1, p. 305.

Theorem 8.3.4. *Suppose that $f \in M(N, k, t, \chi)$ and that τ is a divisor of t, with $N = tt_1$, $t = \tau u$, where $(u, N/N_\chi) = 1$. Then $f|D_\tau^t = 0$ if and only if $f = 0$.*

Proof. It suffices to assume that $f|D_\tau^t = 0$ and prove that either $f = 0$ or that N/N_χ is divisible by one of the primes dividing u.

We consider first the case when $u = p^s$ for some prime p and positive integer s. From (8.3.17) we deduce that $A_r = 0$ whenever $(r, p^s t_1) = 1$. If $p|t_1$, the condition $(r, p^s t_1) = 1$ is equivalent to $(r, t_1) = 1$ and so f must be the zero form. Accordingly, we may assume that $p|t_1$; it then follows that A_r can be non-zero only when $p|r$. Hence

$$f\left(z + \frac{t_1}{p}\right) = f(z) \quad (z \in \mathbb{H}), \tag{8.3.19}$$

Now write

$$U^\alpha = \begin{bmatrix} 1 & \alpha \\ 0 & 1 \end{bmatrix}$$

not only for integral α, but also for any real number α, and put $t_0 = t_1/p$. Then

$$T := U^{xt_0} W^N U^{yt_0} = \begin{bmatrix} 1 + xNt_0 & (1 + yNt_0)xt_0 + yt_0 \\ N & 1 + yNt_0 \end{bmatrix}.$$

By (8.3.19),

$$f|T=f \tag{8.3.20}$$

for any integers x and y. Now $Nt_0 \in \mathbb{Z}$ and therefore $T \in \Gamma_0^0(N)$ provided that we can choose x, for any given $y \in \mathbb{Z}$, to satisfy

$$x(1+yNt_0)+y \equiv 0 \pmod{pt}.$$

This is always possible if

$$(1+yNt_0, pt)=1,$$

which is satisfied whenever

$$(1+yNt_0, N)=1, \tag{8.3.21}$$

since N is divisible by all the primes that divide pt.

Now take any $l \equiv 1 \pmod{N/p}$ satisfying $(l, N)=1$. Then we can find an integer y satisfying (8.3.21) and

$$l \equiv 1+yNt_0 \pmod{N},$$

since, by our assumption, $p \nmid t_1$. It follows that $\chi(T)=\chi(l)$ and so

$$f|T=\chi(T)f.$$

This combined with (8.3.20) shows that $\chi(l)=1$ whenever $l \equiv 1 \pmod{N/p}$ and $(l, N)=1$. From the remark before the theorem it follows that N_χ divides N/p, and therefore $p|(N/N_\chi)$. This completes the proof in the particular case considered.

In the general case we have

$$t=\tau p_1^{m_1}p_2^{m_2}\ldots p_r^{m_r},$$

where $p_1, p_2, \ldots, p_r$ are different primes and $m_i>0$ $(1 \le i \le r)$. Write

$$\tau^{(0)}=\tau, \quad \tau^{(i)}=\tau p_1^{m_1}\ldots p_i^{m_i} \quad (1 \le i \le r).$$

From (8.3.18) we see that $f|D_\tau^t$ can be derived successively from f by mapping it from a space of divisor $t=\tau^{(r)}$ into spaces of divisors $\tau^{(r-1)}, \ldots, \tau^{(1)}, \tau^{(0)}$. If some of the primes $p_1, p_2, \ldots, p_r$ divide N/N_χ, it follows successively that

$$0=f|D_\tau^t=f|D_{\tau^{(1)}}^t=\cdots=f|D_t^t=f,$$

and this completes the proof of the theorem.

Theorem 8.3.5. *Let $f \in M(N, k, \chi, t)$ and suppose that $N = tt_1$ and that τ is the maximum divisor of t that is prime to t_1. Write $t = \tau u$, $\tau_1 = t_1 u$. Then $u^{\frac{1}{2}k} f|J_u = f|D_\tau^t$ and this function is a member of $M(N, k, \chi, \tau)$.*

Proof. This is immediate, since $(r, t_1) = 1$ if and only if $(r, \tau_1) = 1$.

The theorem shows that the operator J_u maps $M(N, k, \chi, t)$ injectively into $M(N, k, \chi, \tau)$. The mapping preserves the Fourier coefficients. Accordingly, in situations where we are concerned only with the arithmetic properties of these coefficients, there is no loss of generality in making the assumption that the divisor t is a *unitary* divisor of N, i.e. that $(t, t_1) = 1$.

8.4. The operator $\boldsymbol{H}_q$. In this section N, N' and q are positive integers such that q divides NN', and we write

$$qq' = NN' \tag{8.4.1}$$

and assume that

$$(q, q') = 1, \tag{8.4.2}$$

so that q is a unitary divisor of NN'.

Let χ be a character modulo NN', so that, by (1.1.11), χ is also a character on the group $\Gamma_0(N, N')$, when we put

$$\chi(T) := \chi(d) \quad \text{for} \quad T \in \Gamma_0(N, N'). \tag{8.4.3}$$

In the next chapter we shall take $N' = t_1$, where $tt_1 = N$, and suppose that χ is a character modulo N, but these assumptions are not required in the present section.

We introduce a family $\boldsymbol{H}_q(N, N')$ of primitive transformations of order q whose elements are matrices of the form

$$H_q = \begin{bmatrix} qx & N'y \\ Nz & qw \end{bmatrix}, \tag{8.4.4}$$

where x, y, z and w are integers satisfying

$$qxw - q'yz = 1. \tag{8.4.5}$$

This is possible by (8.4.2). It is clear that $\boldsymbol{H}_q(N, N')$ is a subset of Ω_q^*; see (4.3.6).

It is easily verified that, when H_q and H'_q belong to $\boldsymbol{H}_q(N, N')$, then both $H_q^{-1}H'_q$ and $H'_q H_q^{-1}$ belong to $\Gamma_0(N, N')$. Further, SH_q

and H_qS belong to $\boldsymbol{H}_q(N, N')$ whenever $S \in \Gamma_0(N, N')$. From these facts it follows that $\boldsymbol{H}_q(N, N')$ is just a single left or right coset (in the group Ω^+ defined in (4.3.7)) of the form

$$\boldsymbol{H}_q(N, N') = H_q\Gamma_0(N, N') = \Gamma_0(N, N')H_q, \tag{8.4.6}$$

where H_q is any member of $\boldsymbol{H}_q(N, N')$. Note also that

$$H_qH_q' = qS,$$

where $S \in \Gamma_0(N, N')$.

Further, if

$$NN' = qq' = rr' = qrs, \tag{8.4.7}$$

say, where

$$(q, r) = (q, q') = (r, r') = 1, \tag{8.4.8}$$

then

$$\boldsymbol{H}_{qr}(N, N') = \boldsymbol{H}_q(N, N')\boldsymbol{H}_r(N, N'). \tag{8.4.9}$$

For it is clear that the right-hand side is contained in the left-hand side; that the converse holds follows from (8.4.6) with $\boldsymbol{H}_q(N, N')$ replaced by $\boldsymbol{H}_{qr}(N, N')$ and H_q replaced by H_qH_r. In particular, we deduce that

$$H_qH_rH_q^{-1}H_r^{-1} \in \Gamma_0(N, N'). \tag{8.4.10}$$

It is convenient for some purposes to choose a *canonical form* for the matrix H_q, namely

$$H_q = \begin{bmatrix} q\lambda & N'\lambda' \\ N & q \end{bmatrix}, \tag{8.4.11}$$

where λ and λ' are the unique integers satisfying

$$q\lambda - q'\lambda' = 1, \quad 0 \leq \lambda' < q. \tag{8.4.12}$$

The members of $\boldsymbol{H}_q(N, N')$ determine an involution on the characters modulo NN', which we now describe. We observe first that every such character χ can be expressed uniquely as

$$\chi = \chi_q\chi_{q'}, \tag{8.4.13}$$

where χ_q and $\chi_{q'}$ are characters modulis q and q', respectively.

Now take a fixed $H_q \in \boldsymbol{H}_q(N, N')$ and any $S \in \Gamma_0(N, N')$. Put

$$S^* = \begin{bmatrix} \alpha^* & \beta^* \\ \gamma^* & \delta^* \end{bmatrix} = H_qSH_q^{-1}, \tag{8.4.14}$$

so that $S^* \in \Gamma_0(N, N')$ and, as is easily shown,

$$\delta^* \equiv q\lambda\delta - q'\lambda'\alpha \pmod{NN'}.$$

Thus

$$\delta^* \equiv \delta \pmod{q'}, \quad \delta^* \equiv \alpha \equiv \delta^{-1} \pmod{q}.$$

If we put

$$\chi^*(S) := \chi^*(\delta) := \chi(\delta^*) = \chi(S^*), \tag{8.4.15}$$

then clearly χ^* is a character modulo NN' and

$$\chi^* = \bar{\chi}_q\chi_{q'}. \tag{8.4.16}$$

Note that χ^* is independent of the choice of H_q in $\boldsymbol{H}_q(N, N')$ and that $(\chi^*)^* = \chi$. Also χ and χ^* have the same conductor. Clearly $\chi^* = \chi$ if and only if χ_q is a real character; we describe this by saying that χ is of *real type modulo q*. It follows that χ is of real type modulo q whenever the conductor N_χ of χ divides the conductor of $\chi_{q'}$; this is the case, in particular, when N_χ divides q'.

The following theorem makes clear the reason for introducing this involution.

Theorem 8.4.1. *Suppose that $f \in M(\Gamma_0(N, N'), k, \chi)$, where $k \in \mathbb{Z}$, χ is a character modulo NN', and (8.4.1, 2) hold. Then*

$$f|H_q \in M(\Gamma_0(N, N'), k, \chi^*),$$

for each $H_q \in \boldsymbol{H}_q(N, N')$, where χ^* is defined by (8.4.13, 16). Moreover, if f is a holomorphic, entire or cusp form, so is $f|H_q$.

Further, if (8.4.7, 8) hold, and if N_χ divides s, then

$$(f|H_q)|H_r = (f|H_r)|H_q$$

for all $H_q \in \boldsymbol{H}_q(N, N')$ and all $H_r \in \boldsymbol{H}_r(N, N')$.

Proof. The first part follows immediately from the fact that, if $S \in \Gamma_0(N, N')$, then

$$(f|H_q)|S = (f|S^*)|H_q,$$

and from theorem 4.3.9(v). The last part follows from (8.4.10), since it is easily verified that

$$\chi(H_qH_rH_q^{-1}H_r^{-1}) = 1$$

when N_χ divides s.

Observe that, for different H_q in $\boldsymbol{H}_q(N, N')$, the functions $f|H_q$ differ only by constant multiples; i.e., if H_q and H'_q are in $\boldsymbol{H}_q(N, N')$, then

$$f|H'_q = \chi(H'_q H_q^{-1}) f|H_q$$

for each $f \in M(\Gamma_0(N, N'), k, \chi)$.

We conclude with two theorems on matrices that we shall require in §9.2 when we consider the interaction of the operators H_q with certain Hecke operators.

Theorem 8.4.2. *Let (8.4.1, 2) hold and suppose that N' divides Nq and that $(n, q) = 1$. Let $H_q^* \in \boldsymbol{H}_q(Nn, N'n)$, so that we may put*

$$H_q^* = \begin{bmatrix} qx & N'ny \\ Nnz & qw \end{bmatrix}, \tag{8.4.17}$$

and write, for any integer r,

$$X_r = \begin{bmatrix} a & r \\ 0 & d \end{bmatrix}, \tag{8.4.18}$$

where $ad = n \in \mathbb{Z}^+$. Then, for any integer ν,

$$X_{qx\nu} H_q^* X_{qw\nu}^{-1} = : H'_q \in \boldsymbol{H}_q(N, N')$$

and, if $S_\nu := H'_q(H_q^*)^{-1}$, then $S_\nu \in \Gamma_0(N, N')$ and the bottom right-hand entry δ_ν in S_ν satisfies

$$\delta_\nu \equiv 1 \ (\mathrm{mod}(q', N)), \quad \delta_\nu \equiv d/a \ (\mathrm{mod}\, q). \tag{8.4.19}$$

The proof is straightforward. Observe that $H_q(Nn, N'n)$ is a subset of $\boldsymbol{H}_q(N, N')$.

Theorem 8.4.3. *Let (8.4.1, 2) hold and suppose that*

$$q = mn, \quad n = p^e = pl, \tag{8.4.20}$$

where p is prime, $e > 0$ and $p \nmid m$. Let H_q be the matrix given in (8.4.11) and write

$$H_q^* = \begin{bmatrix} lmq\lambda & N'\lambda'(1+q\lambda) \\ N & pq \end{bmatrix}, \tag{8.4.21}$$

so that, by (8.4.12), $H_q^* \in \boldsymbol{H}_q(N, N')$. For any positive integers μ and ν, write

$$\mu^* = \mu - \lambda\lambda', \tag{8.4.22}$$

and

$$A(\mu,\nu)=J_nU^{\mu N'}H_qJ_nU^{\nu N'}. \qquad (8.4.23)$$

Then, for any integers μ_0, μ, ν_0, ν satisfying

$$\mu_0^*-\mu^*\equiv q'^2m(\nu-\nu_0)\mu_0^*\mu^* \pmod n, \qquad (8.4.24)$$

we have

$$A(\mu,\nu)\{A(\mu_0,\nu_0)\}^{-1}=:S\in\Gamma_0(N,N'), \qquad (8.4.25)$$

where

$$\delta\equiv 1+(\nu-\nu_0)mq'^2\mu_0^* \pmod{NN'}. \qquad (8.4.26)$$

Further, if $\mu^*=p\rho$, for some integer ρ, then

$$A(\mu,\nu)=pB(\rho,\nu), \qquad (8.4.27)$$

where

$$B(\rho,\nu)=J_lU^{\rho N'}H_q^*J_lU^{\nu N'}. \qquad (8.4.28)$$

Proof. From (8.4.23, 25) we have, for any integers μ_0, μ, ν_0, ν, with $\kappa=\nu-\nu_0$, and noting that

$$\lambda+\mu q'\equiv-\mu^*/\lambda'\equiv q'\mu^* \pmod q,$$

$$\alpha=q\lambda+\mu qq'-q'(\lambda'+\mu q)-\kappa mq'(\lambda+\mu q'),$$

we have

$$\gamma=-\kappa Nq',$$

$$\delta=-q'(\lambda'+\mu_0q)+(q+\kappa mq')(\lambda+\mu_0q')$$

$$\equiv 1+\kappa mq'^2\mu_0^* \pmod{NN'},$$

and

$$\frac{n\beta}{N'}=-(\lambda+\mu q')(\lambda'+\mu_0q)+(\lambda+\mu_0q')(\lambda'+\mu q)$$

$$+\kappa m(\lambda+\mu q')(\lambda+\mu_0q')$$

$$\equiv q'^2\{m\kappa\mu^*\mu_0^*-\lambda'^2(\mu_0^*-\mu^*)\} \pmod q.$$

It follows that, if (8.4.24) holds, then $S\in\Gamma_0(N,N')$ and (8.4.26) holds.

Finally, if $\mu^*=p\rho$ and $B(\rho,\nu)$ is defined by (8.4.21, 28), it follows easily that (8.4.27) holds.

8.5. The operator L_q. In this section we assume that q, q', N and N' are integers satisfying (8.4.1). We no longer assume that (8.4.2) holds, but shall suppose that q divides N and write

$$N = qq_1. \tag{8.5.1}$$

We define an operator $L_q(N, N')$ that acts as a projection operator on the space of modular forms belonging to the group $\Gamma_0(N, N')$. For this purpose we write

$$\Gamma_0(q_1, N') = \Gamma_0(q_1, N'q) \cdot \boldsymbol{Z}_q(N, N'), \tag{8.5.2}$$

where $\boldsymbol{Z}_q$ is any right transversal for which this holds. Then, by (1.4.28),

$$|\boldsymbol{Z}_q(N, N')| = \psi(NN')/\psi(q_1N').$$

When q is a prime number, this number is q or $q+1$ according as q does or does not divide q_1N'. In the former case we may take $\boldsymbol{Z}_q$ to consist of the q matrices

$$U^{\nu N'} \quad (0 \le \nu < q). \tag{8.5.3}$$

(This, in fact, holds even when q is not a prime, provided that every prime divisor of q divides q_1N'.) In the latter case (i.e. when $q \nmid q_1N'$), we may take $\boldsymbol{Z}_q$ to consist of the q matrices in (8.5.3), together with any member S of $J_q^{-1}\boldsymbol{H}_q(N, N')$, such as, for example, $S = U^{\lambda' N'} W^{q_1}$; see (8.4.11). For, since $SU^{-\nu N'} \notin \Gamma_0(q_1, N'q)$, $\boldsymbol{Z}_q$ is in fact a right transversal.

We return to the general case and note that $\boldsymbol{Z}_q$ can be chosen so that, in addition to (8.5.2),

$$\Gamma_0(q_1 s, N'r) = \Gamma_0(q_1 s, N'rq) \cdot \boldsymbol{Z}_q \tag{8.5.4}$$

holds, where r and s are any fixed integers for which $(q, rs) = 1$. For, if $\boldsymbol{Z}_q$ is chosen to satisfy (8.5.4), then

$$|\boldsymbol{Z}_q| = \psi(NN'sr)/\psi(q_1N'sr) = \psi(NN')/\psi(q_1N')$$

and, moreover,

$$\Gamma_0(q_1, N'q)\boldsymbol{Z}_q \subseteq \Gamma_0(q_1, N').$$

Further, if for any two members X and Y of $\boldsymbol{Z}_q$ we have

$$XY^{-1} \in \Gamma_0(q_1, N'q),$$

then

$$XY^{-1} \in \Gamma_0(q_1, N'q) \cap \Gamma_0(q_1 s, N'r) = \Gamma_0(q_1 s, N'qr),$$

so that $X = Y$. Thus (8.5.2) holds with this choice of $\boldsymbol{Z}_q$.

Now let χ be any character modulo NN' whose conductor N_χ divides q_1N', and define, for each $f \in M(\Gamma_0(N, N'), k, \chi)$ and positive integer q,

$$f|L_q^\chi(N, N') = \frac{\psi(q_1N')}{\psi(NN')} \sum_{T \in \boldsymbol{Z}_q} \bar{\chi}(T) f|J_q T J_q^{-1}, \tag{8.5.5}$$

where $\boldsymbol{Z}_q$ is any transversal satisfying (8.5.2). When no confusion can arise, we shall write $f|L_q^\chi$ in place of $f|L_q^\chi(N, N')$.

We observe first that (8.5.5) is independent of the choice of the transversal $\boldsymbol{Z}_q$. For let $T' = ST$, where $S \in \Gamma_0(q_1, N'q)$ and $T \in \Gamma_0(q_1, N')$. Then $J_q S J_q^{-1} \in \Gamma_0(N, N')$ and

$$\chi(T') = \chi(S)\chi(T),$$

since $N_\chi | q_1N'$. Hence

$$\begin{aligned} \bar{\chi}(T') f|J_q T' J_q^{-1} &= \bar{\chi}(S)\bar{\chi}(T) f|J_q S J_q^{-1} \,.\, J_q T J_q^{-1} \\ &= \bar{\chi}(T) f|J_q T J_q^{-1}. \end{aligned}$$

We now show that $f|L_q^\chi(N, N')$ belongs to the group $\Gamma_0(N, N')$ and that $f|L_q^\chi(N, N')J_q$ belongs to $\Gamma_0(q_1, N')$. Take any $S \in \Gamma_0(q_1, N')$ and note that, by theorem 8.3.2, $f|J_q$ belongs to $M(\Gamma_0(q_1, N'q), k, \chi)$. As T runs through $\boldsymbol{Z}_q$, TS runs through a right transversal also. Since

$$\bar{\chi}(T) = \chi(S)\bar{\chi}(TS),$$

(8.5.5) shows that $f|L_q^\chi(N, N')J_q \in M(\Gamma_0(q_1, N'), k, \chi)$. It follows from theorem 8.3.2 that $f|L_q^\chi(N, N') \in M(\Gamma_0(N, N'), k, \chi)$.

When f belongs to $\Gamma_0(N, N')$ it belongs also to the subgroups $\Gamma_0(Ns, N'r)$, and (8.5.4) shows that

$$f|L_q^\chi(Ns, N'r) = f|L_q^\chi(N, N'), \quad \text{when} \quad (q, rs) = 1. \tag{8.5.6}$$

Thus, in situations of this kind it is safe to write $f|L_q^\chi$ in place of $f|L_q^\chi(N, N')$.

Further, if $(q, r) = 1$, we may, by (8.5.4), choose the transversal $\boldsymbol{Z}_q$ in (8.5.2) to satisfy

$$\Gamma_0(q_1r, N') = \Gamma_0(q_1r, N'q) \cdot \boldsymbol{Z}_q.$$

It then follows that

$$\Gamma_0(q_1, N'r) = \Gamma_0(q_1, N'qr) \cdot J_r^{-1} \boldsymbol{Z}_q J_r,$$

so that, by (8.5.4),

$$\Gamma_0(q_1, N') = \Gamma_0(q_1, N'q) \cdot J_r^{-1} \boldsymbol{Z}_q J_r,$$

and therefore

$$f|L_q^\chi(N,N')=f|J_r^{-1}L_q^\chi(Nr,N')J_r. \tag{8.5.7}$$

Here, of course, χ is any character whose conductor divides q_1N'.

For different values of q the operators L_q^χ commute with each other. In fact, we show that, if

$$N=qq_1=rr_1=qrs, \tag{8.5.8}$$

where

$$(q,r)=1, \quad N_\chi|N's, \tag{8.5.9}$$

then

$$f|L_{qr}^\chi(N,N')=(f|L_q^\chi(N,N'))|L_r^\chi(N,N'), \tag{8.5.10}$$

whenever $f\in M(\Gamma_0(N,N'),k,\chi)$.

To prove this choose transversals $\boldsymbol{Z}_q$ and $\boldsymbol{Z}_r$ satisfying

$$\Gamma_0(s,N'q)=\Gamma_0(s,N'qr)\cdot\boldsymbol{Z}_r, \quad \Gamma_0(s,N'r)=\Gamma_0(s,N'qr)\cdot\boldsymbol{Z}_q,$$

so that $\Gamma_0(s,N')=\Gamma_0(s,N'r)\cdot\boldsymbol{Z}_r$ and therefore

$$\Gamma_0(s,N')=\Gamma_0(s,N'qr)\cdot\boldsymbol{Z}_q\cdot\boldsymbol{Z}_r.$$

We note also that, if $f\in M(\Gamma_0(N,N'),k,\chi)$, we have, by theorem 8.3.2, that $f|J_r$ belongs to the group $\Gamma_0(r_1,N'r)$ and, by what we have already proved and (8.5.6), $f|L_q^\chi(N,N'r)J_q$ belongs to the group $\Gamma_0(q_1,N'q)$. Hence

$$\begin{aligned}
f|L_{qr}^\chi(N,N') &= \frac{\psi(sN')}{\psi(NN')}\sum_{T_r\in\boldsymbol{Z}_r}\sum_{T_q\in\boldsymbol{Z}_q}\bar{\chi}(T_qT_r)f|J_r\,.\,J_qT_qJ_q^{-1}\,.\,J_qT_rJ_{qr}^{-1}\\
&= \frac{\psi(sN')}{\psi(q_1N')}\sum_{T_r\in\boldsymbol{Z}_r}\bar{\chi}(T_r)f|J_rL_q^\chi(r_1,N'r)\,.\,J_qT_rJ_{qr}^{-1}\\
&= \frac{\psi(sN')}{\psi(q_1N')}\sum_{T_r\in\boldsymbol{Z}_r}\bar{\chi}(T_r)f|L_q^\chi(N,N'r)J_r\,.\,J_qT_rJ_{qr}^{-1}\\
&= \frac{\psi(sN')}{\psi(q_1N')}\sum_{T_r\in\boldsymbol{Z}_r}\bar{\chi}(T_r)f|L_q^\chi(N,N'r)J_q\,.\,J_rT_rJ_r^{-1}\,.\,J_q^{-1}\\
&= f|L_q^\chi(N,N'r)J_qL_q^\chi(q_1,N'q)J_q^{-1}\\
&= f|L_q^\chi(N,N'r)L_r^\chi(N,N'q)\\
&= f|L_q^\chi(N,N')L_r^\chi(N,N'),
\end{aligned}$$

by (8.5.6).

In certain cases the operator L_q^χ leaves a function invariant. For example, suppose that $f \in M(\Gamma_0(N, N'), k, \chi)$ and that

$$f|J_qU^{N'} = f|J_q. \tag{8.5.11}$$

Then $f|J_q$ belongs to the group generated by $U^{N'}$ and $\Gamma_0(q_1, N'q)$, by theorem 8.3.2, and so, by theorem 1.4.5, $f|J_q$ belongs to $\Gamma_0(q_1, N')$. It then follows from (8.5.5) that

$$f|L_q^\chi(N, N') = f. \tag{8.5.12}$$

In particular, since $\boldsymbol{Z}_qU^{N'}$ is a transversal in (8.5.2) whenever $\boldsymbol{Z}_q$ is, we deduce that, for all $f \in M(\Gamma_0(N, N'), k, \chi)$,

$$(f|L_q^\chi)|L_q^\chi = f|L_q^\chi. \tag{8.5.13}$$

We summarize these results in

Theorem 8.5.1. *Let χ be a character modulo NN' whose conductor divides q_1N'. Then*

(i) *the linear operator $L_q^\chi(N, N')$ is well defined by (8.5.2) and maps $M(\Gamma_0(N, N'), k, \chi)$ into itself, while $L_q^\chi(N, N')J_q$ maps this space into $M(\Gamma_0(q_1, N'), k, \chi)$.*

(ii) *The operator L_q^χ is a projection operator in the sense that (8.5.13) holds.*

(iii) *(8.5.7) holds whenever $(q, r) = 1$.*

(iv) *If f is a holomorphic, entire or a cusp form, so is $f|L_q^\chi$.*

(v) *If (8.5.11) holds, then so does (8.5.12).*

(vi) *The operators L_q^χ commute in the sense that (8.5.10) holds under the assumptions (8.5.8, 9).*

We now show that, for coprime q and r, the operators L_q^χ and H_r commute.

Theorem 8.5.2. *Suppose that*

$$N = qq_1, \quad NN' = rr',$$

where $(q, r) = (r, r') = 1$ and let $\chi \mapsto \chi^*$ be the involution corresponding to the divisor r. Then, if $f \in M(\Gamma_0(N, N'), k, \chi)$ and $H_r \in \boldsymbol{H}_r(N, N')$,

$$f|L_q^\chi H_r = f|H_rL_q^{\chi^*}.$$

In particular, if χ is of real type modulo r, then the operators L_q^χ and H_r commute.

Proof. We take a transversal $\boldsymbol{Z}_q$ satisfying (8.5.2) and any $H_r \in \boldsymbol{H}_r(N, N')$. Write $K_r := J_q^{-1} H_r J_q$, so that

$$K_r \in \boldsymbol{H}_r(q_1, N'q) \subseteq \boldsymbol{H}_r(q_1, N'),$$

by (8.4.4). Also put

$$\boldsymbol{Z}_q^* := K_r^{-1} \boldsymbol{Z}_q K_r.$$

Then $\boldsymbol{Z}_q^*$ is also a right transversal. For $|\boldsymbol{Z}_q^*| = |\boldsymbol{Z}_q|$ and, if $S \in \boldsymbol{Z}_q$, then $S \in \Gamma_0(q_1, N')$, so that

$$K_r^{-1} S K_r \in \Gamma_0(q_1, N'),$$

by the remarks preceding (8.4.6).

Further, if for

$$X^* = K_r^{-1} X K_r \in \boldsymbol{Z}_q^*, \quad Y^* = K_r^{-1} Y K_r \in \boldsymbol{Z}_q^*$$

we had

$$T = X^*(Y^*)^{-1} \in \Gamma_0(q_1, N'q),$$

then it would follow that

$$XY^{-1} = K_r T K_r^{-1} \in \Gamma_0(q_1, N'q),$$

again by the remarks preceding (8.4.6). This implies that $X = Y$, since X and Y belong to $\boldsymbol{Z}_q$. Thus $\boldsymbol{Z}_q^*$ is a transversal.

Finally, for $S \in \boldsymbol{Z}_q$, write $S^* = K_r^{-1} S K_r$ and note that, by (8.4.15),

$$\chi(S) = \chi^*(S^*).$$

Accordingly,

$$\begin{aligned} f|L_q^\chi(N, N')H_q &= \frac{\psi(q_1 N')}{\psi(NN')} \sum_{S \in \boldsymbol{Z}_q} \bar{\chi}(S) f|J_q S J_q^{-1} H_r \\ &= \frac{\psi(q_1 N')}{\psi(NN')} \sum_{S^* \in \boldsymbol{Z}_q^*} \bar{\chi}^*(S) f|H_r J_q S^* J_q^{-1} \\ &= f|H_r L_q^{\chi^*}(N, N'), \end{aligned}$$

as required.

8.6. The conjugate linear map *K*. In this section we consider a map K that is closely connected with the matrix J introduced in

(1.2.23). If f is any function defined on $\mathbb{H}$ as domain, we define the function $f|K$ on $\mathbb{H}$ by

$$f^K(z) := f(z)|K = \overline{f(-\bar{z})}. \tag{8.6.1}$$

The mapping K is conjugate linear in the sense that, if f and g are two functions defined on $\mathbb{H}$ and λ, μ are complex constants, then

$$(\alpha f + \beta g)|K = \bar{\alpha}(f|K) + \bar{\beta}(g|K).$$

Now let L be any member of Ω_0; i.e. L is a matrix with integral entries and positive determinant. We examine the interaction of the operators K and L. It is easily verified that

$$L(z)|K =: -L^*(z) \tag{8.6.2}$$

and

$$(L:z)|K = (L^*:z), \tag{8.6.3}$$

where

$$L^* = J^{-1}LJ. \tag{8.6.4}$$

Now let $f \in M(\Gamma, k, v)$, where Γ is a subgroup of $\Gamma(1)$, $k \in \mathbb{Z}$ and v is a multiplier system of weight k for Γ. Then, by (4.3.21) and (8.6.2, 3),

$$\begin{aligned}\{f(z)|L\}|K &= (\det L)^{\frac{1}{2}k}(L^*:z)^{-k} f^K(L^*(z)) \\ &= f^K(z)|L^*,\end{aligned}$$

which we may write as

$$f|LK = f|KL^*. \tag{8.6.5}$$

If we take $L = T \in \Gamma$ in (8.6.5), we obtain

$$\overline{v(T)} f^K = f^K|T^*,$$

which shows that

$$f^K \in M(\Gamma^*, k, v^*), \tag{8.6.6}$$

where

$$\Gamma^* = J^{-1}\Gamma J = \Gamma^J \tag{8.6.7}$$

and

$$v^*(S) = \overline{v(J^{-1}SJ)} \quad (S \in \Gamma^*). \tag{8.6.8}$$

Since $(f^K)^K = f$, it is clear that the mapping K is an involution. Further, if for sufficiently large $y = \text{Im } z > 0$,

$$f(z) = \sum_{m=N}^{\infty} a_m e^{2\pi imz/n},$$

then

$$f^K(z) = \sum_{m=N}^{\infty} \bar{a}_m e^{2\pi imz/n},$$

and it is clear that f^K is a holomorphic, entire or cusp form according as f is.

In particular, if $f \in M(\Gamma_0(N, N'), k, \chi)$, where $k \in \mathbb{Z}$ and χ is a character modulo NN', then

$$f|K \in M(\Gamma_0(N, N'), k, \bar{\chi})$$

by (8.6.8). If q and q' satisfy (8.4.1, 2), it follows that $f|H_q$ and $f|KH_{q'}$ are forms with the same character χ^*, given by (8.4.16). In particular, $f|K$ and $f|H_{NN'}$ have the same character $\bar{\chi}$.

We note finally that the operator K maps Poincaré series into Poincaré series. For, by means of (8.6.2, 3, 4, 7, 8), we deduce from (5.1.9) that

$$G_L(z; m; \Gamma, k, v)|K = G_{L^*}(z; m; \Gamma^*, k, v^*). \qquad (8.6.9)$$

This holds even when k is not an integer. Both G_L and $G_L|K$ have the same cusp width and cusp parameter.

It may be recalled that we have already, in the proof of theorem 6.2.1, obtained and used (8.6.9) in the case when $\Gamma = \Gamma(1)$.

8.7. Historical remarks. The operators R_χ, D_t, $D_t^{(\varepsilon)}$ and K all originate in the works of Hecke, even when, as in the case of D_t, they do not appear explicitly as operators. The classification of forms by divisor and character is also due to him, as are many of the more ingenious arguments, such as that employed in the proof of theorem 8.3.4.

Operators similar to H_q appear in various forms in the literature, especially in the simplest case when $N' = t = 1$, in which case we may take $q = N'$, $q' = 1$, $x = w = 0$, $y = -1$, $z = 1$, so that

$$H_N = \begin{bmatrix} 0 & -1 \\ N & 0 \end{bmatrix};$$

see for example Hurwitz's 1882 paper (Hurwitz (1932), vol. 1, p. 98), and Klein and Fricke (1892), *passim*. However, it is in the papers of Hecke that this operator appears first to have been fully exploited; see also Atkin and Lehner (1970).

The contents of §§8.4 and 8.5 are largely based on the work of Atkin and Lehner just mentioned, although developed in a more formal way; these authors restricted their attention to the case when $t_1 = 1$ and χ is the principal character modulo N. In particular, the operator L_q^χ appears to have been employed first (in the particular case mentioned) by Atkin and Lehner. Both R^χ and L_q^χ (or more precisely $L_q^\chi J_q$) are examples of the trace operators defined in (4.2.9, 10).

Theorem 8.4.3 is a generalization of a result of Ogg (1969*a*).

The results obtained in this chapter become very simple when the level N is 1, so that the weight k is even. Characters are then irrelevant and the spaces $M(N, k, \chi, t)$ reduce to the single space $M(1, k)$.

When $N = 2$ we have again restricted the weight k to be even. Here there is only one character, namely the principal character χ_0, and $M(N, k, \chi, t)$ is either $M(2, k, \chi_0, 1)$ or $M(2, k, \chi_0, 2)$.

The reason why we have assumed k to be even in these two cases is, of course, because there exists no constant multiplier system on $\Gamma(1)$ or $\Gamma(2)$ when k is odd. It may be noted, however, that a form of level 1 or 2 and odd weight can be regarded as a form of level $N > 2$ (e.g. $N = 4$), and so is included in the theory provided that its MS is constant on $\Gamma(N)$.

Alternatively, when $N = 2$ and k is odd, we could develop the theory in a similar way to that done by considering the space $M(\tilde{\Gamma}(2), k, \tilde{v})$, where $\tilde{\Gamma}(2)$ and $\tilde{v}$ are defined in (3.3.19, 20). Characters are not relevant here, but forms of divisors 1 and 2 can be defined as in §8.1. For $\tilde{v}^U = \tilde{v}$ since $\tilde{\Gamma}(2)$ is normal in $\Gamma(1)$.

9: Hecke operators and congruence groups

9.1. Double coset modules. In this section we study double cosets of the form $\Gamma(N)L\Gamma(N)$, where L belongs to the set Ω_n of matrices of order $n \in \mathbb{Z}^+$. We then introduce transversals $\boldsymbol{T}_n$, that satisfy $\Omega_n = \Gamma(1) \cdot \boldsymbol{T}_n$ and have other desirable properties in relation to the integer N. In §9.2 this will be used to define a family of linear ('Hecke') operators acting on modular forms associated with congruence groups of level N.

As indicated, we write, for each positive integer n,

$$\Omega_n := \{T: a, b, c, d \in \mathbb{Z}, \det T = n\}, \tag{9.1.1}$$

so that Ω_n is the set of all matrices of order n. For each $T \in \Omega_n$ we define

$$\operatorname{div} T = (a, b, c, d), \tag{9.1.2}$$

the right-hand side being the highest factor of the entries of T. The subset of primitive matrices of order n is then, as in (4.3.6),

$$\Omega_n^* = \{T \in \Omega_n : \operatorname{div} T = 1\}. \tag{9.1.3}$$

We also write

$$\Omega_0 = \bigcup_{n=1}^{\infty} \Omega_n, \quad \Omega_0^* = \bigcup_{n=1}^{\infty} \Omega_n^*. \tag{9.1.4}$$

Then Ω_0 is a semigroup under matrix multiplication and is a subset of the group Ω^+ defined in (4.3.7); Ω_0^* is not a semigroup, since the product of primitive matrices need not be primitive.

When $T \in \Omega_0$ and $\operatorname{div} T = h$ we have

$$T = hT^*, \tag{9.1.5}$$

where $\operatorname{div} T^* = 1$ and $T^* \in \Omega_0^*$; we call T^* the primitive matrix associated with T. The matrices T and T^* determine the same bilinear transformation. Note that

$$\det T^* = (\det T)/(\operatorname{div} T)^2. \tag{9.1.6}$$

Lemma. *Let S and T belong to Ω_0. Then*

$$\operatorname{div} ST = \operatorname{div} S \operatorname{div} T \operatorname{div} S^*T^* \tag{9.1.7}$$

and

$$\operatorname{div} S^*T^* \quad \textit{divides} \quad (\det S^*, \det T^*). \tag{9.1.8}$$

Further, if $S \in \Gamma(1)$, then

$$\operatorname{div} ST = \operatorname{div} TS = \operatorname{div} T. \tag{9.1.9}$$

Proof. The equation (9.1.7) is obvious. To prove (9.1.8) we may assume that S and T are primitive. We write $h = \operatorname{div} ST$, so that we have to prove that $h|(s, t)$, where $s = \det S$, $t = \det T$. Write

$$T' = \begin{bmatrix} d & -b \\ -c & a \end{bmatrix}.$$

Since h divides each entry in ST, it divides each entry in $STT' = tS$. Hence $h|t$; the proof that $h|s$ is similar.

If $S \in \Gamma(1)$, then $S = S^*$ and (9.1.8) shows that $\operatorname{div} ST^* = 1$. It follows that $\operatorname{div} ST = \operatorname{div} T$ and, similarly, $\operatorname{div} TS = \operatorname{div} T$.

From now on we assume that N is a fixed positive integer. Take $L \in \Omega_l$, where $l \in \mathbb{Z}^+$, and write

$$l^* := \det L^* = l/\lambda^2, \tag{9.1.10}$$

where

$$\lambda = \operatorname{div} L. \tag{9.1.11}$$

As in (4.3.14) (with $\Gamma = \Gamma(N)$), we write

$$\Gamma_L = \Gamma(N) \cap L^{-1}\Gamma(N)L, \tag{9.1.12}$$

and note that

$$\Gamma_L = \Gamma_{L^*}. \tag{9.1.13}$$

Theorem 9.1.1. *If $L \in \Omega_l$, then Γ_L is conjugate in $\Gamma(1)$ to the subgroup*

$$\Gamma(N, l^*) := \Gamma(N) \cap \Gamma^0(l^*N), \tag{9.1.14}$$

which has index

$$\psi_N(l^*) := \psi(Nl^*)/\psi(N) \tag{9.1.15}$$

in $\Gamma(N)$.

Proof. In view of (9.1.13), we may assume that L is primitive. By theorem 4.3.6, there exist $S_1, S_2 \in \Gamma(1)$ such that

$$L = S_1 J_l S_2.$$

It follows from (9.1.12) that

$$\Gamma_L = S_2^{-1}\{\Gamma(N) \cap J_l^{-1}\Gamma(N)J_l\}S_2.$$

We may therefore, without loss of generality, take L to be J_l.

Clearly $T \in \Gamma(N) \cap J_l^{-1}\Gamma(N)J_l$ if and only if

$$a \equiv d \equiv 1 \pmod N, \quad b \equiv 0 \pmod{Nl} \quad \text{and} \quad c \equiv 0 \pmod N,$$

so that

$$\Gamma(N) \cap J_l^{-1}\Gamma(N)J_l = \Gamma(N) \cap \Gamma^0(Nl) = \Gamma(N, l).$$

It remains to show that

$$[\Gamma(N) : \Gamma(N, l)] = \psi_N(l),$$

as defined by (9.1.15). Now

$$\psi_N(l) = [\Gamma^0(N) : \Gamma^0(Nl)]$$

and the required result will follow from the second isomorphism theorem if we show that

$$\Gamma(N)\Gamma^0(Nl) = \Gamma^0(N).$$

This follows immediately from theorem 1.4.4 on taking $n = Nl$.

The relationship between these groups is shown in fig. 11.

We now define

$$(L) := \Gamma(N)L\Gamma(N) \quad (L \in \Omega_0) \tag{9.1.16}$$

and write $\mathcal{D}$ for the set of all such double cosets (L) $(L \in \Omega_0)$.

Theorem 9.1.2. *For $L \in \Omega_0$ let*

$$\Gamma(N) = \Gamma_L \cdot \boldsymbol{R}_L. \tag{9.1.17}$$

Then

$$(L) = \Gamma(N) \cdot L\boldsymbol{R}_L. \tag{9.1.18}$$

Moreover, it is possible to find a common transversal $\mathcal{T}_L$ such that

$$(L) = \Gamma(N) \cdot \boldsymbol{T}_L = \boldsymbol{T}_L \cdot \Gamma(N).$$

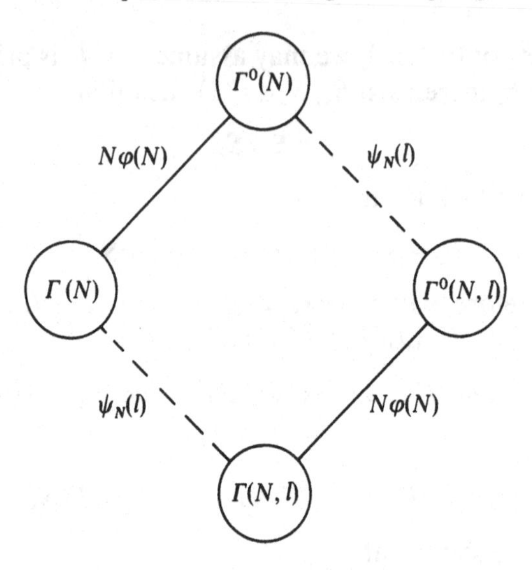

Fig. 11. Groups between $\Gamma(N, l)$ and $\Gamma^0(N)$.

Proof. For S_1 and S_2 in $\Gamma(N)$ we have

$$\Gamma(N)LS_1 = \Gamma(N)LS_2$$

if and only if $S_2S_1^{-1} = L^{-1}SL$ for some $S \in \Gamma(N)$, i.e. if and only if $S_2S_1^{-1} \in \Gamma_L$.

To prove the last part let A and B be any sets of matrices for which

$$(L) = \Gamma(N) \cdot LA = BL \cdot \Gamma(N).$$

Clearly $|A| = |B| = n$, say. Let A_i, B_i $(1 \le i \le n)$ be the members of A and B, respectively. Then the set

$$\boldsymbol{T}_L = \{B_iLA_i : i = 1, 2, \ldots, n\}$$

has the required properties.

For every $L \in \Omega_0$ we define

$$[L] := \{T \in \Omega_0 : T \equiv L \ (\text{mod}\ N), \det T = \det L, \operatorname{div} T = \operatorname{div} L\}. \tag{9.1.19}$$

Then we prove

Theorem 9.1.3. *For every $L \in \Omega_0^*$, $[L] = (L)$.*

Proof. Take $L \in \Omega_l^*$ and consider the set

$$G_l = \{(S_1, S_2) \in \Gamma(1) \times \Gamma(1) \colon S_1 J_l \equiv J_l S_2 \pmod{N}\}. \tag{9.1.20}$$

It is easily verified that G_l is a subgroup of $\Gamma(1) \times \Gamma(1)$ containing $\Gamma(N) \times \Gamma(N)$ as a normal subgroup.

Let $\theta(l)$ be the number of incongruent sets modulo N into which Ω_l^* splits. We denote elements of $\Gamma(1)$ by S_ν, with different suffixes ν. Any element of Ω_l^* can, by theorem 4.3.6, be expressed in the form $S_1 J_l S_2^{-1}$ and

$$S_1 J_l S_2^{-1} \equiv S_3 J_l S_4^{-1} \pmod{N}$$

if and only if

$$(S_1, S_2) \in (S_3, S_4) G_l.$$

Thus $\theta(l)$ is equal to the number of left cosets of G_l in $\Gamma(1) \times \Gamma(1)$ and so

$$\theta(l) = [\Gamma(1) \times \Gamma(1) \colon G_l] = \mu^2(N)/[G_l \colon \Gamma(N) \times \Gamma(N)], \tag{9.1.21}$$

where, as usual, $\mu(N) = [\Gamma(1) \colon \Gamma(N)]$.

We now evaluate $[G_l \colon \Gamma(N) \times \Gamma(N)]$, which is the number of incongruent pairs (S_1, S_2) satisfying the congruence

$$S_1 J_l \equiv J_l S_2 \pmod{N}. \tag{9.1.22}$$

This congruence is, in an obvious notation, equivalent to

$$\alpha_1 \equiv \alpha_2, \quad \beta_1 l \equiv \beta_2, \quad \gamma_1 \equiv \gamma_2 l, \quad \delta_1 l \equiv \delta_2 l \pmod{N}. \tag{9.1.23}$$

Write

$$N_1 = (N, l), \quad N_2 = N/N_1.$$

Then $\beta_2 \equiv 0 \pmod{N_1}$, so that necessarily $S_2 \in \Gamma^0(N_1)$. For each $S_2 \in \Gamma^0(N_1)$ we can choose α_1, $\gamma_1 \pmod{N}$ and $\beta_1 \pmod{N_2}$ to satisfy the first three congruences in (9.1.23). We then have, for any choice of δ_1,

$$\begin{aligned} \alpha_1\delta_1 - \beta_1\gamma_1 &\equiv \alpha_2\delta_1 - \beta_1\gamma_2 l \equiv \alpha_2\delta_1 - \beta_2\gamma_2 \pmod{N} \\ &\equiv 1 + \alpha_2(\delta_1 - \delta_2) \pmod{N}. \end{aligned}$$

Since $\alpha_2\delta_2 \equiv 1 \pmod{N_1}$, $(\alpha_2, N_1)=1$. In order to satisfy (9.1.22) we therefore have to choose δ_1 to satisfy

$$\delta_1-\delta_2 \equiv 0 \pmod{N_2}, \quad \delta_1-\delta_2 \equiv 0 \pmod{N/(N,\alpha_2)}.$$

Now the least common multiple of N_2 and $N/(N, \alpha_2)$ is N so that $\delta_1 \equiv \delta_2 \pmod N$. Since there are N_1 choices of $\beta_1 \pmod N$, it follows that, for each $S_2 \in \Gamma^0(N_1)$ we can find N_1 incongruent matrices $S_1 \pmod N$ to satisfy (9.1.24) and therefore

$$[G_l : \Gamma(N)\times\Gamma(N)] = N_1[\Gamma^0(N_1):\Gamma(N)] = \frac{N_1\mu(N)}{\psi(N_1)}.$$

Accordingly, by (9.1.21),

$$\theta(l) = \mu(N)\psi(N_1)/N_1.$$

Now write

$$\Omega_l^* = \Gamma(1)\cdot \boldsymbol{T}_l,$$

so that, by theorem 4.3.6, $|\boldsymbol{T}_l| = \psi(l)$. If $\Gamma(1)=\Gamma(N)\cdot \boldsymbol{R}$, we have

$$\Omega_l^* = \Gamma(N)\cdot\boldsymbol{R}\cdot\boldsymbol{T}_l,$$

where $|\boldsymbol{R}| = \mu(N)$. Thus Ω_l^* is a union of $\mu(N)\psi(l)$ right cosets of $\Gamma(N)$. Now Ω_l^* is also a union of double cosets (L) for $L\in\Omega_l^*$ and, by (9.1.18), each double coset (L) is a union of

$$|\boldsymbol{R}_L| = [\Gamma(N):\Gamma_L] = \psi_N(l)$$

right cosets of $\Gamma(N)$. It follows that Ω_l^* is a union of

$$\frac{\mu(N)\psi(l)}{\psi_N(l)} = \frac{\mu(N)\psi(l)\psi(N)}{\psi(Nl)}$$

double cosets (L). But this number is just $\theta(l)$, since it is easily verified that

$$N_1\psi(l)\psi(N) = \psi(N_1)\psi(Nl).$$

Now $(L)\subseteq[L]$ and, since $\theta(l)$ is the number of different sets $[L]$, the theorem follows.

Corollary 9.1.3. (i) *Let $L\in\Omega_0$ and suppose that* $(\operatorname{div} L, N)=1$. *Then* $[L]=(L)$.

(ii) *Let L_1, L_2 belong to Ω_0 and suppose that* $L_1\equiv L_2 \pmod N$, $\det L_1 = \det L_2$, $\operatorname{div} L_1 = \operatorname{div} L_2$ *and that* $(\operatorname{div} L_1, N)=1$. *Then* $(L_1)=(L_2)$.

Proof. (i) In the notation of (9.1.10, 11),

$$[L]=\{T=\lambda T^*: T^*\in \Omega_{l^*}^*, \lambda T^*\equiv\lambda L^* \pmod N\}.$$

Since $(\lambda, N)=1$, the last congruence reduces to $T^*\equiv L^* \pmod N$ and so

$$[L]=\lambda(L^*)=(\lambda L^*)=(L).$$

Part (ii) is immediate since $[L_1]=[L_2]$.

We now denote by $\mathcal{M}$ the free left-module of $\mathcal{D}$ over $\mathbb{Z}$. Each element of $\mathcal{M}$ is a sum of the form

$$\sum_{(L)\in\mathcal{D}} c_L(L),$$

where $c_L\in\mathbb{Z}$, and only a finite number of the coefficients c_L are non-zero. We wish to make $\mathcal{M}$ into a ring by defining multiplication. Note that, in this ring, $q(L)$ and (qL) have different meanings.

For this purpose, let (L) and (M) be any two members of $\mathcal{D}$ and take transversals $\boldsymbol{T}_L$, $\boldsymbol{T}_M$ such that

$$(L)=\Gamma(N)\cdot\boldsymbol{T}_L, \quad (M)=\Gamma(N)\cdot\boldsymbol{T}_M.$$

For each $X\in\Omega_0$ consider the equation

$$\Gamma(N)L_iM_j=\Gamma(N)X, \tag{9.1.24}$$

where $L_i\in\boldsymbol{T}_L$, $M_j\in\boldsymbol{T}_M$. Here i and j run through the positive integers not exceeding $|\boldsymbol{T}_L|$ and $|\boldsymbol{T}_M|$, respectively. The number of pairs (i,j) satisfying (9.1.24) we denote by $c(L,M;X)$, so that $c(L,M;X)$ is a non-negative integer.

Moreover, $c(L,M;X)$ is independent of the choice of the transversal $\boldsymbol{T}_L$, since left multiplication of L_i by a member of $\Gamma(N)$ does not alter (9.1.24). It is also independent of the choice of $\boldsymbol{T}_M$. For take any $S\in\Gamma(N)$. Then, since $L_iS\in(L)$, there exists a unique integer $k=\pi_S(i)$ such that

$$L_iS=S_iL_k,$$

where $S_i\in\Gamma(N)$. The mapping π_S is easily seen to be a permutation of the integers in the interval $[1,|\boldsymbol{T}_L|]$. When M_j is replaced by SM_j the equation (9.1.24) becomes

$$\Gamma(N)L_kM_j=\Gamma(N)X,$$

and it follows that $c(L,M;X)$ is unaltered by this replacement.

Similar arguments show that

$$c(L, M; S_1XS_2) = c(L, M; X)$$

for any S_1, S_2 in $\Gamma(N)$. Hence $c(L, M; X)$ depends only on the double coset (X) in which X lies.

We now define

$$(L)(M) := \sum_{(X)\in\mathcal{D}} c(L, M; X)(X). \tag{9.1.25}$$

The product is well defined and $c(L, M; X)$ is non-zero for a finite number of double cosets (X) only, since, by (9.1.24), $(X) = (L_iM_j)$.

The definition of multiplication is now extended to the whole of $\mathcal{M}$ in the obvious way, by distributivity. Thus, for any $L^{(i)}, M^{(j)}$ in $\mathcal{D}$,

$$\left\{\sum_{i=1}^{n} a_i(L^{(i)})\right\}\left\{\sum_{j=1}^{q} b_j(M^{(j)})\right\} = \sum_{i=1}^{n}\sum_{j=1}^{q} a_ib_j(L^{(i)})(M^{(j)}).$$

That multiplication is associative follows from the fact that

$$\sum_{(X)\in\mathcal{D}} c(L, M; X)c(X, R; Y) = \sum_{(X)\in\mathcal{D}} c(M, R; X)c(L, X; Y).$$

For each side is the number of triplets (i, j, k) satisfying

$$\Gamma(N)L_iM_jR_k = \Gamma(N)Y,$$

in an obvious notation. It follows that

$$(L)(M).(R) = (L).(M)(R).$$

In the next theorem we collect together some further properties of the coefficients $c(L, M; X)$.

Theorem 9.1.4. (i) *If L and M belong to Ω_0 and $L^* \in \Gamma(1)$, then*

$$(L)(M) = (LM) = L(M) \quad \textit{and} \quad (M)(L) = (ML) = (M)L,$$

so that

$$c(L, M; LM) = c(M, L; ML) = 1.$$

(ii) *For any L, M, $X \in \Omega_0$,*

$$c(L, M; X) = c(M', L'; X'), \tag{9.1.26}$$

where dashes denote transposes.

(iii) *Let* $L \in \Omega_l$, $M \in \Omega_m$. *Then*

$$\sum_{(X)\in\mathcal{D}} c(L, M; X)\psi_N(x^*) = \psi_N(l^*)\psi_N(m^*), \qquad (9.1.27)$$

where l^*, m^* *and* x^* *are the determinants of the primitive matrices* L^*, M^* *and* X^* *associated with* L, M *and* X, *respectively.*

Proof. (i) This is an immediate consequence of the normality of $\Gamma(N)$ in $\Gamma(1)$ and (9.1.24).

(ii) By theorem 9.1.2 we may take common transversals $\boldsymbol{T}_L$, $\boldsymbol{T}_M$ such that

$$(L) = \boldsymbol{T}_L \cdot \Gamma(N) = \Gamma(N) \cdot \boldsymbol{T}_L, \quad (M) = \boldsymbol{T}_M \cdot \Gamma(N) = \Gamma(N) \cdot \boldsymbol{T}_M$$

and put

$$\boldsymbol{T}_L = \{L_i : 1 \le i \le \psi_N(l^*)\}, \quad \boldsymbol{T}_M = \{M_j : 1 \le j \le \psi_N(m^*)\},$$

where $l = \det L$, $m = \det M$. Then $c(L, M; X)$ is the number of solutions (i, j) of the equation $X = SL_iM_j$ for some $S \in \Gamma(N)$. By arguments of the type used earlier, this equation is equivalent to

$$X = L_{\lambda(i)}M_{\mu(j)}T,$$

where $T \in \Gamma(N)$ and λ, μ are permutations depending on S.

Write $q = \lambda(i)$, $r = \mu(j)$. Then this is equivalent to

$$X' = T'M_r'L_q',$$

where we note that $T' \in \Gamma(N)$. Now

$$(L') = \Gamma(N) \cdot \boldsymbol{T}_L', \quad (M') = \Gamma(N) \cdot \boldsymbol{T}_M',$$

where $\boldsymbol{T}_L'$ and $\boldsymbol{T}_M'$ consist of the transpose matrices L_i' and M_j', respectively. From this (9.1.26) follows.

Finally, (iii) follows from (9.1.25) by enumerating the number of right cosets of $\Gamma(N)$ on each side of the equation.

We now particularize by taking

$$L = J_l, \quad M = J_m,$$

where l and m are any two positive integers. Take transversals $\boldsymbol{R}_l$, $\boldsymbol{R}_m$ such that

$$(J_l) = \Gamma(N) \cdot J_l\boldsymbol{R}_l, \quad (J_m) = \Gamma(N) \cdot J_m\boldsymbol{R}_m,$$

as in (9.1.18). If $c(J_l, J_m; X) \neq 0$ for any $X \in \Omega_0$ we must have, by (9.1.24),

$$X \in (J_l S J_m), \tag{9.1.28}$$

for some $S \in \Gamma(N)$, so that $X \in \Omega_{lm}$ and

$$\operatorname{div} X = \operatorname{div}(J_l S J_m).$$

Thus $\operatorname{div} X$ divides (l, m), by the lemma. Moreover, since the first entry in the matrix $J_l S J_m$ is α and $\alpha \equiv 1 \pmod N$, it follows that

$$(\operatorname{div} X, N) = 1.$$

We therefore take any integer q satisfying

$$q|(l, m), \quad (q, N) = 1$$

and investigate which double cosets (X) with $\operatorname{div} X = q$ can arise for which (9.1.28) holds.

By corollary 9.1.3(i) we have, since $X \equiv J_{lm} \pmod N$,

$$(J_l S J_m) = (X) = [X] = \{T \in \Omega_{lm} : T \equiv J_{lm} \pmod N, \operatorname{div} T = q\}.$$

Write $T = qT^*$. To solve the congruence $T \equiv J_{lm} \pmod N$ we make use of the matrices $R_q \in \Gamma(1)$ for which (see (8.1.5))

$$R_q \equiv \begin{bmatrix} q^{-1} & 0 \\ 0 & q \end{bmatrix} \pmod N. \tag{9.1.29}$$

Here, of course, q^{-1} denotes any residue r modulo N for which $qr \equiv 1 \pmod N$. Although R_q is not uniquely determined, $\Gamma(N)R_q$ and (R_q) are. Then $T \equiv J_{lm} \pmod N$ is equivalent to

$$T^* \equiv R_q J_{lm/q^2} \pmod N$$

and therefore there is a unique double coset (X) with $\operatorname{div} X = q$ satisfying (9.1.28), namely

$$(X) = (qR_q J_{lm/q^2}) = (qR_q)(J_{lm/q^2}).$$

Observe that the product on the right is independent of the order of the factors. For, by corollary 9.1.3(ii) and theorem 9.1.4(i),

$$(R_q)(J_n) = (R_q J_n) = (J_n R_q) = (J_n)(R_q) \tag{9.1.30}$$

for any $n \in \mathbb{Z}^+$ and q prime to N; further, by theorem 9.1.4(i), (qR_q) may be replaced by qR_q.

We now write $c(l, m; q)$ in place of $c(J_l, J_m; X)$, for this double coset (X). We have therefore proved the first part of the following theorem.

Theorem 9.1.5. *Let l and m be positive integers and, for each q prime to N take a matrix $R_q \in \Gamma(1)$ satisfying (9.1.29). Then there exist non-negative integers $c(l, m; q)$ such that*

$$(J_l)(J_m) = \sum_{\substack{q|(l,m)\\(q,N)=1}} c(l, m; q)(qR_q)(J_{lm/q^2}) \tag{9.1.31}$$

and

$$\psi_N(l)\psi_N(m) = \sum_{\substack{q|(l,m)\\(q,N)=1}} c(l, m; q)\psi_N(lm/q^2). \tag{9.1.32}$$

On the right-hand side of (9.1.31), $(qR_q)(J_{lm/q^2})$ can be replaced by $(J_{lm/q^2})(qR_q)$ and the double coset (qR_q) may be replaced by the matrix qR_q. Moreover,

$$c(l, m; q) = c(m, l; q), \tag{9.1.33}$$

so that

$$(J_l)(J_m) = (J_m)(J_l). \tag{9.1.34}$$

Further, if $(l, m) = 1$ or if (l, m) is composed entirely of primes dividing N, then $c(l, m; 1) = 1$ and

$$(J_l)(J_m) = (J_{lm}). \tag{9.1.35}$$

Finally, if p is a prime, $l = p$ and $m = p^\mu$ $(\mu > 1)$, then

$$(J_p)(J_{p^\mu}) = (J_{p^{\mu+1}}) + c_\mu(pR_p)(J_{p^{\mu-1}}), \tag{9.1.36}$$

where

$$c_\mu = 0 \quad \text{if } p|N, \quad c_\mu = \psi(p^\mu)/\psi(p^{\mu-1}) \quad \text{if } p \nmid N. \tag{9.1.37}$$

Proof. (9.1.32) follows from (9.1.31) by enumerating right cosets, as in the proof of theorem 9.1.4(iii). Also (9.1.33) follows from (9.1.26), since $J_l' = J_l$, $J_m' = J_m$ and since

$$\begin{aligned}(\{qR_qJ_{lm/q^2}\}') &= [\{qR_qJ_{lm/q^2}\}'] = [J'_{lm/q^2}qR'_q]\\ &= [J_{lm/q^2}qR_q] = [qR_qJ_{lm/q^2}]\\ &= (qR_qJ_{lm/q^2}).\end{aligned}$$

From (9.1.31, 33) we obtain (9.1.34).

Since (9.1.24) is always satisfied with $L_i = L$, $M_j = M$ and $X = LM$ we have $c(L, M; LM) \geq 1$. If (9.1.35) did not hold under the conditions stated, we should have, by (9.1.32),

$$\psi_N(l)\psi_N(m) > \psi_N(lm)$$

whereas, as is easily verified, equality holds in these circumstances; in particular, $c(l, m; 1) = 1$.

In view of this, it remains to prove (9.1.36) when $p \nmid N$. We have, from (9.1.32),

$$\psi_N(p)\psi_N(p^\mu) = \psi_N(p^{\mu+1}) + c(p, p^\mu; p)\psi_N(p^{\mu-1}).$$

Solving for $c_\mu := c(p, p^\mu; p)$ we obtain (9.1.37).

We now recall that, in the definition of $[L]$ in (9.1.19), the members of the set were restricted to having the same divisor as L. We now relax this restriction and put, for any $L \in \Omega_0$,

$$[\![L]\!] := \{T \in \Omega_0 : T \equiv L \pmod N, \det T = \det L\} \quad (9.1.38)$$

so that

$$(L) \subseteq [L] \subseteq [\![L]\!].$$

$[\![L]\!]$ is clearly a union of double cosets and so can be regarded as an element of the module $\mathcal{M}$.

Theorem 9.1.6. *Let $L \in \Omega_0^*$, so that there exist S_1, S_2 in $\Gamma(1)$ such that $L = S_1 J_l S_2$, where $l = \det L$. Then*

$$[\![L]\!] = \sum_{\substack{q^2 | l \\ (q,N)=1}} (S_1 q R_q J_{l/q^2} S_2). \quad (9.1.39)$$

In particular,

$$[\![J_l]\!] = \sum_{\substack{q^2 | l \\ (q,N)=1}} (qR_q)(J_{l/q^2}) = \sum_{\substack{q^2 | l \\ (q,N)=1}} (J_{l/q^2})(qR_q). \quad (9.1.40)$$

Moreover, for any m prime to N,

$$[\![J_l]\!] R_m = [\![J_l]\!](R_m) = (R_m)[\![J_l]\!] = R_m [\![J_l]\!]. \quad (9.1.41)$$

Further

$$[\![L]\!] = \Gamma(N) \cdot \boldsymbol{T}_L^+, \quad (9.1.42)$$

where

$$|\boldsymbol{T}_L^+| = \sigma(l; N) := \sum_{\substack{q|l \\ (N,l/q)=1}} q. \tag{9.1.43}$$

Moreover, for any positive integers l and m,

$$[\![J_l]\!][\![J_m]\!] = [\![J_m]\!][\![J_l]\!] \tag{9.1.44}$$

$$= \sum_{\substack{d|(l,m) \\ (d,N)=1}} d(dR_d)[\![J_{lm/d^2}]\!]. \tag{9.1.45}$$

In particular, for $l = p$, $m = p^{\mu}$, where p is a prime, $p \nmid N$, $\mu \geq 1$,

$$[\![J_p]\!][\![J_{p^\mu}]\!] = [\![J_{p^{\mu+1}}]\!] + p(pR_p)[\![J_{p^{\mu-1}}]\!]. \tag{9.1.46}$$

Proof. We note that in (9.1.39) the summation is carried out over all positive integers q such that $q^2|l$ and $(q, N) = 1$.

Take any $T \in [\![L]\!]$ and write $q = \text{div } T$. Then $q^2|l$ and

$$qT^* \equiv L \pmod{N}.$$

Clearly $(q, N) = 1$ as otherwise some prime divisor of both q and N would divide all the entries in L. Since

$$q^{-1}J_l \equiv R_q J_{l/q^2} \pmod{N},$$

we have

$$T^* \equiv S_1 R_q J_{l/q^2} S_2 \pmod{N}$$

and so

$$(T^*) = (S_1 R_q J_{l/q^2} S_2).$$

Since $[\![L]\!]$ is a union of all the sets $[T] = [qT^*]$ where $q_2|l$ and $(q, N) = 1$ and since $[T^*] = (T^*)$, (9.1.39) follows.

If $[\![L]\!] = \Gamma(N) \cdot \boldsymbol{T}_L^+$ it follows that

$$|\boldsymbol{T}_L^+| = \sum_{\substack{q^2|l \\ (q,N)=1}} \psi_N(l/q^2) = \sigma(l; N),$$

say. Clearly $\sigma(lm; N) = \sigma(l, N)\sigma(m; N)$ when $(l, m) = 1$, and $\sigma(l, N) = 1$ when l is composed entirely of primes dividing N. If $l = p^\lambda$, where p is prime and does not divide N, then

$$\sigma(p^\lambda; N) = \sum_{0 \leq \alpha \leq \frac{1}{2}\lambda} \psi(p^{\lambda - 2\alpha}).$$

By considering the cases when λ is odd and even separately we easily verify that

$$\sigma(p^\lambda; N) = 1 + p + p^2 + \cdots + p^\lambda$$

and (9.1.43) follows.

We now prove (9.1.46). By (9.1.40), $[\![J_p]\!] = (J_p)$ and

$$[\![J_{p^2}]\!] = (J_{p^2}) + (pR_p) = (J_p)(J_p) - p(pR_p),$$

by (9.1.36, 37). This gives (9.1.46) for $\mu = 1$. If $\mu > 1$ we have, by (9.1.40) and (9.1.36),

$$\begin{aligned}[\![J_p]\!][\![J_{p^\mu}]\!] &= \sum_{2\alpha \le \mu} (J_p)(pR_p)^\alpha (J_{p^{\mu-2\alpha}}) = \sum_{2\alpha \le \mu} (pR_p)^\alpha (J_p)(J_{p^{\mu-2\alpha}}) \\ &= \sum_{2\alpha \le \mu} (pR_p)^\alpha (J_{p^{\mu+1-2\alpha}}) + \sum_{2\alpha < \mu} c_{\mu-2\alpha} (pR_p)^{\alpha+1} (J_{p^{\mu-1-2\alpha}}).\end{aligned}$$

Now $c_{\mu-2\alpha} = p$ except in the case when μ is odd and $\alpha = \frac{1}{2}(\mu - 1)$, in which case $c_{\mu-2\alpha} = p + 1$, so that in the second summation on the right-hand side there is an extra term, namely $(pR_p)^{\frac{1}{2}(\mu+1)}$, which can be added as a final term to the first series arising from $\alpha = \frac{1}{2}(\mu + 1)$. Thus, in all cases,

$$\begin{aligned}[\![J_p]\!][\![J_{p^\mu}]\!] &= \sum_{2\alpha \le \mu+1} (pR_p)^\alpha (J_{p^{\mu+1-2\alpha}}) + p(pR_p) \sum_{2\alpha \le \mu-1} (pR_p)^\alpha (J_{p^{\mu-1-2\alpha}}) \\ &= [\![J_{p^{\mu+1}}]\!] + p(pR_p)[\![J_{p^{\mu-1}}]\!],\end{aligned}$$

which is (9.1.46).

From (9.1.30, 34, 40) we deduce (9.1.41, 44) and, further, by (9.1.35), that

$$[\![J_l]\!][\![J_m]\!] = [\![J_{lm}]\!] \quad \text{when} \quad (l, m) = 1.$$

It follows that we need only prove (9.1.45) when l and m are powers of the same prime p and $p \nmid N$.

Now define, for $q \ge 0$, $r \ge 1$

$$F(q, r) = [\![J_{p^q}]\!][\![J_{p^r}]\!] - [\![J_{p^{q+1}}]\!][\![J_{p^{r-1}}]\!].$$

Then, by applying (9.1.46) with $\mu = q$ and $\mu = r - 1$, we have for $r \ge 2$

$$F(q, r) = p(pR_p)F(q - 1, r - 1).$$

Now take $l = p^\lambda$, $m = p^\mu$, where $1 \le \nu \le \lambda \le \mu$; we deduce that

$$\begin{aligned}F(\lambda + \mu - \nu, \nu) &= p^{\nu-1}(pR_p)^{\nu+1} F(\lambda + \mu - 2\nu + 1, 1) \\ &= p^\nu (pR_p)^\nu [\![J_{p^{\lambda+\mu-2\nu}}]\!].\end{aligned}$$

We sum this for $\nu = 1, 2, \ldots, \lambda$ and deduce that

$$[\![J_{p^\lambda}]\!][\![J_{p^\mu}]\!] = \sum_{\nu=0}^{\lambda} p^\nu (pR_p)^\nu [\![J_{p^{\lambda+\mu-2\nu}}]\!],$$

which is (9.1.45).

Now write, for any $n \in \mathbb{Z}^+$,

$$[\![J_n]\!] = \Gamma(N) \cdot \boldsymbol{T}_n^+. \tag{9.1.47}$$

We conclude by obtaining canonical expressions for the right transversal $\boldsymbol{T}_n^+$.

We assume first of all that $(n, N) = 1$. Then, by theorem 4.3.6,

$$\Omega_n^* = \Gamma(1) \cdot \boldsymbol{T}_n^*,$$

where we may take

$$\boldsymbol{T}_n^* = \{T \in \Omega_n^* : c = 0, a > 0, ad = n, b = \nu N (0 \leq \nu < d)\}. \tag{9.1.48}$$

Here we have chosen b to be divisible by N, which is possible since $(d, N) = 1$. Only values of ν for which $(a, \nu, d) = 1$ are allowed, since div $T = 1$, but they do not necessarily have to be taken between 0 and $d - 1$. Now define

$$\boldsymbol{T}_n^0 := \{S : S = R_a T, T \in \boldsymbol{T}_n^*\}. \tag{9.1.49}$$

Then, if $S \in \boldsymbol{T}_n^0$, we have $S \equiv J_n \pmod N$. Moreover,

$$\Omega_n^* = \Gamma(1) \cdot \boldsymbol{T}_n^0$$

and therefore

$$\Gamma(N)\boldsymbol{T}_n^0 = \Gamma(N) \cdot \boldsymbol{T}_n^0.$$

But

$$\Gamma(N)\boldsymbol{T}_n^0 \subseteq [J_n] = (J_n)$$

and (J_n) is a disjoint union of $\psi_N(n) = \psi(n)$ right cosets of $\Gamma(N)$. Since $|\boldsymbol{T}_n^0| = \psi(n)$, it follows that

$$(J_n) = \Gamma(N) \cdot \boldsymbol{T}_n^0. \tag{9.1.50}$$

Now take any q with $q^2 | n$. Then, if $S = R_a T \in \boldsymbol{T}_{n/q^2}^0$, we have

$$qR_q S \equiv R_{aq} \,.\, qT \pmod N.$$

Hence, replacing aq by a, we have from (9.1.40) that (9.1.47) holds with

$$\boldsymbol{T}_n^+ = \{S : S = R_a T,\ T \in \Omega_n,\ c = 0,\ a > 0,\ ad = n,\ b = \nu N (0 \le \nu < d)\}. \qquad (9.1.51)$$

Here all values of ν between 0 and $d-1$ are allowed. This is the canonical form required.

If we write, for each positive divisor a of n,

$$\boldsymbol{A}_n^a = \{T \in \Omega_n : c = 0,\ b = \nu N (0 \le \nu < a)\} \qquad (9.1.52)$$

and

$$\boldsymbol{D}_n^a = \{T \in \Omega_n : c = 0,\ b = \nu N (0 \le \nu < d)\}, \qquad (9.1.53)$$

then we have

$$\boldsymbol{T}_n^+ = \bigcup_{a|n} R_a \boldsymbol{D}_n^a \quad \text{and put} \quad {}^+\boldsymbol{T}_n := \bigcup_{a|n} \boldsymbol{A}_n^a R_a. \qquad (9.1.54)$$

It can be shown, similarly, that ${}^+\boldsymbol{T}_n$ is a left transversal of $[\![J_n]\!]$ with respect to $\Gamma(N)$; i.e.

$$[\![J_n]\!] = \Gamma(N) \cdot \boldsymbol{T}_n^+ = {}^+\boldsymbol{T}_n \cdot \Gamma(N). \qquad (9.1.55)$$

Moreover,

$$\Omega_n = \Gamma(1) \cdot \boldsymbol{T}_n^+ = \boldsymbol{T}_n^+ \cdot \Gamma(1) = \Gamma(1) \cdot {}^+\boldsymbol{T}_n = {}^+\boldsymbol{T}_n \cdot \Gamma(1). \qquad (9.1.56)$$

For, if $S_1 T_1 = S_2 T_2$ where $S_r \in \Gamma(1)$, $T_r \in \boldsymbol{T}_n^+$ $(r = 1, 2)$, then

$$S_1 J_n \equiv S_2 J_n \pmod N, \quad \text{so that} \quad S_1 \equiv S_2 \pmod N,$$

since $(n, N) = 1$. Hence $S_2^{-1} S_1 \in \Gamma(N)$ and, by (9.1.47), the equation

$$S_2^{-1} S_1 \,.\, T_1 = I \,.\, T_2$$

implies that $S_1 = S_2$ and $T_1 = T_2$. Hence $\Gamma(1)\boldsymbol{T}_n^+ = \Gamma(1) \cdot \boldsymbol{T}_n^+$. Further, since by (9.1.43),

$$|\boldsymbol{T}_n^+| = \sigma(n; N) = \sigma_1(n),$$

it follows that $\Omega_n = \Gamma(1) \cdot \boldsymbol{T}_n^+$. The remaining parts of (9.1.56) are proved similarly or are obvious.

Finally, assume that n is composed entirely of primes dividing N. We denote this by

$$n \,|\, N^\infty. \qquad (9.1.57)$$

Then, as in the proof of theorem 9.1.1, for $L = J_n$,

$$\Gamma_L = \Gamma(N) \cap \Gamma^0(Nn).$$

The $n = \psi_N(n)$ matrices $U^{\nu N} (0 \le \nu < n)$ belong to different right cosets of Γ_L in $\Gamma(N)$. Hence we can take

$$\boldsymbol{T}_n^+ = \{S: S = J_n U^{\nu N} \ (0 \le \nu < n)\} \tag{9.1.58}$$

in (9.1.47).

Finally, we note that in (9.1.51, 53) the values of ν may run through any complete set of residues modulo d and need not necessarily lie in the range $0 \le \nu < d$. For example, we can replace ν by νx where x is any fixed integer prime to n. Similar remarks apply to (9.1.52, 58).

9.2. Definition and properties of Hecke operators. We continue the study of modular forms of level N begun in chapter 8 and make the same assumptions regarding N, the weight k and the multiplier system v as done in (8.1.1, 2). We define a family of Hecke operators T_n ($n \in \mathbb{Z}^+$) on $M(N, k)$ in the following way.

For any $n \in \mathbb{Z}^+$ we write

$$[\![J_n]\!] = \Gamma(N) \cdot \boldsymbol{T}_n^+, \tag{9.2.1}$$

as in (9.1.47), for any choice of the transversal $\boldsymbol{T}_n^+$. If $f \in M(N, k)$, we define

$$f|T_n := n^{\frac{1}{2}k-1} \sum_{T \in \boldsymbol{T}_n^+} f|T. \tag{9.2.2}$$

Thus,

$$f(z)|T_n = n^{k-1} \sum_{T \in \boldsymbol{T}_n^+} (T:z)^{-k} f(Tz). \tag{9.2.3}$$

Since, for any $S \in \Gamma(N)$ and $T \in \boldsymbol{T}_n^+$,

$$f|ST = (f|S)|T = f|T,$$

it is clear that $f|T_n$ is well defined and independent of the particular choice of the right transversal $\boldsymbol{T}_n^+$.

Theorem 9.2.1. *For all* $n \in \mathbb{Z}^+$,

$$M(N, k)|T_n \subseteq M(N, k). \tag{9.2.4}$$

Moreover, each of the subspaces $H(N, k)$, $\{N, k\}$ *and* $\{N, k\}_0$ *is also invariant under the operators* T_n.

The operators T_n $(n \in \mathbb{Z}^+)$ *commute with each other and with the operators* R_m *and* R^χ, *where* $(m, N) = 1$ *and* χ *is a character modulo* N. *Further, for any positive integers* l *and* m, *and any* $f \in M(N, k)$,

$$(f|T_l)|T_m = (f|T_m)|T_l = \sum_{\substack{d|(l,m)\\(d,N)=1}} d^{k-1}(f|R_d)|T_{lm/d^2}. \qquad (9.2.5)$$

Suppose that $f(z)$ *has the Fourier expansion (8.1.13) for* $y = \operatorname{Im} z > y_0 \geq 0$. *Then* $f|T_n$ *has the Fourier expansion*

$$f(z)|T_n = \sum_{r=-\infty}^{\infty} a_r(n)\, e^{2\pi i r z/N}, \qquad (9.2.6)$$

where

$$a_r(n) = a_{rn} \quad \text{when} \quad n|N^\infty. \qquad (9.2.7)$$

If $f \in M(N, k, \chi)$, *where* χ *is a character modulo* N *satisfying (8.1.7), then*

$$a_r(n) = \sum_{d|(r,n)} \chi(d) d^{k-1} a_{rn/d^2} \quad (n \in \mathbb{Z}^+). \qquad (9.2.8)$$

Proof. Take any $n \geq 1$ and $S \in \Gamma(N)$. Then, since

$$\Gamma(N) \cdot \boldsymbol{T}_n^+ S = [\![J_n]\!] S = [\![J_n]\!] = \Gamma(N) \cdot \boldsymbol{T}_n^+,$$

it follows from (9.2.2) that

$$(f|T_n)|S = f|T_n S = f|T_n.$$

The first part of the theorem now follows from theorem 4.3.9.

More generally, if $S \in \Gamma_0^0(N)$, then $S = S_1 R_m$ for some $S_1 \in \Gamma(N)$ and some m prime to N. Then, by (9.1.41),

$$[\![J_n]\!] S = [\![J_n]\!] R_m = R_m [\![J_n]\!],$$

so that

$$\Gamma(N) \cdot \boldsymbol{T}_n^+ S = \Gamma(N) \cdot R_m \boldsymbol{T}_n^+ = \Gamma(N) \cdot S \boldsymbol{T}_n^+.$$

Accordingly,

$$(f|T_n)|S = (f|S)|T_n.$$

It follows, in particular, that the operators T_n and R_m commute; so therefore, by (8.1.10), do T_n and R^χ.

Now take any positive integers l and m. Then, for any $f \in M(N, k)$,

$$(lm)^{1-\frac{1}{2}k}(f|T_l)|T_m = \sum_{S\in \boldsymbol{T}_l^+} \sum_{T\in \boldsymbol{T}_m^+} f|ST, \tag{9.2.9}$$

where

$$[\![J_l]\!] = \Gamma(N)\cdot \boldsymbol{T}_l^+, \quad [\![J_m]\!] = \Gamma(N)\cdot \boldsymbol{T}_m^+.$$

The product $[\![J_l]\!][\![J_m]\!]$ is a sum of double cosets each of which is a union of right cosets. From the definition of multiplication in the module $\mathcal{M}$ it is seen that these right cosets are precisely the right cosets $\Gamma(N)ST$, where $S\in \boldsymbol{T}_l^+$ and $T \in \boldsymbol{T}_m^+$; these $|\boldsymbol{T}_l^+||\boldsymbol{T}_m^+|$ right cosets need not, of course, all be distinct. Since, by (9.1.45),

$$[\![J_l]\!][\![J_m]\!] = \sum_{\substack{d|(l,m)\\ (d,N)=1}} d[\![J_{lm/d^2}]\!](dR_d),$$

the same is true of the sum on the right, which shows how many times each coset occurs; the double coset (dR_d) on the right can be replaced by the matrix dR_d. Now the sum on the right of (9.2.9) can be regarded as a sum taken over the right cosets $\Gamma(N)ST$ and is therefore equal to

$$\sum_{\substack{d|(l,m)\\ (d,N)=1}} d\{(d^2/lm)^{\frac{1}{2}k-1}(f|T_{lm/d^2})\}|(dR_d);$$

here, of course, the (dR_d) on the right denotes the matrix dR_d. Since $F|(dR_d) = F|R_d$, (9.2.5) follows.

It remains to prove (9.2.7, 8). For this purpose, assume in the first instance that $n|N^\infty$ and take $\boldsymbol{T}_n^+$ as in (9.1.58). Then

$$\begin{aligned} f(z)|T_n &= n^{-1}\sum_{\nu=0}^{n-1} f\{(z+\nu N)/n\} \\ &= n^{-1}\sum_{r=-\infty}^{\infty}\sum_{\nu=0}^{n-1} \exp\left\{\frac{2\pi i r}{Nn}(z+\nu N)\right\} \\ &= \sum_{r\equiv 0 \,(\mathrm{mod}\, n)} a_r e^{2\pi i r z/Nn}, \end{aligned}$$

which gives (9.2.7).

Now let $f \in M(N, k, \chi)$ and assume in the first instance that $(n, N) = 1$. We take $\boldsymbol{T}_n^+$ in the form (9.1.51) and obtain

$$f(z)|T_n = n^{k-1} \sum_{d|n} \sum_{\nu=1}^{d} \chi(a) d^{-k} f\{(az+\nu N)/d\}$$

$$= n^{k-1} \sum_{d|n} \chi(a) d^{-k} \sum_{r=-\infty}^{\infty} a_r \sum_{\nu=1}^{d} \exp\left\{\frac{2\pi i r}{Nd}(az+\nu N)\right\}$$

$$= \sum_{d|n} \chi(a) a^{k-1} \sum_{s=-\infty}^{\infty} a_{sd} e^{2\pi i s a z/N},$$

where, of course, $ad = n$. This gives (9.2.8).

Thus we have proved (9.2.8) when $(n, N) = 1$ and when $n|N^\infty$. From these two particular cases we easily deduce (9.2.8) in the general case, since, if $n = lm$, where $(m, N) = 1$ and $l|N^\infty$,

$$f(z)|T_n = (f(z)|T_l)|T_m.$$

We now examine the relationship between the Hecke operators T_n and the operators D_t introduced in §8.1. As in that section, t is any divisor of N and we write, for any subspace M of $M(N, k)$,

$$M_t = M|D_t, \quad M^\chi = M|R^\chi.$$

Theorem 9.2.2. *The operators D_t and T_n commute on $M(N, k)$ whenever $(n, N) = 1$.*

Let M be a subspace of $M(N, k)$ such that, for some $n \in \mathbb{Z}^+$, $M|T_n \subseteq M$. Then we have the following results:

(i) *For each $t|N$,*

$$M_t|T_n = 0 \quad \text{if} \quad (n, N) \nmid t, \tag{9.2.10}$$

while

$$M_t|T_n \subseteq \bigoplus_{d|(n,N)} M_{t/d} \quad \text{if} \quad (n, N)|t. \tag{9.2.11}$$

In particular,

$$M_t|T_n \subseteq M_t \quad \text{if} \quad (n, N) = 1 \tag{9.2.12}$$

and

$$M_1|T_n = 0 \quad \text{if} \quad (n, N) > 1. \tag{9.2.13}$$

(ii) *For all characters χ modulo N,*

$$M^\chi|T_n \subseteq M^\chi.$$

(iii) *For all $t|N$ and all χ modulo N,*

$$M_t^\chi|T_n = 0 \quad \textit{if} \quad (n, N) \nmid t, \tag{9.2.14}$$

while

$$M_t^\chi|T_n \subseteq \bigoplus_{d|(n,N)} M_{t/d}^\chi \quad \textit{if} \quad (n, N)|t. \tag{9.2.15}$$

(iv) *If $f \in M^{(t)}(N, k)$, where $t|N$, and ε is any character modulo t_1, then*

$$(f|D_t^{(\varepsilon)})|T_n = \varepsilon(n)(f|T_n)|D_t^{(\varepsilon)} \quad \textit{if} \quad (n, N) = 1. \tag{9.2.16}$$

In particular, the operators T_n for $(n, N) = 1$ map each of the spaces $M(N, k, \chi)$, $M(N, k, \chi, t)$, $M^{(t)}(N, k, \chi)$ and $M^{(t)}(N, k)$ into itself.

Proof. We assume, in the first place, that $f \in M(N, k)$ and that $(n, N) = 1$. Since, for any $l \in \mathbb{Z}^+$,

$$U^l J_n = J_n U^{ln},$$

it follows from (9.1.38) and (9.2.1) that

$$\Gamma(N) \cdot U^l \boldsymbol{T}_n^+ = \Gamma(N) \cdot \boldsymbol{T}_n^+ U^{ln}$$

and so

$$(f|U^l)|T_n = (f|T_n)|U^{ln}.$$

We therefore deduce from (8.1.16) and one of the remarks following (8.1.20) that

$$(f|D_t)|T_n = (f|T_n)|D_t \quad \text{when} \quad (n, N) = 1, \tag{9.2.17}$$

so that D_t and T_n commute in this case.

(i) Now suppose that M is a subspace for which $M|T_n \subseteq M$ and that $f \in M_t$. Then $f = f|D_t$ and it follows from (9.2.17) that $f|T_n \in M_t$ when $(n, N) = 1$; thus (9.2.12) holds.

In view of (9.2.5, 12) we need only prove (9.2.10, 11, 13) in the case when $n|N^\infty$, which we therefore assume. If $(n, N) \nmid t$, then $(rn, N) \neq t$ for all $r \in \mathbb{Z}$. Since, by (9.2.6, 7),

$$f(z)|T_n = \sum_{r=-\infty}^{\infty} a_{rn} e^{2\pi i r z/N}$$

and a_m is non-zero only when $(m, N) = t$, it follows that $a_{rn} = 0$ for all $r \in \mathbb{Z}$. Hence $f|T_n = 0$, so that (9.2.10, 13) follow.

Accordingly, we now assume that $(n, N)|t$. If $f|T_n \neq 0$, then, for some $r \in \mathbb{Z}$, $(rn, N)=t$ and so $(r, N)=t/d$, say. Since

$$t=(rn, N) \quad \text{divides } (r, N)(n, N),$$

it follows that d divides (n, N) and (9.2.11) follows.

(ii) Let $f \in M^\chi$ and take any m prime to N. Then

$$(f|T_n)|R_m=(f|R_m)|T_n=\chi(m)f|T_n,$$

by theorem 9.2.1, from which it follows that $f|T_n$ belongs to M^χ.

(iii) Since $M^\chi|T_n \subseteq M^\chi$, the results stated follow by applying part (i) to M^χ in place of M.

(iv) Suppose that $f \in M^{(t)}(N, k)$. We apply the argument of part (i) of the proof to the operator $D_t^{(\varepsilon)}$ defined in (8.2.14). We have, in the notation of (8.1.13) and (9.2.6),

$$\begin{aligned}
(f(z)|D_t^{(\varepsilon)})|T_n &= \frac{1}{t_1}\sum_{l=1}^{t_1}\sum_{s=1}^{t_1}\varepsilon(s)\,e^{-2\pi i l s/t_1}f(z)|T_nU^{ln} \\
&= \frac{1}{t_1}\sum_{r=-\infty}^{\infty}a_{rt}(n)\sum_{s=1}^{t_1}\varepsilon(s)\sum_{l=1}^{t_1}e^{2\pi i(rz+rln-ls)/t_1} \\
&= \sum_{r=-\infty}^{\infty}a_{rt}(n)\sum_{\substack{s=1\\ s\equiv rn\,(\mathrm{mod}\ t_1)}}^{t_1}\varepsilon(s)\,e^{2\pi i r z/t_1} \\
&= \varepsilon(n)\sum_{r=-\infty}^{\infty}a_{rt}(n)\varepsilon(r)\,e^{2\pi i r z/t_1} \\
&= \varepsilon(n)(f(z)|T_n)|D_t^{(\varepsilon)}.
\end{aligned}$$

We note that $f|D_t^{(\varepsilon)}$ belongs to $M^{(t)}(N, k)$.

The last part follows from (8.1.23) and (9.2.12, 15).

We now examine the effect of the Hecke operator T_n, when $(n, N)=1$, on functions belonging to a group Γ lying between $\Gamma(N)$ and $\Gamma(1)$ and make use of the results of theorem 8.3.1.

Theorem 9.2.3. *Let $(n, N)=1$, $n \geq 1$ and let the groups Γ and Γ_n be as in (8.3.7, 8). Then, in the notation of (9.1.54),*

$$\Gamma\boldsymbol{T}_n^+=\Gamma\cdot\boldsymbol{T}_n^+={}^+\boldsymbol{T}_n\cdot\Gamma_n.$$

Let the multiplier systems u and u_n be as described in theorem

8.3.1. Then, if $f \in M(\Gamma, k, u)$, and $S \in \Gamma(1)$,

$$(f|S)|T_n = (f|T_n)|S_n, \tag{9.2.18}$$

where S_n is defined by (8.3.3), and

$$f|T_n \in M(\Gamma_n, k, u_n). \tag{9.2.19}$$

Proof. From (9.1.56) we see that

$$\Gamma \boldsymbol{T}_n^+ = \Gamma \cdot \boldsymbol{T}_n^+ = {}^+\boldsymbol{T}_n \cdot \Delta,$$

say, where Δ is some subset of $\Gamma(1)$ containing $\Gamma(N)$. If $S \in \Gamma$, $T \in \boldsymbol{T}_n^+$, there exist unique $S' \in \Delta$, $T' \in {}^+\boldsymbol{T}_n$ such that $ST = T'S'$ and it follows that

$$J_n S_n \equiv S J_n \equiv ST \equiv T'S' \equiv J_n S' \pmod N,$$

so that $S' \equiv S_n \pmod N$ and therefore $S' \in \Gamma_n$. This proves that

$$\Gamma \cdot \boldsymbol{T}_n^+ \subseteq {}^+\boldsymbol{T}_n \cdot \Gamma_n$$

and the converse relation is proved similarly.

Finally, for any $L \in \Gamma(1)$, by (9.1.38) and (8.3.4),

$$[\![J_n]\!] S_n = [\![J_n S_n]\!] = [\![S J_n]\!] = S[\![J_n]\!]$$

for any $S \in \Gamma(1)$; (9.2.18) follows immediately from this and (9.2.1, 2). In particular, if $S \in \Gamma$, we have, when $f \in M(\Gamma, k, u)$,

$$(f|T_n)|S_n = (f|S)|T_n = u(S) f|T_n = u_n(S_n) f|T_n,$$

from which (9.2.19) follows.

The foregoing results have taken their simplest form when $(n, N) = 1$. When this condition does not hold the situation becomes more complicated and we note, for example, that T_n need not leave spaces of forms of divisor t invariant. We meet this situation by introducing a modified 'conjugate' operator $T_n^{(t)}$ which operates not on $M(N, k)$ but on its subspace $M^{(t)}(N, k)$.

Accordingly, we define, for each positive divisor t of N and each positive integer n, an operator

$$T_n^{(t)} := J_t T_n J_t^{-1} \tag{9.2.20}$$

acting on the space $M^{(t)}(N, k)$. Observe that $T_n^{(1)} = T_n$.

In the following theorems we show that $T_n^{(t)}$ is well defined and that its properties are similar to those of T_n but simpler in so far as forms of divisor t are concerned. We shall assume that the form $f \in M^{(t)}(N, k)$ has the Fourier expansion

$$f(z) = \sum_{r=-\infty}^{\infty} A_r e^{2\pi i r z/t_1} \tag{9.2.21}$$

for $y = \operatorname{Im} z \geq y_0 > 0$, where t_1 is the codivisor of t; the series contains, of course, only a finite number of non-zero terms with negative r. If we also write $f(z)$ in the form (8.1.13), as we have done earlier in this section, we observe that

$$A_r = a_{rt}, \quad a_r = A_{r/t} \quad (r \in \mathbb{Z}), \tag{9.2.22}$$

where, in the second equation we make the convention that $A_{r/t}$ denotes zero when r/t is not an integer.

Theorem 9.2.4. *Suppose that, for some fixed positive divisor t of N, $f \in M^{(t)}(N, k)$ and that $f(z)$ has the expansion (9.2.21). Then, for each $n \in \mathbb{Z}^+$, the operator $T_n^{(t)}$ is well-defined and*

$$f|T_n^{(t)} \in M^{(t)}(N, k). \tag{9.2.23}$$

Thus $M^{(t)}(N, k)$ is invariant under the operators $T_n^{(t)}$ and so are its subspaces contained in $H(N, k)$, $\{N, k\}$ and $\{N, k\}_0$.

The operators $T_n^{(t)} (n \in \mathbb{Z}^+)$ commute with each other and with the operators R_m, where $(m, N) = 1$. Further, for any positive integers l and m,

$$(f|T_l^{(t)})|T_m^{(t)} = (f|T_m^{(t)})|T_l^{(t)} = \sum_{\substack{d|(l,m)\\(d,N)=1}} d^{k-1}(f|R_d)|T_{lm/d^2}. \tag{9.2.24}$$

Also

$$f|T_n^{(t)} = f|T_n^{(t/q)}, \quad \textit{whenever} \quad q|t \quad \textit{and} \quad (q, n) = 1; \tag{9.2.25}$$

in particular,

$$f|T_n^{(t)} = f|T_n \quad \textit{if} \quad (n, t) = 1. \tag{9.2.26}$$

When $n|N^\infty$, we may take

$$f|T_n^{(t)} = n^{\frac{1}{2}k-1} \sum_{\nu=0}^{n-1} f|J_n U^{\nu t_1}. \tag{9.2.27}$$

Finally, $f|T_n^{(t)}$ has the Fourier expansion

$$f(z)|T_n^{(t)}=\sum_{r=-\infty}^{\infty} A_r(n)\,e^{2\pi irz/t_1}\quad (y\geq y_0),\qquad (9.2.28)$$

where

$$A_r(n)=A_{rn},\quad \text{when}\quad n|N^\infty.\qquad (9.2.29)$$

If $f\in M^{(t)}(N,k,\chi)$, where χ is a character modulo N satisfying (8.1.7), then

$$A_r(n)=\sum_{d|(r,n)}\chi(d)d^{k-1}A_{rn/d^2}.\qquad (9.2.30)$$

Proof. Write $[\![J_n]\!]=\Gamma(N)\cdot \boldsymbol{T}_n^+$, for any choice of the transversal $\boldsymbol{T}_n^+$. Then

$$(f|J_t)|\sum_{T\in \boldsymbol{T}_n^+} T$$

is independent of the choice of $\boldsymbol{T}_n^+$, by theorem 8.3.3, since $f|J_t\in M(N,k)$. Thus $T_n^{(t)}$ is well defined. From (9.2.20) and (9.1.58) we derive (9.2.27), since

$$J_tJ_nU^{\nu N}J_t^{-1}=J_nU^{\nu t_1}.$$

That the operators $T_n^{(t)}$ commute ($n\in\mathbb{Z}$) and that (9.2.24) holds follow immediately from the fact that J_t and R_d commute on $M^{(t)}(N,k)$.

To prove (9.2.25) assume that $q|t$ and that $(q,n)=1$. As ν runs through a complete set of residues modulo n so does νq. This and (9.2.27) prove (9.2.25) in the case when $n|N^\infty$. By (9.2.24) it therefore suffices to prove that

$$f|T_n^{(t)}=f|T_n\quad \text{when}\quad (n,N)=1.\qquad (9.2.31)$$

Since J_t and T_a commute, we have, by (9.1.54), when $(n,N)=1$,

$$f|T_n^{(t)}=n^{\frac{1}{2}k-1}\sum_{a|n}\sum_{T\in \boldsymbol{D}_n^a} f|J_tR_aTJ_t^{-1}$$

$$=n^{\frac{1}{2}k-1}\sum_{a|n}\sum_{T\in \boldsymbol{D}_n^a} f|R_aJ_tTJ_t^{-1}.$$

Now we may take $\boldsymbol{D}_n^a$ to consist of the matrices

$$T=\begin{bmatrix} a & \nu tN\\ 0 & d\end{bmatrix}\quad (0\leq \nu<d),$$

since νt runs through a complete residue system modulo d. From the fact that

$$J_tTJ_t^{-1}=\begin{bmatrix} a & \nu N\\ 0 & d\end{bmatrix}$$

(9.2.31) follows.

We have not yet proved (9.2.23). By (9.2.31), we may restrict ourselves to the case $n|N^\infty$ and, by making use of (9.2.24), assume further that $n=p$ where p is a prime dividing N. We suppose that $f\in M^{(t)}(N, k, \chi)=M(\Gamma_0(N, t_1), k, \chi)$ and note that $M^{(t)}(N, k)$ is a vector sum of these spaces, for different characters χ. Then, by theorems 8.3.2 or 8.3.3, $f|J_t\in M(\Gamma_0(N, N), k, \chi)=M(N, k, \chi)$ and so $f|J_tT_p\in M(N, k, \chi)$, by theorem 9.2.2(ii). Thus, by theorem 8.3.3,

$$f|T_p^{(t)}\in M(\Gamma_0(Np, t_1), k, \chi).$$

Since $\Gamma_0(N, t_1)$ is generated by W^N and $\Gamma_0(Np, t_1)$, it suffices to prove that

$$f|T_p^{(t)}W^N=f|T_p^{(t)}. \tag{9.2.32}$$

Write, for any integer ν,

$$S_\nu=U^{\nu t_1}W^NU^{-\nu t_1}=\begin{bmatrix}1+N\nu t_1 & -N\nu^2t_1^2\\ N & 1-N\nu t_1\end{bmatrix},$$

so that $J_pS_\nu J_p^{-1}\in\Gamma_0(N, t_1)$ and $\chi(J_pS_\nu J_p^{-1})=1$. Since

$$f|J_pU^{\nu t_1}W^N=f|J_pS_\nu J_p^{-1}\,.\,J_pU^{\nu t_1}=f|J_pU^{\nu t_1},$$

(9.2.32) follows, and therefore $f|T_p^{(t)}\in M^{(t)}(N, k, \chi)$, so that (9.2.23) holds. The subspaces of $M^{(t)}(N, k)$ contained in $H(N, k)$, $\{N, k\}$ and $\{N, k\}_0$ are clearly invariant under the operations $T_n^{(t)}$.

To show that $T_n^{(t)}$ commutes with R_m, when $(m, N)=1$, we may, by what we have proved, choose the entries of R_m so that $R_m\in\Gamma_0^0(N^2)$. Then $[J_t^{-1}, R_m]\in\Gamma(N)$ and so we have

$$\begin{aligned}f|T_n^{(t)}R_m&=f|J_tT_nJ_t^{-1}R_m=f|J_tT_nR_mJ_t^{-1}\\&=f|R_mT_n^{(t)},\end{aligned}$$

since R_m commutes with T_n and J_t.

It remains to prove (9.2.29, 30) and this is done exactly as in the proof of Theorem 9.2.1.

Theorem 9.2.2 can be extended to the operators $T_n^{(t)}$; however, in the next theorem we give only those properties that we shall use later.

Theorem 9.2.5. *Suppose that M is a subspace of $M^{(t)}(N, k)$, where $t|N$, and that $M|T_n^{(t)} \subseteq M$ for some $n \in \mathbb{Z}^+$. Then*

$$M_t|T_n^{(t)} \subseteq M_t, \quad M^\times|T_n^{(t)} \subseteq M^\times, \quad M_t^\times|T_n^{(t)} \subseteq M_t^\times \qquad (9.2.33)$$

and

$$M_t|T_n^{(t)} = M_t^\times|T_n^{(t)} = 0 \quad \text{if} \quad (n, t_1) = 1. \qquad (9.2.34)$$

In particular, the operators $T_n^{(t)}$ $(n \in \mathbb{Z}^+)$ map each of the spaces $M^{(t)}(N, k)$, $M^{(t)}(N, k, \chi)$, $M_t(N, k)$, $M(N, k, \chi, t)$ into itself.

Finally, if $f \in M^{(t)}(N, k)$ and ε is any character modulo t_1, then

$$(f|D_t^{(\varepsilon)})|T_n^{(t)} = \varepsilon(n)(f|T_n^{(t)})|D_t^{(\varepsilon)}. \qquad (9.2.35)$$

In particular, the right-hand side is zero when $(n, t_1) > 1$.

Proof. If $(n, N) = 1$, the theorem follows directly from theorem 9.2.2. It is therefore only necessary to prove the results stated when $n|N^\infty$. The arguments required to prove (9.2.33, 34) are exactly similar to those used in the proof of theorem 9.2.2 (i), (ii), (iii) for the case $t = 1$, and we omit them.

To prove (9.2.35) assume that f is given by (9.2.21). Since $U^l J_n = J_n U^{nl}$, we have, by (8.2.14) and (9.2.28, 29),

$$(f(z)|D_t^{(\varepsilon)})|T_n^{(t)} = \frac{n^{\frac{1}{2}k-1}}{t_1} \sum_{l=1}^{t_1} \sum_{s=1}^{t_1} \varepsilon(s)\, e^{-2\pi i l s/t_1} f(z)|U^l J_n \sum_{\nu=1}^{n} U^{\nu t_1}$$

$$= \frac{1}{t_1} \sum_{l=1}^{t_1} \sum_{s=1}^{t_1} \varepsilon(s)\, e^{-2\pi i l s/t_1} \sum_{r=-\infty}^{\infty} A_{rn}\, e^{2\pi i r(z+nl)/t_1}$$

$$= \frac{1}{t_1} \sum_{r=-\infty}^{\infty} A_{rn}\, e^{2\pi i r z/t_1} \sum_{s=1}^{t_1} \varepsilon(s) \sum_{l=1}^{t_1} e^{2\pi i l(nr-s)/t_1}$$

$$= \sum_{r=-\infty}^{\infty} A_{rn} \varepsilon(nr)\, e^{2\pi i r z/t_1},$$

from which (9.2.35) follows, since $\varepsilon(nr) = \varepsilon(n)\varepsilon(r)$.

We now show that the operators $T_n^{(t)}$ and K (see §8.6) commute.

Theorem 9.2.6. *Let $f \in M^{(t)}(N, k)$, where t divides N. Then*

$$(f|T_n^{(t)})|K = (f|K)|T_n^{(t)} \tag{9.2.36}$$

for all $n \in \mathbb{Z}^+$.

Proof. Suppose first that $(n, N) = 1$. It is clear from (9.1.38) that

$$J^{-1}[\![J_n]\!]J = [\![J_n]\!].$$

Since $J^{-1}\Gamma(N)J = \Gamma(N)$, it is clear that $J^{-1}\boldsymbol{T}_n^+ J$ is a right transversal satisfying (9.1.47) when $\boldsymbol{T}_n^+$ is. Accordingly, by (9.2.2) and (8.6.4, 5), it follows that

$$(f|T_n)K = (f|K)|T_n$$

and T_n can be replaced by $T_n^{(t)}$ since $(n, N) = 1$.

If $n|N^\infty$, the result follows similarly from (9.2.27) since

$$J^{-1}(J_n U^{\nu t_1})J = J_n U^{-\nu t_1}.$$

Subject to certain conditions the operators H_q and $T_n^{(t)}$ commute. More precisely, we have

Theorem 9.2.7. *Suppose that $N = tt_1$ and that $Nt_1 = qq'$, where $(q, q') = 1$. Let $f \in M^{(t)}(N, k, \chi)$, where χ is a character modulo N and $\chi = \chi_q \chi_{q'}$, where χ_q and $\chi_{q'}$ are characters modulis (q, N) and (q', N), respectively. Then, for any $H_q \in \boldsymbol{H}_q(N, t_1)$ and for every positive integer n prime to q,*

$$(f|T_n^{(t)})|H_q = \chi_q(n)(f|H_q)|T_n^{(t)}. \tag{9.2.37}$$

Finally, if $q|N$ and f has divisor t, so has $f|H_q$.

Proof. Suppose, in the first instance, that $(n, N) = 1$ and take any $H_q \in \boldsymbol{H}_q(N, t_1)$. Let H_q^* be the matrix defined by (8.4.17) (with $N' = t_1$), so that $H_q = SH_q^*$, where $S \in \Gamma_0(N, t_1)$. As right transversals for defining the operator $T_n = T_n^{(t)}$ we may take the matrices

$$R_a X_{\nu N x q} \quad \text{or} \quad R_a X_{\nu N w q} \quad (ad = n, \nu \bmod d),$$

where X_r is defined by (8.4.18), since $(xq, d)=(wq, d)=1$. Hence, by theorem 8.4.2,

$$\begin{aligned}(f|T_n)|H_q &= \chi(S)f|T_nH_q^* \\ &= n^{\frac{1}{2}k-1}\chi(S)\sum_{d|n}\chi(n/d)\sum_{\nu \pmod d} f|X_{\nu Nxq}H_q^* \\ &= n^{\frac{1}{2}k-1}\chi(S)\sum_{d|n}\chi(n/d)\sum_{\nu \pmod d} f|H_q'X_{\nu Nwq} \\ &= n^{\frac{1}{2}k-1}\chi(S)\sum_{d|n}\chi(n/d)\sum_{\nu \pmod d} f|S_\nu S^{-1}H_qX_{\nu Nwq} \\ &= n^{\frac{1}{2}k-1}\sum_{d|n}\chi(n/d)\sum_{\nu \pmod d}\chi(S_\nu)f|X_{\nu Nwq}.\end{aligned}$$

Since, by (8.4.16, 19),

$$\chi(n/d)\chi(S_\nu)=\chi(n/d)\chi_q(d^2)\bar{\chi}_q(n)=\chi_q(n)\chi^*(n/d),$$

(9.2.37) follows.

To complete the proof in the general case it suffices to prove (9.2.37) when $n|N^\infty$ and $(n, q)=1$. We use (9.2.27) and apply theorem 8.4.2 with $a=1$ and $d=n$. We obtain, as before,

$$\begin{aligned}(f|T_n^{(t)})|H_q &= \chi(S)f|T_n^{(t)}H_q^* \\ &= n^{\frac{1}{2}k-1}\chi(S)\sum_{\nu=0}^{n-1} f|X_{\nu t_1xq}H_q^* \\ &= n^{\frac{1}{2}k-1}\chi(S)\sum_{\nu=0}^{n-1} f|S_\nu S^{-1}H_qX_{\nu t_1wq} \\ &= n^{\frac{1}{2}k-1}\sum_{\nu=0}^{n-1}\chi(S_\nu)f|H_qX_{\nu t_1wq}.\end{aligned}$$

Since $\chi(S_\nu)=\chi_q(n)$, the required result follows.

To prove the last part assume that $q|N$ and that f has divisor t, so that $f|D_t=f$. Take H_q in the form (8.4.4) (with $N'=t_1$) and write, for any $l\in\mathbb{Z}$,

$$H_q' := U^{lx}H_qU^{-lw}.$$

Then it is easily seen that $H_q'\in \boldsymbol{H}_q(N, t_1)$ and that

$$\chi\{H_q(H_q')^{-1}\}=1.$$

Thus

$$f|H_q U^{lw} = f|H_q' U^{lw} = f|U^{lx} H_q$$

and it follows, from (8.1.15) and the remarks before theorem 8.1.2, that

$$(f|H_q)|D_t = (f|D_t)|H_q = f|H_q.$$

Thus $f|H_q$ has divisor t.

Theorem 9.2.8. *Suppose that $f \in M^{(t)}(N, k, \chi)$ and that its Fourier expansion is given by (9.2.21). Let d be a divisor of N satisfying*

$$d|t_1(N/d, N/N_\chi), \quad d \text{ squarefree}. \tag{9.2.38}$$

Then

$$f|T_d^{(t)} J_d^{-1} = d^{\frac{1}{2}k} f|L_d^\chi(N, t_1). \tag{9.2.39}$$

Moreover, for any squarefree divisor u of N, write

$$f_u = \sum_{d|u} \mu(d) d^{-\frac{1}{2}k} f|T_d^{(t)} J_d^{-1}. \tag{9.2.40}$$

Then $f_u \in M(\Gamma_0(Nu, t_1), k, \chi)$ and

$$f_u(z) = \sum_{\substack{r=-\infty \\ (r,u)=1}}^{\infty} A_r e^{2\pi i r z/t_1}. \tag{9.2.41}$$

In particular, if $f \in M(N, k, \chi, t)$, then $f_u \in M(Nu, k, \chi, tu)$.

Finally, if d is any divisor of t and $d = qr$, where $(q, r) = 1$, then

$$f|T_d^{(t)} J_d^{-1} = f|T_q^{(t)} J_q^{-1} \,.\, T_r^{(tq)} J_r^{-1}. \tag{9.2.42}$$

Proof. Let d satisfy (9.2.38). Then N_χ divides $t_1 N/d$, so that the operator $L_d^\chi(N, t_1)$ can be defined as in (8.5.5). By the remark following (8.5.3), the transversal $\boldsymbol{Z}_d(N, t_1)$ can be taken to consist of the d matrices $U^{\nu t_1}$ $(0 \le \nu < d)$. Then (9.2.39) follows from (9.2.27) and (8.5.5); we note also that, by theorem 8.5.1, $f|T_d^{(t)} J_d^{-1}$ belongs to the group $\Gamma_0(N, t_1)$.

On the other hand, for any divisor d of N, $f|T_d^{(t)} J_d^{-1}$ belongs to the group $\Gamma_0(Nd, t_1)$, by theorem 8.3.2, and it follows that $f_u \in M(\Gamma_0(Nu, t_1), k, \chi)$; the remark at the end of the last paragraph enables us to reduce the level Nu in certain cases, but we shall not require this.

The expansion (9.2.41) follows immediately from (9.2.21, 29, 40), and the following sentence is an immediate consequence.

To prove (9.2.42) we note that, by (9.2.25),

$$\begin{aligned} f|T_d^{(t)}J_d^{-1} &= f|T_q^{(t)}T_r^{(t)}J_d^{-1} = f|T_q^{(t)}T_r^{(t/q)}J_d^{-1} \\ &= r^{\frac{1}{2}k-1}\sum_{\nu=0}^{r-1} f|T_q^{(t)}J_rU^{\nu q t_1}J_d^{-1} \\ &= r^{\frac{1}{2}k-1}\sum_{\nu=0}^{r-1} f|T_q^{(t)}J_q^{-1}\,.\,J_rU^{\nu t_1}J_r^{-1} \\ &= f|T_q^{(t)}J_q^{-1}\,.\,T_r^{(tq)}J_r^{-1}, \end{aligned}$$

as required. Observe that the last Hecke operator is $T_r^{(tq)}$ and not $T_r^{(t)}$, as would appear to be the case at first sight. For $f|T_q^{(t)}J_q^{-1}$ belongs to the group $\Gamma_0(Nq, t_1)$ rather than $\Gamma_0(N, q)$; this is connected with the fact, made clear by (9.2.27), that the Hecke operator depends on the codivisor t_1 (which is the same in each case) rather than tq or t.

Theorem 9.2.9. *Let $f\in M^{(t)}(N, k, \chi)$ and let p be a prime dividing $(t, N/N_\chi)$ but not t_1. Suppose also that n is a positive integer not divisible by p. Then*

$$(f|L_p^\chi(N, t_1))|T_n^{(t)} = (f|T_n^{(t)})|L_p^\chi(N, t_1). \qquad (9.2.43)$$

Moreover, if q and d are squarefree divisors of $(t, N/N_\chi)$ and prime to t_1, then

$$(f|L_q^\chi(N, t_1))|T_d^{(t)}J_d^{-1} = (f|T_d^{(t)}J_d^{-1})|L_q^\chi(dN, t_1), \qquad (9.2.44)$$

when $(d, q) = 1$.

Finally, if d satisfies the conditions stated (for $q = 1$) and $f\in M(N, k, \chi, t)$, then both $f|L_d^\chi(N, t_1)$ and $f|L_d^\chi(N, t_1)J_d$ also belong to this space; in fact the latter function belongs to the subspace $M(N/d, k, \chi, t/d)$.

Proof. Suppose that p satisfies the conditions stated and write $\varepsilon_p = 0$ or 1 according as p^2 does or does not divide N. In the latter case the set $\boldsymbol{H}_p(N, t_1)$ (see §8.4) has a meaning, and we denote by H_p any member of this set. By (8.5.3), we can take $\boldsymbol{Z}_p(N, t_1)$ to consist of the p matrices $U^{\nu t_1}(0\le\nu<p)$, when $\varepsilon_p = 0$, to which we add

$J_p^{-1}H_p$, when $\varepsilon_p = 1$. Accordingly, by (8.5.5),

$$f|L_p^\chi(N, t_1) = \frac{\psi(Nt_1/p)}{\psi(Nt_1)}\left\{\sum_{\nu=0}^{p-1} f|J_pU^{\nu t_1}J_p^{-1} + \varepsilon_p\bar{\chi}(J_p^{-1}H_p)f|H_pJ_p^{-1}\right\}.$$

Hence, by (9.2.27),

$$f|T_p^{(t)}J_p^{-1} = A_pf|L_p^\chi(N, t_1) - \varepsilon_pp^{\frac{1}{2}k-1}\bar{\chi}(J_p^{-1}H_p)f|H_pJ_p^{-1}, \tag{9.2.45}$$

where

$$A_p = \frac{p^{\frac{1}{2}k-1}\psi(Nt_1)}{\psi(Nt_1/p)}.$$

It follows from (9.2.45) that $f|L_p^\chi J_p$ is a linear combination of $f|T_p^{(t)}$ and $f|H_p$ (the latter only arising when $\varepsilon_p = 1$). By theorems 9.2.4 and 9.2.7, the operator $T_n^{(t)}$ commutes with $T_p^{(t)}$ and H_p, since χ_p is the principal character modulo p in this case. It follows that

$$f|L_p^\chi J_pT_n^{(t)} = f|T_n^{(t)}L_p^\chi J_p$$

and so, by (9.2.25),

$$\begin{aligned} f|L_p^\chi T_n^{(t)} &= f|L_p^\chi \,.\, J_pT_n^{(t/p)}J_p^{-1} = f|L_p^\chi J_p \cdot T_n^{(t)}J_p^{-1} \\ &= f|T_n^{(t)}J_p^\chi, \end{aligned}$$

which is (9.2.43).

Now suppose that p and q satisfy the conditions stated and that $p \nmid q$. By theorem 8.3.2, $f|T_p^{(t)}J_p^{-1} \in M(\Gamma_0(Np, t_1), k, \chi)$, and, by theorem 8.4.1, $f|H_pJ_p^{-1}$ also belongs to this space (when $\varepsilon_p = 1$). Accordingly, we may operate on these functions by $L_q^\chi(Np, t_1)$. We obtain, by (8.5.6, 7) and theorems 8.5.1(vi) and 8.5.2 (with $r = p$),

$$\begin{aligned} f|T_p^{(t)}J_p^{-1}L_q^\chi(Np, t_1) &= A_pf|L_p^\chi(N, t_1)L_q^\chi(N, t_1) \\ &\quad - \varepsilon_pp^{\frac{1}{2}k-1}\bar{\chi}(J_p^{-1}H_p)f|H_pJ_p^{-1}L_q^\chi(Np, t_1) \\ &= A_pf|L_q^\chi(N, t_1)L_p^\chi(N, t_1) - \varepsilon_pp^{\frac{1}{2}k-1}\bar{\chi}(J_p^{-1}H_p)f|H_pL_q^\chi(N, t_1)J_p^{-1} \\ &= f|L_q^\chi(N, t_1)\{A_pL_p^\chi(N, t_1) - \varepsilon_pp^{\frac{1}{2}k-1}\bar{\chi}(J_p^{-1}H_p)H_pJ_p^{-1}\} \\ &= f|L_q^\chi(N, t_1)T_p^{(t)}J_p^{-1}. \end{aligned}$$

This is (9.2.44) with d replaced by p. We obtain (9.2.44) in the general case by expressing d as a product of primes and by repeated use of (9.2.42).

Finally, suppose that $f \in M(N, k, \chi, t)$ and that d is a squarefree divisor of $(t, N/N_\chi)$ and is prime to t_1. If p is any prime divisor of d, then, by (9.2.45), $f|L_p^\chi J_p$ is a linear combination of $f|T_p^{(t)}$ and $\varepsilon_p f|H_p$ and so, by theorems 9.2.5 and 9.2.7, $f|L_p^\chi J_p \in M(N, k, \chi, t)$.

Now, if $d = qr$, where $(q, r) = 1$, we have, by (8.5.6, 10),

$$\begin{aligned} f|L_{qr}^\chi(N, t_1)J_{qr} &= f|L_q^\chi(N, t_1)L_r^\chi(N, t_1)J_{qr} \\ &= f|L_q^\chi(N, t_1)L_r^\chi(Nq, t_1)J_qJ_r \\ &= f|L_q^\chi(N, t_1)J_qL_r^\chi(N, t_1)J_r. \end{aligned}$$

By repeated use of this and the previous result we deduce that

$$f|L_d^\chi(N, t_1)J_d \in M(N, k, \chi, t).$$

By considering the Fourier series of $f|L_d^\chi$, we see that this function also has divisor t. Since, by theorem 8.5.1, $f|L_d^\chi J_d$ also belongs to $M(\Gamma_0(N/d, t_1), k, \chi)$ this function belongs to $M(N/d, k, \chi, t/d)$.

Theorem 9.2.10. *Let $N = qq_1 = tt_1$, where $(q, q_1t_1) = 1$, and suppose that, for some prime p,*

$$q = mn, \quad \text{where} \quad p \nmid m \quad \text{and} \quad n = p^e = pl \quad (e > 0).$$

Let χ be a character modulo N such that $p|N_\chi$ and let $f \in M^{(t)}(N, k, \chi)$. Then for any $H_q \in \boldsymbol{H}_q(N, t_1)$,

$$f|T_n^{(t)}H_qT_n^{(t)} = p^{k-1}\chi_{q'}(p)f|T_l^{(t)}H_qT_l^{(t)}, \tag{9.2.46}$$

where $\chi_{q'}$ is defined as in theorem 9.2.7 and $q' = q_1t_1$.

Proof. Since both $f|T_n^{(t)}$ and $f|T_l^{(t)}$ belong to $M^{(t)}(N, k, \chi)$ and since H_q occurs on each side of (9.2.46), there is no loss of generality in assuming that H_q is given by (8.4.11), with $N' = t_1$. We now apply theorem 8.4.3, with $N' = t_1$, and note that

$$\chi(H_q^*H_q^{-1}) = \chi\{1 + q\lambda(p-1)\} = \chi_{q'}(p). \tag{9.2.47}$$

Let Σ denote the set of n^2 ordered pairs (μ, ν), where

$$0 \le \mu - \lambda\lambda' = \mu^* < n, \quad 0 \le \nu < n,$$

and let Σ_0 and Σ_1 be the subsets of Σ consisting of pairs (μ, ν) for which $p|\mu^*$ and $p\nmid\mu^*$, respectively. Then, in the notation of (8.4.23),

$$f|T_n^{(t)}H_qT_n^{(t)}=n^{k-2}\sum_{(\mu,\nu)\in\Sigma}f|A(\mu,\nu)=f_0+f_1,$$

say, where f_i denotes the contribution to the sum from pairs $(\mu_i,\nu_i)\in\Sigma_i$ $(i=0,1)$. By (8.4.28),

$$f_0=n^{k-2}\sum_{\rho=0}^{l-1}\sum_{\nu=0}^{n-1}f|\{pJ_lU^{\rho t_1}H_q^*J_lU^{\nu t_1}\}$$

$$=p^{k-1}f|T_l^{(t)}H_q^*T_l^{(t)}=p^{k-1}\chi_{q'}(p)f|T_l^{(t)}H_qT_l^{(t)},$$

by (9.2.47).

We now show that $f_1=0$. We define a relation $\sim$ on the $n(n-l)$ members of Σ_1 as follows. Write $(\mu,\nu)\sim(\mu_0,\nu_0)$ if and only if

$$\mu_0^*-\mu^*\equiv mq'^2(\nu-\nu_0)\mu_0^*\mu^*\ (\mathrm{mod}\ n).$$

It is easily verified that this is an equivalence relation. Further, if $(\mu,\nu)\sim(\mu_0,\nu_0)$, then $\mu\equiv\mu_0\ (\mathrm{mod}\ n)$ if and only if $\mu=\mu_0$ and $\nu=\nu_0$, and ν is uniquely determined by μ, μ_0 and ν_0. Since μ can take $n-l$ different values, each equivalence class consists of $n-l$ pairs, so that there are n equivalence classes.

If we regard the above congruence as one in which μ is the unknown, we see that it is necessary that

$$1+mq'^2(\nu-\nu_0)\mu_0^*\not\equiv 0\ (\mathrm{mod}\ p)$$

and this determines uniquely the $n-l$ different values taken by ν modulo n for given (μ_0,ν_0).

The function f_1, being a sum over Σ_1, is a sum over the n equivalence classes. The contribution from the equivalence class containing (μ_0,ν_0) is, by (8.4.23, 25, 26),

$$n^{k-2}\sum_{\nu}f|A(\mu,\nu)=n^{k-2}\sum_{\nu}f|\{A(\mu,\nu)A^{-1}(\mu_0,\nu_0)\}A(\mu_0,\nu_0)$$

$$=n^{k-2}f|A(\mu_0,\nu_0)\sum_{\nu}\chi\{1+mq'^2(\nu-\nu_0)\mu_0^*\}.$$

We can write $\chi=\chi'\chi_n$, where χ' and χ_n are characters modulis N/n and n, respectively, so that

$$\chi\{1+mq'^2(\nu-\nu_0)\mu_0^*\}=\chi_n\{1+mq'^2(\nu-\nu_0)\mu_0^*\}.$$

Now $1+mq'^2(\nu-\nu_0)\mu_0^*$ runs through the $n-l$ residues modulo n

that are prime to n, and χ_n is not the principal character modulo n, since $p|N_\chi$. Accordingly, the sum over each equivalence class is zero, and so $f_1=0$. This completes the proof.

9.3. The effect of Hecke operators on Poincaré series. In this section we study the action of Hecke operators on Poincaré series. This leads to interesting results concerning these functions and their Fourier coefficients. From these properties we deduce theorem 9.3.1, from which it can be inferred that the operators T_n are 'normal operators' when $(n, N)=1$. A separate proof of this important result, independent of the theory of Poincaré series, is also given.

As in §9.2, we assume that the integer k and the MS v satisfy (8.1.1, 2) and we suppose that Γ is a subgroup of the modular group satisfying

$$\bar{\Gamma}(N)\subseteq\Gamma\subseteq\Gamma(1). \tag{9.3.1}$$

We shall suppose, as in theorem 8.3.1, that u is a MS of weight k on Γ, whose restriction to $\bar{\Gamma}(N)$ is v, and that n is a positive integer prime to N. The group Γ_n is then defined as in (8.3.8) and has the MS u_n defined by (8.3.9). In the most important applications Γ will be $\Gamma_0^0(N)$ or $\Gamma_0(N, t_1)$ and u is then a character modulo N satisfying (8.1.6).

Write

$$q=n(\infty,\Gamma),\quad \kappa=\kappa(\infty,\Gamma,u) \tag{9.3.2}$$

and, for any positive divisor a of n, put, as usual

$$n=ad. \tag{9.3.3}$$

Then

$$n(R_a^{-1}\infty,\Gamma_n)=q, \tag{9.3.4}$$

since it is the least positive integer r with the property that

$$R_a^{-1}U^rR_a\in\Gamma_n.$$

This, by (8.2.11), is equivalent to $U^{ra^2}\in\Gamma_n$, i.e. to $U^{n'ra^2}\in\Gamma$, where $nn'\equiv1 \pmod N$; hence $r=q$. The number q is a divisor of N and we write, as usual,

$$qq_1=N. \tag{9.3.5}$$

Accordingly, since $e^{2\pi i\kappa}=u(U^q)$ and $u(U^N)=1$,

$$\kappa=p/q,\quad\text{where}\quad 0\leq p\leq q_1. \tag{9.3.6}$$

Similarly, if we write

$$\kappa_{n,d} = \kappa(R_a^{-1}\infty, \Gamma_n, u_n), \tag{9.3.7}$$

we have

$$\kappa_{n,d} = p_{n,d}/q_1 \quad (0 \le p_{n,d} < q_1), \tag{9.3.8}$$

where

$$\begin{aligned} e^{2\pi i \kappa_{n,d}} &= u_n(R_a^{-1}\, U^q\, R_a) = u(U^{n'qa^2}) \\ &= e^{2\pi i n' a^2 p/q_1}, \end{aligned}$$

so that

$$p_{n,d} \equiv n' a^2 p \pmod{q_1}. \tag{9.3.9}$$

If m is any integer satisfying

$$mq_1 + p \equiv 0 \pmod{d} \tag{9.3.10}$$

and if

$$m_{n,d} := \left[\frac{(m+\kappa)a}{d}\right] = \left[\frac{(mq_1+p)a}{dq_1}\right], \tag{9.3.11}$$

it is easily verified from (9.3.8, 9, 10) that

$$\frac{(m+\kappa)a}{d} = m_{n,d} + \kappa_{n,d}. \tag{9.3.12}$$

We are now in a position to state and prove

Theorem 9.3.1. *For any integers m, n, k with $n > 0$, $(n, N) = 1$ and $k \ge 2$,*

$$G_I(z; m; \Gamma, k, u)|T_n = \sum_{d|(n, mq_1+p)} (n/d)^{k-1} G_{R_a}(z; m_{n,d}; \Gamma_n, k, u_n), \tag{9.3.13}$$

where $ad = n$. More generally, for any $L \in \Gamma(1)$ with $L_n \in \theta_n(\Gamma(N)L)$ (as in theorem 8.3.1),

$$G_L(z; m; \Gamma, k, u)|T_n = \sum_{d|(n, mq_1+p)} (n/d)^{k-1} G_{R_a L_n}(z; m_{n,d}; \Gamma_n, k, u_n). \tag{9.3.14}$$

Note that, in (9.3.14), $m_{n,d}$ is defined by (9.3.11), but with κ replaced by the cusp parameter appropriate to G_L, namely $\kappa(L^{-1}\infty, \Gamma, u)$, and also

$$mq_1 + p = q_1(m+\kappa) = N(m+\kappa)/n(L^{-1}\infty, \Gamma).$$

Observe also that the modular forms on the right of (9.3.13, 14) belong to $M(\Gamma_n, k, u_n)$ as we should expect from theorem 9.2.3.

Proof. We require the result

$$\Gamma_{U^q} \cdot \boldsymbol{D}_n^a = \boldsymbol{A}_n^a \cdot \Gamma_{U^q}, \tag{9.3.15}$$

where q is any division of N and where $\boldsymbol{A}_n^a$ and $\boldsymbol{D}_n^a$ are defined by (9.1.52, 53). Take a fixed $r \in \mathbb{Z}$ and $T \in \boldsymbol{D}_n^a$, so that, for any $s \in \mathbb{Z}$,

$$U^{qr}TU^{-qs} = \begin{bmatrix} a & \nu N + q(rd - sa) \\ 0 & d \end{bmatrix}.$$

Now choose s_0 to satisfy

$$s_0 a \equiv rd \pmod{q_1},$$

which is possible since $(a, q_1) = 1$. Write

$$s_0 a = rd + hq_1, \quad \text{where} \quad h \in \mathbb{Z},$$

and put $s = s_0 + h'q_1$. Then

$$\nu N + q(rd - sa) = N(\nu - h - h'a) = N\mu,$$

say, and the integer h' can be chosen so that $0 \leq \mu < a$. It follows that

$$U^{qr}T = T_1 U^{qs},$$

where $T_1 \in \boldsymbol{A}_h^a$. But T_1 and s are uniquely determined by T and r, by (9.1.56). It follows that

$$\Gamma_{U^q} \cdot \boldsymbol{D}_n^a \subseteq \boldsymbol{A}_n^a \cdot \Gamma_{U^q}$$

and the reverse inclusion is proved similarly.

Write

$$\Gamma = \Gamma_{U^q} \cdot \boldsymbol{B}, \quad \Gamma_n = R_a^{-1} \Gamma_{U^q} R_a \cdot \boldsymbol{B}_n^a, \tag{9.3.16}$$

where, as previously, a is any positive divisor of n. Then, by (9.2.17) and (9.3.15, 16),

$$\begin{aligned} \Gamma \cdot \boldsymbol{T}_n^+ = {}^+\boldsymbol{T}_n \cdot \Gamma_n &= \bigcup_{a|n} \boldsymbol{A}_n^a \cdot R_a \Gamma_n \\ &= \bigcup_{a|n} \boldsymbol{A}_n^a \cdot \Gamma_{U^q} \cdot R_a \boldsymbol{B}_n^a \\ &= \Gamma_{U^q} \cdot \bigcup_{a|n} \boldsymbol{D}_n^a R_a \cdot \boldsymbol{B}_n^a. \end{aligned} \tag{9.3.17}$$

It follows that, if $S \in \Gamma$ and $T \in \mathbf{T}_n^+$, then there exist unique integers $r = 0, 1$, $s \in \mathbb{Z}$ and $a|n$, and unique $S_n' \in \mathbf{B}_n^a$, $T' \in \mathbf{D}_n^a R_a$ such that

$$ST = (-I)^r U^{sq} T' S_n'. \tag{9.3.18}$$

Since $T \equiv T' \equiv J_n \pmod{N}$, we have

$$SJ_n \equiv (-I)^r U^{sq} J_n S_n' \equiv (-I)^r U^{sq} S' J_n \pmod{N},$$

where S' and S_n' are related as in theorem 8.3.1. Hence, since $(n, N) = 1$,

$$S \equiv (-I)^r U^{sq} S' \pmod{N}$$

and so

$$\begin{aligned} u(S) &= u^r(-I) u(U^{sq}) u(S') \\ &= (-1)^{kr} e^{2\pi i \kappa s} u_n(S_n'). \end{aligned} \tag{9.3.19}$$

Now, by (5.1.9), if $k > 2$,

$$G_I(z; m; \Gamma, k, u) = \sum_{S \in \mathbf{B}} \frac{\exp\left\{\dfrac{2\pi i}{q}(m+\kappa) Sz\right\}}{u(S)(S : z)^k}.$$

We abbreviate by writing

$$G_I(z, m) = G_I(z; m; \Gamma, k, u).$$

Then, by (9.2.3),

$$G_I(z, m)|T_n = n^{k-1} \sum_{S \in \mathbf{B}} \sum_{T \in \mathbf{T}_n^+} \frac{\exp\left\{\dfrac{2\pi i}{q}(m+\kappa) STz\right\}}{u(S)(ST : z)^k}.$$

In this sum we use (9.3.18, 19), observing that

$$STz = sq + T'S_n'z, \quad (ST:\ z) = (-1)^r (T'S_n' : z),$$

so that we have, by (9.3.17),

$$G_I(z, m)|T_n = n^{k-1} \sum_{a|n} \sum_{S_n' \in \mathbf{B}_n^a} \sum_{T' \in \mathbf{D}_n^a R_a} \frac{\exp\left\{\dfrac{2\pi i}{q}(m+\kappa) T'S_n'z\right\}}{u_n(S_n')(T'S_n' : z)^k}.$$

Now, by (9.1.53), $(T'S'_n: z) = d(R_aS'_n: z)$, and, by (9.3.6, 12),

$$\sum_{T' \in \boldsymbol{D}_n^a R_a} \exp\left\{\frac{2\pi i}{q}(m+\kappa)T'S'_n z\right\} = \sum_{\nu=0}^{d-1} \exp\{2\pi i(mq_1+p) \times (aR_aS'_n z + \nu N)/Nd\}$$

$$= d\exp\left\{\frac{2\pi i}{q}(m_{n,d}+\kappa_{n,d})R_aS'_n z\right\},$$

when $d|(mq_1+p)$, and is zero otherwise. Accordingly,

$$G_I(z, m)|T_n = \sum_{d|(n, mq_1+p)} (n/d)^{k-1} \sum_{S'_n \in \boldsymbol{B}_n^a} \frac{\exp\left\{\dfrac{2\pi i}{q}(m_{n,d}+\kappa_{n,d})R_aS'_n z\right\}}{u_n(S'_n)(R_aS'_n: z)^k},$$

from which (9.3.13) follows. Because of the absolute convergence of the Poincaré series, the various processes carried out above are all justified.

To derive (9.3.14) for any $L \in \Gamma(1)$, we use the fact that

$$G_L(z; m; \Gamma, k, u) = G_I(z; m; \Gamma^M, k, u^M)|L.$$

This follows from (5.1.10) and we have put $M = L^{-1}$. We obtain the required result by theorem 9.2.3, since, by theorem 8.3.1,

$$((\Gamma^M)_n)^{L_n} = \Gamma_n, \quad ((u^M)_n)^{L_n} = u_n,$$

$$n(\infty, \Gamma^M) = n(L^{-1}\infty, \Gamma) \quad \text{and} \quad \kappa(\infty, \Gamma^M, u^M) = \kappa(L^{-1}\infty, \Gamma, u).$$

The corresponding results for $k = 2$ can be obtained by similar methods using theorem 5.1.5 and the definitions and limiting processes employed in §5.7. It is, of course, necessary first to study the effect of T_n on the series $G_L(z; m, N; s)$ defined in (5.7.1) for $s > 0$ and then let $s \to 0+$. For this purpose the definition of T_n given in (9.2.3) must be modified by replacing $n^{k-1}(T: z)^{-k}$ by

$$n^{1+2s}(T: z)^{-2}|T: z|^{-2s}.$$

This will give the results required for $\Gamma = \bar{\Gamma}(N)$ and the corresponding results for general Γ can be obtained in a similar way or by applying theorem 5.1.5.

As particular cases of the theorem we have the results stated in

Corollary 9.3.1. *Let $(n, N) = 1$ and let k be an integer satisfying $k \geq 2$.* (i) *For any $L \in \Gamma(1)$ and $m \in \mathbb{Z}$,*

$$G_L(z; m; \bar{\Gamma}(N), k, v)|T_n = \sum_{d|(m,n)} (n/d)^{k-1} G_{R_a L_n}(z; mn/d^2; \bar{\Gamma}(N), k, v), \qquad (9.3.20)$$

where $ad = n$. In particular, for $m = 0$, whenever $(C, D, N) = 1$,

$$E_k(z; C, D; N)|T_n = \sum_{d|n} (n/d)^{k-1} E_k(z; Cd^{-1}, Da; N). \qquad (9.3.21)$$

Here E_k is defined by (5.5.7) *and the inverse d^{-1} is taken modulo N.*

(ii) *Let χ be any character modulo N satisfying* (8.1.7). *Then, for any $m \in \mathbb{Z}$ and t_1 dividing N,*

$$G_I(z; m; \Gamma_0(N, t_1), k, \chi)|T_n = \sum_{d|(m,n)} (n/d)^{k-1} \chi(n/d) G_I(z; mn/d^2; \Gamma_0(N, t_1), k, \chi). \qquad (9.3.22)$$

Theorem 9.3.2. *Let $\Gamma = \bar{\Gamma}(N)$ or $\Gamma_0^0(N)$ and $k > 2$. Then the Poincaré series $G_I(z; m; \Gamma, k, v)$, where $v = \chi$ when $\Gamma = \Gamma_0^0(N)$, has divisor (m, N). Accordingly, the spaces $H_t(N, k)$ and $H(N, k, \chi, t)$ are spanned by the functions $G_I(z; m; \Gamma, k, v)$ for which $(m, N) = t$, in the cases $\Gamma = \bar{\Gamma}(N)$ and $\Gamma = \Gamma_0^0(N)$, respectively. In particular, the spaces $\{N, k\}_0|D_t$ and $\{N, k, \chi, t\}_0$ are spanned by the corresponding functions with $m > 0$ and $(m, N) = t$, and this holds also for $k = 2$. Finally, the spaces $\{N, k\}_0|D_t$, and also the spaces $\{N, k, \chi, t\}_0$, are mutually orthogonal for $t|N$.*

Proof. We shall write, for convenience (and for fixed k and N),

$$G(z, m) = G_I(z; m; \bar{\Gamma}(N), k, v), \qquad (9.3.23)$$

and

$$G(z, m, \chi) = G_I(z; m; \Gamma_0^0(N), k, \chi). \qquad (9.3.24)$$

If $\bar{\Gamma}(N) = \Gamma_{U^N} \cdot \boldsymbol{B}$, then, for any $l \in \mathbb{Z}$,

$$\Gamma_{U^N} \cdot \boldsymbol{B} U^l = \bar{\Gamma}(N) U^l = U^l \bar{\Gamma}(N) = \Gamma_{U^N} \cdot U^l \boldsymbol{B},$$

so that

$$G(z, m)|U^l = \sum_{S \in B} (S:z)^{-k} \exp\left\{\frac{2\pi i m}{N} U^l S z\right\}$$
$$= e^{2\pi i l m/N} G(z, m).$$

It now follows from (8.1.15) that, for any $t|N$,

$$G(z, m)|D_t = \delta G(z, m),$$

where $\delta = 1$ or 0 according as $(m, N) = t$ or not.

Further, by theorem 5.1.5,

$$G(z, m, \chi) = \sum_a \bar{\chi}(a) G(z, m)|R_a,$$

where the summation is over a prime to N for which the matrices R_a run through a transversal of $\bar{\Gamma}(N)$ in $\Gamma_0^0(N)$. By (8.2.11) and the following remark, the operators D_t and R_a commute, so that

$$G(z, m, \chi)|D_t = \delta G(z, m, \chi).$$

The last part of the theorem follows from theorem 5.2.2, or from theorem 8.1.3.

We now study the Fourier coefficients of Poincaré series and introduce for this purpose a more convenient notation. We confine our attention to cusp forms (so that $m > 0$) and write

$$g(z, m) := m^{k-1} G(z, m) \tag{9.3.25}$$

$$=: \sum_{r=1}^{\infty} c(r, m)\, e^{2\pi i r z/N}. \tag{9.3.26}$$

We have already in §6.2 used a similar notation for the case when $N = 1$. Similarly, for any character χ modulo N, satisfying (8.1.7), write

$$g(z, m, \chi) := m^{k-1} G(z, m, \chi) \tag{9.3.27}$$

$$=: \sum_{r=1}^{\infty} c(r, m; \chi)\, e^{2\pi i r z/N}. \tag{9.3.28}$$

As usual $k = 2$ is allowed as a limiting case. Write also, for any $n \in \mathbb{Z}_N$,

$$g(z, m)|T_n = \sum_{r=1}^{\infty} C(r, m, n)\, e^{2\pi i r z/N} \tag{9.3.29}$$

and

$$g(z, m, \chi)|T_n = \sum_{r=1}^{\infty} C(r, m, n; \chi)\, e^{2\pi i r z/N}. \qquad (9.3.30)$$

We observe that, by (8.6.9),

$$g(z, m)|K = g(z, m),\ g(z, m, \chi)|K = g(z, m, \bar{\chi}). \quad (9.3.31)$$

The following theorem shows that the Fourier coefficients just defined have interesting symmetric properties; some of these have been obtained previously in theorem 6.2.1 in the case when $N = 1$.

Theorem 9.3.3. *Let r, m and n be positive integers with $(n, N) = 1$ and suppose that χ is a character modulo N. Then $c(r, m)$ and $c(r, m; \chi)$ are non-zero only when $(r, N) = (m, N)$, and $c(r, m)$ is real. Further,*

$$c(m, r) = c(r, m), \qquad (9.3.32)$$

$$c(m, r; \chi) = \overline{c(r, m; \chi)} = c(r, m; \bar{\chi}), \qquad (9.3.33)$$

and

$$C(r, m, n; \chi) = \chi(n)\overline{C(m, r, n; \chi)} = \chi(n) C(m, r, n; \bar{\chi}). \qquad (9.3.34)$$

Finally,

$$g(z, m, \chi)|T_n = \sum_{d|(m,n)} \chi(n/d) d^{k-1} g(z, mn/d^2, \chi). \qquad (9.3.35)$$

Proof. That $c(r, m)$ and $c(r, m; \chi)$ are non-zero only when $(r, N) = (m, n)$ follows from theorem 9.3.2. That $c(r, m)$ is real follows from (9.3.26, 31). By theorem 5.2.2,

$$(g(z, m), g(z, r)) = Ac(r, m),$$

where A is a non-zero number depending only on k and N. Equation (9.3.32) follows from this and the properties of the inner product. We obtain (9.3.33) in a similar way from (9.3.28, 31) and theorem 5.2.2.

From (9.2.8) we have

$$C(r, m, n; \chi) = \sum_{d|(r,n)} \chi(d) d^{k-1} c(rn/d^2, m; \chi), \qquad (9.3.36)$$

while, by (9.3.22), (9.3.35) follows and this gives

$$C(r, m, n; \chi) = \sum_{d|(m,n)} \chi(n/d)d^{k-1}c(r, mn/d^2; \chi)$$

$$= \chi(n) \sum_{d|(m,n)} \bar{\chi}(d)d^{k-1}c(mn/d^2, r; \bar{\chi})$$

$$= \chi(n)C(m, r, n; \bar{\chi}),$$

by (9.3.33). Finally, by (9.3.36), the conjugate complex number to $C(r, m, n; \chi)$ is $C(r, m, n; \bar{\chi})$, and this completes the proof of the theorem.

Theorem 9.3.4. *Let f and g be members of $\{N, k, \chi\}_0$, where χ is a character modulo N and k is a positive integer. Then, for all $n \in \mathbb{Z}_N^+$,*

$$(f|T_n, g) = \chi(n)(f, g|T_n). \tag{9.3.37}$$

Proof. Assume that $k > 2$. Since the forms $g(z, m, \chi)$ span $\{N, k, \chi\}_0$, it suffices to prove that

$$(g(z, r, \chi)|T_n, g(z, m, \chi)) = \chi(n)(g(z, r, \chi), g(z, m, \chi|T_n)),$$

whenever r and m are positive and $(n, N) = 1$. This, by theorem 5.2.2, is equivalent to

$$C(m, r, n; \chi) = \chi(n)\overline{C(r, m, n; \chi)},$$

which follows from (9.3.34).

We can obtain the result for $k = 2$ as a limiting case. However, because of the importance of the result, we now give an independent proof valid for any positive integer k.

Because of the commutative properties of the Hecke operators, it is clearly sufficient to prove (9.3.37) when n is a prime p that does not divide N. In this case $[\![J_p]\!] = (J_p)$ (by (9.1.40), for example).

Now, by theorem 9.1.2, we can find a common transversal $\boldsymbol{C}_p$ such that

$$(J_p) = \Gamma(n) \cdot \boldsymbol{C}_p = \boldsymbol{C}_p \cdot \Gamma(N).$$

Write

$$\boldsymbol{C}_p^* := \{L \in \Omega_p^* : L = R_p(pT^{-1}), T \in \boldsymbol{C}_p\}.$$

Note that $pT^{-1} \in \Omega_p^*$ and that $L \equiv J_p \pmod{N}$ for $T \in \boldsymbol{C}_p$ and $L \in \boldsymbol{C}_p^*$. It follows that

$$\Gamma(N)\boldsymbol{C}_p^* \subseteq (J_p).$$

We show that

$$(J_p)=\Gamma(N)\cdot \boldsymbol{C}_p^*.$$

For this purpose it is enough to take any $L\in(J_p)$ and prove that $L\in\Gamma(N)\boldsymbol{C}_p^*$.

We have $L=S_1T_1$, where S_1 (and later S_2, S_3) belongs to $\Gamma(N)$ and $T_1\in\boldsymbol{C}_p$. Write

$$L_1=(pT_1^{-1})S_1^{-1}R_p,$$

so that $L_1\in(J_p)$ and so $L_1=T_2S_2$, where $T_2\in\boldsymbol{C}_p$. Hence

$$L=S_1T_1=pR_pS_2^{-1}T_2^{-1}=S_3R_p(pT_2^{-1}),$$

and so $L\in\Gamma(N)\boldsymbol{C}_p^*$.

Accordingly, by theorem 5.2.1(v),

$$(f|T_p,g)=p^{\frac{1}{2}k-1}\sum_{T\in\boldsymbol{C}_p}(f|T,g)=p^{\frac{1}{2}k-1}\sum_{T\in\boldsymbol{C}_p}(f,g|T^{-1}).$$

Since

$$g|T^{-1}=\bar{\chi}(p)g|R_pT^{-1}=\bar{\chi}(p)g|R_p(pT^{-1}),$$

we deduce that

$$\begin{aligned}(f|T_p,g)&=\chi(p)p^{\frac{1}{2}k-1}\sum_{L\in\boldsymbol{C}_p^*}(f,g|L)\\&=\chi(p)(f,g|T_p),\end{aligned}$$

as required.

9.4. Eigenforms. In this section we confine our attention to cusp forms. We have seen that the space $\{N,k\}_0$ of cusp forms of level N and weight k is the direct sum of subspaces

$$\{N,k,\chi,t\}_0:=\{N,k\}_0|D_tR^{\chi} \tag{9.4.1}$$

of fixed character χ and divisor t; see (8.2.12). Theorem 9.2.6 shows that each subspace $\{N,k,\chi,t\}_0$ is invariant under the family of operators $\{T_n^{(t)}:n\in\mathbb{Z}^+\}$, and it is our object to try to find a basis for $\{N,k,\chi,t\}_0$ consisting of eigenfunctions for these operators; for it turns out that the Fourier coefficients of such cusp forms have interesting multiplicative properties.

A function f in $\{N,k,\chi\}_0^{(t)}$ or $\{N,k,\chi,t\}_0$ is said to be *primitive* if it has a simple zero at ∞, i.e. if, in its expansion (9.2.21), $A_1\neq 0$; we have, of course, $A_r = 0$ for $r \leq 0$. Otherwise f is said to be *imprimitive*; in particular, the zero form is imprimitive.

We begin by stating a general theorem in algebra concerning normal operators and finite-dimensional Hilbert (i.e. inner product) spaces; see Gantmacher (1960), p. 291 (theorem 11).

Theorem 9.4.1. *Let M be a finite-dimensional Hilbert space over the complex field and let $\mathcal{T}$ be a family of pairwise commuting normal operators on M, under which M is invariant. Then there exists an orthogonal basis B for M, whose elements are eigenvectors of all the operators in the family.*

We note that each operator $T \in \mathcal{T}$ determines a unique adjoint operator T^*, with the property that

$$(f, g|T^*) = (f|T, g)$$

for all f and g in M, and recall that T is defined to be normal when it commutes with T^*. In particular, if

$$(f|T, g) = c_T(f, g|T)$$

for all f and g in M, where c_T depends only on T, then T^* is just a constant multiple of T and so T is normal.

The basis B has the property that, for each $f \in B$,

$$f|T = \lambda(f, T)f \quad \text{for all } T \in \mathcal{T},$$

where the eigenvalue $\lambda(f, T)$ is a complex number depending only on T and on the basis vector f. Further, $(f, g) = 0$ for any two different elements f and g of B. It is clear that each member of B may be replaced by a constant non-zero multiple of itself, and this allows us to 'normalize' B in various ways. A standard way of doing this is to choose B to be an 'orthonormal' basis, so that $(f, f) = 1$ for each $f \in B$; however, for our purposes, this is not a convenient method of normalization and we shall later on, and in certain special cases, adopt a different method.

We call the elements of B *eigenforms*, rather than eigenvectors, since in our application of the theorem they will be modular forms. More generally, any element of M, that is an eigenvector for all the operators in $\mathcal{T}$, will be called an eigenform.

We make our first application of the theorem to the space

$$M = \{\Gamma, k, u\}_0 \cap M_0,$$

where (i) M_0 is any subspace of $M(N, k)$ invariant under the operators T_n with $(n, N) = 1$, (ii) Γ is a group satisfying

$$\Gamma_0^0(N) \subseteq \Gamma \subseteq \Gamma(1),$$

and (iii) u is a MS of positive integral weight k on Γ whose restriction to $\Gamma_0^0(N)$ is the character χ modulo N. Accordingly, M is a subspace of $\{N, k, \chi\}_0$ and, by theorem 9.2.3, is invariant under the family of operators T_n where $n \in \mathbb{Z}_\Gamma^+$; see (8.3.12). Moreover, these operators are normal, by theorem 9.3.4. Hence M possesses an orthogonal basis of eigenforms for the operators T_n, where $n \in \mathbb{Z}_\Gamma^+$.

There are various possible choices for the space M_0, such as $\{N, k, \chi\}_0$, $\{N, k, \chi\}_0^{(t)}$ and $\{N, k, \chi, t\}_0$. Suppose that M_0 is one of the latter two spaces and that f is an eigenform of the kind described. Its Fourier expansion is then given, as in (9.2.21), by

$$f(z) = \sum_{r=1}^{\infty} A_r e^{2\pi irz/t_1}. \tag{9.4.2}$$

Let

$$f(z)|T_n = \lambda_f(n) f(z) \quad (n \in \mathbb{Z}_\Gamma^+). \tag{9.4.3}$$

It follows from (9.2.28, 30) that

$$\lambda_f(n) A_r = \sum_{d|(r,n)} \chi(d) d^{k-1} A_{rn/d^2} \quad (r \geq 1, n \in \mathbb{Z}_\Gamma^+). \tag{9.4.4}$$

In particular,

$$\lambda_f(n) A_1 = A_n \quad (n \in \mathbb{Z}_\Gamma^+). \tag{9.4.5}$$

Accordingly, if f is primitive (so that $A_1 \neq 0$), the eigenvalue $\lambda_f(n)$ is just a constant multiple of the Fourier coefficient A_n. In this case we may replace the basis form f by f/A_1, so obtaining an eigenform with first Fourier coefficient 1; such an eigenform we call a *normalized eigenform*.

However, if $A_1 = 0$, this is not possible and we take m to be the least positive integer such that $A_m \neq 0$; thus $m > 1$. It is easy to see that m is not divisible by any integer n in $\mathbb{Z}_\Gamma^+$ apart from $n = 1$. For, if $m = nr$ where $n \in \mathbb{Z}_\Gamma^+$ and $n > 1$, then $r < m$ and

$$\lambda_f(n) A_r = \sum_{d|(r,n)} \chi(d) d^{k-1} A_{m/d^2} = A_m,$$

by (9.4.4). This leads to the contradictory conclusion that $A_r \neq 0$. It follows that m is composed entirely of factors belonging to $\mathbb{Z}^+ - \mathbb{Z}_\Gamma^+$. Moreover, if f has divisor t, then $(m, t_1) = 1$.

By applying (9.3.37) to $g = f$ we obtain a formula relating eigenvalues to their complex conjugates, namely

$$\lambda_f(n) = \chi(n)\overline{\lambda_f(n)} \quad (n \in \mathbb{Z}_\Gamma^+). \tag{9.4.6}$$

This shows, in particular, that $\lambda_f(n)$ is real when χ is the principal character.

We now restrict our attention to the space $\{N, k, \chi, t\}_0$, i.e. to the case when $\Gamma = \Gamma_0(N, t_1)$; here, of course, $\Gamma_l = \Gamma$ for all l prime to N and $\mathbb{Z}_\Gamma = \mathbb{Z}_N$. We therefore have the following

Theorem 9.4.2. *For each divisor t of N and character χ modulo N the space $\{N, k, \chi, t\}_0$ possesses an orthogonal basis (which is empty if the space has zero dimension) of eigenforms of all the Hecke operators T_n, where $(n, N) = 1$. If f is such a non-zero eigenform, with eigenvalues $\lambda_f(n)$ for T_n, and if f has the Fourier expansion (9.4.2), then*

$$\lambda_f(n)A_r = \sum_{d|(r,n)} \chi(d)d^{k-1}A_{rn/d^2} \quad \text{for } r \geq 1, (n, N) = 1 \tag{9.4.7}$$

and, in particular,

$$\lambda_f(n)A_1 = A_n \quad \text{for } (n, N) = 1. \tag{9.4.8}$$

If m is the smallest positive integer such that $A_m \neq 0$, then $m|t^\infty$ and $(m, t_1) = 1$; further, if $m > 1$, then $A_n = 0$ whenever $(n, N) = 1$.

The last sentence follows immediately from (9.4.8).

Observe that, if we know in addition that f is an eigenform of the operators $T_n^{(t)}$ (for $n|N^\infty$), with eigenvalues $\lambda_f(n)$, then (9.4.8) holds for all positive integers n.

We now study in greater detail the subset E of the space

$$M = \{N, k, \chi, t\}_0 \tag{9.4.9}$$

consisting of cusp forms that are eigenforms for all the operators T_n $(n \in \mathbb{Z}_N)$. Let $f \in E$, so that

$$f|T_n = \lambda_f(n)f,$$

say, for all $n \in \mathbb{Z}_N$.

We denote by $[f]$ the subset of E consisting of all eigenforms having the same eigenvalues $\lambda_f(n)$ as f $(n \in \mathbb{Z}_N)$. We include the zero form in $[f]$, so that $[f]$ is a subspace of M; we call $[f]$ a *class* of eigenforms, or an *eigenclass* (in M). Two different classes clearly have only the zero form in common. It is also convenient to regard the set consisting of the zero form only as a class and to denote it by $[0]$; we refer to this class as the *trivial* class. The vector space M, being spanned by E, can be expressed as a vector sum of different classes.

Theorem 9.4.3. *Let C be a class of eigenforms in $\{N, k, \chi, t\}_0$ with eigenvalues $\lambda(n)$ $(n \in \mathbb{Z}_N)$. Then C is invariant under the operators (i) $T_n^{(t)}$ $(n \in \mathbb{Z}^+)$, (ii) $H_q \in \boldsymbol{H}_q(N, t_1)$, where $qN_\chi|N, (q, Nt_1/q) = 1$, and (iii) $L_p^\chi(N, t_1)$, where p is a prime dividing $(t, N/N_\chi)$ but not t_1. Moreover, if C is not the trivial class, C contains a normalized eigenform F with Fourier expansion*

$$F(z) = \sum_{r=1}^{\infty} \lambda(r)\, e^{2\pi i r z/t_1} \quad (z \in \mathbb{H}), \tag{9.4.10}$$

and

$$\lambda(n)\lambda(r) = \sum_{d|(r,n)} \chi(d) d^{k-1} \lambda(rn/d^2), \tag{9.4.11}$$

whenever $r \geq 1$ and $(n, N) = 1$.

Proof. We note that the coefficient $\lambda(r)$ of F is the eigenvalue of T_r, when $(r, N) = 1$; in particular, $\lambda(1) = 1$. We do not claim that $\lambda(r)$ is an eigenvalue when $(r, N) > 1$.

Suppose that $f \in C$, and that its Fourier series is given by (9.4.2). Then, if $(n, N) = 1$ and $q \geq 1$,

$$(f|T_q^{(t)})|T_n = (f|T_n)|T_q^{(t)} = \lambda(n) f|T_q^{(t)},$$

so that $f|T_q^{(t)}$, which is a member of M, belongs also to C. Similar arguments, using theorems 9.2.7, 9, show that $f|H_q$ and $f|L_p^\chi$ belong to C, under the conditions stated; for χ_q is the principal character modulo (q, N) when N_χ divides N/q.

If C is not the trivial class, we may assume that $f(z)$ has a smallest non-vanishing Fourier coefficient A_m, where, by theorem 9.4.2,

$m|t^\infty$. Then, by (9.2.29), function

$$F = A_m^{-1} f | T_m^{(t)}$$

is a normalized eigenform in C and has the stated properties.

We now study the space M of (9.4.9) in greater detail and remove from it certain forms of smaller level, for which, by an inductive argument, we assume the problem of finding eigenforms of *all* the T_n $(n \in \mathbb{Z})$ to have been solved.

Let t_χ be the greatest factor of $(t, N/N_\chi)$ that is prime to t_1. In particular, $t_\chi = (t, N/N_\chi)$ whenever t is a unitary divisor of N (see end of §8.3). We write

$$M(d) := \{N/d, \chi, t/d\}_0 \quad (d|t_\chi), \tag{9.4.12}$$

so that $M = M(1)$. Note that χ is a character modulo N/d, since N_χ divides N/d. The space $M(d)$ has level N/d, but has the same character and codivisor as M.

Moreover, for each δ dividing d, the space $M(d)|J_\delta^{-1}$ is a subspace of M. For, if $g \in M(d)$, then, by theorem 8.3.2, $g|J_\delta^{-1} \in \{N, k, \chi\}^{(t)}$. But, since $(d, t_1) = 1$, it follows from a study of the Fourier expansion of $g|J_\delta^{-1}$ that it has divisor t. Hence $g|J_\delta^{-1} \in M$.

We now define M^- to be the vector sum of all the spaces $M(d)|J_\delta^{-1}$, where

$$d|t_\chi, d > 1, \delta|d; \tag{9.4.13}$$

it may of course happen that no integers d and δ satisfy (9.4.13), in which case we take M^- to consist of the zero form only; this happens, for example, when $t_\chi = 1$. Finally, we define M^+ to be the orthogonal complement of M^- in M, so that

$$M = M^- \oplus M^+. \tag{9.4.14}$$

We shall apply Hecke operators to the spaces $M(d)$, for $d|t_\chi$. Since $M(d)$ has level N/d, we shall write $T_n(N/d)$ to make it clear how the operator is defined; we need not display the other parameters involved, since all the spaces $M(d)$ have the same codivisor t_1 and character χ.

When $(n, N) = 1$, the operators $T_n(N/d)$ and $T_n(N)$ both act on $M(d)$ with identical effect; this is perhaps most easily seen from (9.2.30). In particular, it follows that $T_n(N)$ maps $M(d)$ into itself. Moreover, $M(d)|J_\delta^{-1}$ is mapped into itself by $T_n(N)$, whenever $\delta|d$.

For, if $g \in M(d)$, then

$$g|J_\delta^{-1}T_n(N) = g|J_\delta^{-1}T_n^{(\delta)}(N) = g|T_n(N)J_\delta^{-1}. \qquad (9.4.15)$$

We deduce that M^- is invariant under the operators T_n, for $(n, N) = 1$.

The space M^+ is also invariant under T_n for $(n, N) = 1$. For, if $f \in M^+$ and $g \in M^-$, then $g|T_n \in M^-$, so that $(f, g|T_n) = 0$. It follows from theorem 9.3.4 that $(f|T_n, g) = 0$, and so $f|T_n \in M^+$.

We can form the subspaces $M(d)^+$ and $M(d)^-$ in exactly the same way for each divisor d of t_χ. They are invariant under the operators $T_n(N/d)$, whenever $(n, N/d) = 1$.

Theorem 9.4.4.

$$M^- = \oplus\{M(d)^+|J_\delta^{-1}\},$$

where the vector sum is over all

$$d|t_\chi,\ d > 1,\ \delta|d. \qquad (9.4.16)$$

Proof. Denote by $D(t_\chi)$ the set of all (d, δ) satisfying (9.4.16). We prove that, for all τ dividing t_χ,

$$M(t_\chi/\tau)^- = \oplus\{M(t_\chi d/t)^+|J_\delta^{-1}\}, \qquad (9.4.17)$$

where the summation is over all $(d, \delta) \in D(\tau)$. This is true when $\tau = 1$, since it then states that $M(t_\chi)^- = [0]$.

It is enough to assume the truth of (9.4.17) whenever τ has m or fewer prime factors and prove it for a divisor of t_χ with $m+1$ prime factors; here each repeated prime factor is to be counted separately. Accordingly, suppose that τ has m prime factors and that p is a prime dividing t_χ/τ. Then, if $\tau' = p\tau$, τ' has $m+1$ prime factors and

$$M(t_\chi/\tau')^- = \oplus\{M(t_\chi d/\tau')|J_\delta^{-1}\}, \qquad (9.4.18)$$

where the summation is over all $(d, \delta) \in D(\tau')$. Write

$$M(t_\chi d/\tau') = M(t_\chi d/\tau')^- \oplus M(t_\chi d/\tau')^+.$$

We can now apply our assumption to $M(t_\chi d/\tau')^-$, since $\tau'/d = p\tau/d$ and has m or fewer prime factors. On substitution in (9.4.18) we obtain the required result. Thus (9.4.17) holds and we deduce the theorem by taking $\tau = t_\chi$.

Theorem 9.4.5. *Let $f \in M = \{N, k, t, \chi\}_0$ and suppose that, in its Fourier expansion* (9.4.2), *$A_r = 0$ whenever $(r, \tau) = 1$, where τ is some fixed divisor of t. Then $f \in M^-$.*

Moreover, if $A_r = 0$ whenever $(r, t) = 1$, then

$$f = \sum f_d,$$

where

$$f_d \in M(d)|J_d^{-1} \subseteq M$$

and the summation is over all squarefree divisors d of t_χ with $d > 1$. If f belongs to an eigenclass C in M, then so does each f_d.

Proof. Without loss of generality we may suppose that τ is squarefree. Let Q be the product of all primes that divide τ but not t_1 and put

$$Q_0 = (Q, N/N_\chi), \quad Q = Q_0 Q_1. \tag{9.4.19}$$

We prove that

$$f \Big| \sum_{d|Q_0} \mu(d) L_d^\chi(N, t_1) = 0 \tag{9.4.20}$$

where the operator L_d^χ is defined in §8.5.

By (8.3.17), $f|D'_{t/Q} = 0$. Hence, if $Q_0 = 1$, it follows from theorem 8.3.4 (with $u = Q, \tau = t/Q$) and (9.4.19) that $f = 0$; i.e. (9.4.20) holds when $Q_0 = 1$.

We now prove (9.4.20) in general, by induction. We assume that it holds for any τ dividing t for which Q_0 is a product of n different primes ($n \geq 0$) and for any function $f \in M$ for which $A_r = 0$ whenever $(r, \tau) = 1$. Now suppose that τ is a divisor of t for which Q_0 is a product of $n+1$ different primes and that f belongs to M and is such that $A_r = 0$ whenever $(r, \tau) = 1$. We may write

$$Q_0 = q q_0,$$

where q is prime and q_0 is a product of n different primes, each different from q. Put

$$g := f \Big| \sum_{d|q_0} \mu(d) d^{-\frac{1}{2}k} T_d^{(t)} J_d^{-1},$$

so that, by the second part of theorem 9.2.8 (with $u = q_0$),

$$g \in \{N q_0, k, \chi, t q_0\}_0$$

and

$$g(z) = \sum_{\substack{r=1\\ r, q_0 t_1}}^{\infty} A_r{}^{2\pi irz/t_1}.$$

Now write

$$G := f \Big| \textstyle\sum_{d|Q_0} \mu(d) d^{-\frac{1}{2}k} T_d^{(t)} J_d^{-1},$$

so that, similarly, $G \in \{NQ_0, k, \chi, tq_0\}$ and

$$G(z) = \sum_{\substack{r=1\\ (r, Q_0 t_1)=1}}^{\infty} A_r e^{2\pi irz/t_1}.$$

By (8.3.17), $G|D'_{t/Q_1} = 0$ and, since $(Q_1, NQ_0/N_\chi) = 1$, it follows from theorem 8.3.4 (with $u = Q_1$, $\tau = t/Q_1$) that $G = 0$. But, by (9.2.42) and (9.2.25),

$$\begin{aligned} G &= f \Big| \sum_{d|q_0} \sum_{\delta|q} \mu(d\delta)(d\delta)^{-\frac{1}{2}k} T_d^{(t)} J_d^{-1} T_\delta^{(td)} J_\delta^{-1} \\ &= f \Big| \sum_{d|q_0} \sum_{\delta|q} \mu(d) d^{-\frac{1}{2}k} T_d^{(t)} J_d^{-1} \mu(\delta) \delta^{-\frac{1}{2}k} T_\delta^{(tq_0)} J_\delta^{-1} \\ &= g|\{I - q^{-\frac{1}{2}k} T_q^{(tq_0)} J_q^{-1}\}, \end{aligned}$$

for $f|T_d^{(t)} J_d^{-1} \in M(\Gamma_0(Nd, t_1), k, \chi)$ and $(q_0/d, \delta) = 1$. Accordingly,

$$g(z) = g(z)|q^{-\frac{1}{2}k} T_q^{(tq_0)} J_q^{-1} = \sum_{\substack{r=1\\ (r, q_0 t_1)=1}}^{\infty} A_{rq}\, e^{2\pi irqz/t_1}$$

and so

$$g|J_q U^{t_1} = g|J_q.$$

We now apply theorem 8.5.1(v), with N replaced by Nq_0 and N' by t_1. The condition on N_χ is that it divides $Nq_0 t_1/q$ and this is satisfied by (9.4.19). Accordingly,

$$g|L_q^\chi(Nq_0, t_1) = g. \tag{9.4.21}$$

Since, by (8.5.6), $f|L_q^\chi(Nd, t_1) = f|L_q^\chi(Nq_0, t_1)$ for any $d|q_0$, we have, by theorem 9.2.9,

$$\{f - f|L_q^\chi(N, t_1)\}\bigg|\sum_{d|q_0} \mu(d)d^{-\frac{1}{2}k}T_d^{(t)}J_d^{-1}$$

$$= \{f\bigg|\sum_{d|q_0} \mu(d)d^{-\frac{1}{2}k}T_d^{(t)}J_d^{-1}\}|\{I - L_q^\chi(Nq_0, t_1)\}$$

$$= g - g|L_q^\chi(Nq_0, t_1)$$

$$= 0, \tag{9.4.22}$$

by (9.4.21).

Now write

$$F = f - f|L_q^\chi(N, t_1), \tag{9.4.23}$$

so that $F \in \{N, k, \chi, t\}_0$ by theorem 9.2.9. Write

$$F(z) = \sum_{r=1}^{\infty} b_r e^{2\pi i r z/t_1}.$$

From (9.4.22) and theorem 9.2.8, we deduce that $b_r = 0$ whenever $(r, q_0 t_1) = 1$, and therefore whenever $(r, \tau/q) = 1$. We now apply our induction hypothesis to F, with τ replaced by τ/q. Then Q_0 is replaced by q_0, which has n prime factors and so

$$F\bigg|\sum_{d|q_0} \mu(d)L_d^\chi(N, t_1) = 0.$$

It follows from (9.4.23) and theorem 8.5.1(vi) that (9.4.20) holds. Since we have shown that it holds when $Q_0 = 1$, it follows that (9.4.20) holds for any function f satisfying the conditions of the theorem.

But (9.4.20) may be written as

$$f = \sum_{\substack{d|Q_0 \\ d>1}} f_d,$$

where

$$f_d = -\mu(d)f|L_d^\chi(N, t_1).$$

Hence, by theorem 9.2.9, both f_d and $f_d|J_d$ belong to M; further, by the same theorem,

$$f_d|J_d \in \{N/d, k, \chi, t/d\}_0 = M(d).$$

(Note that, although t has become t/d, the codivisor t_1 is unaltered.) Accordingly, $f_d \in M^-$ $(d>1)$, and therefore $f \in M^-$.

The last part of the theorem follows by taking τ to be the squarefree kernel of t, and by theorem 9.4.3(iii) and 8.5.1(vi).

Theorem 9.4.6. *Let C be a non-trivial class of eigenforms in $M = \{N, k, \chi, t\}_0$, with eigenvalues $\lambda(n)$ for $(n, N) = 1$. Then either* (i) $C \subseteq M^+$, *or* (ii) $C \subseteq M^-$.

If (i) *holds, then* $\dim C = 1$ *and each form in C is a scalar multiple of a unique normalized eigenform F, which we call a* newform *in M (of character χ and divisor t).*

Finally, (i) *always holds when $t_\chi | t_1^\infty$; in particular, this occurs when $t | t_1^\infty$.*

Proof. Let d be any divisor of t_χ and consider the spaces $C \cap M(d)^+$ and $C \cap M(d)^-$. Each of these spaces is invariant under the operators $T_m(N/d)$, where $(m, N/d) = 1$. Since $M(d)^+$ and $M(d)^-$ are invariant under these operators, we need only show that, if $f \in C \cap M(d)^+$, for example, then $f | T_m(N/d) \in C$. For we have, if $(n, N) = 1$,

$$f | T_m(N/d) T_m(N) = f | T_m(N/d) T_n(N/d) = f | T_n(N/d) T_m(N/d)$$
$$= f | T_n(N) T_m(N/d) = \lambda(n) f | T_m(N/d).$$

It now follows that each of the spaces $C \cap M(d)^+$, $C \cap M(d)^-$ is a vector sum of eigenclasses of the operators $T_m(N/d)$ for $(m, N/d) = 1$. If $C(d)$ is one of these eigenclasses, then $C(d) \subseteq C$ and every form in $C(d)$ is an eigenform for the operators $T_m(N/d)$ and has the eigenvalues $\lambda(m)$, say, for these operators, where $(m, N/d) = 1$.

Moreover,

$$C \cap M(d) = \{C \cap M(d)^+\} \oplus \{C \cap M(d)^-\}. \qquad (9.4.24)$$

It is enough to prove that, if $f \in C \cap M(d)$, then f belongs to the right-hand side of (9.4.24). Since $f \in M(d)$, we may write $f = f_1 + f_2$, where $f_1 \in M(d)^-$ and $f_2 \in M(d)^+$; we have to show that both f_1 and f_2 belong to C.

For any $\phi \in M(d)^-$ and $(n, N) = 1$,

$$(\lambda(n) f_1 - f_1 | T_n, \phi) = (\lambda(n) f - f_1 | T_n, \phi) = (f_2 | T_n, \phi) = 0,$$

so that $\lambda(n) f_1 | T_n$, which belongs to $M(d)^-$, belongs also to $M(d)^+$.

Accordingly,

$$f_1|T_n = \lambda(n)f_1,$$

and we can show, similarly, that $f_2|T_n = \lambda(n)f_2$; thus f_1 and f_2 are in C.

We now observe that every imprimitive form in C lies in M^-. For, if f is such a form, with expansion given by (9.4.2), then $A_n = 0$, whenever $(n, N) = 1$, by (9.4.8). It follows from theorem 9.4.5 that $f \in M^-$.

For the remainder of the proof we assume that

$$C \cap M^+ \neq [0].$$

Then every non-zero form in the vector space $C \cap M^+$ is primitive and it follows that $C \cap M^+$ has dimension 1 and is generated by a unique primitive normalized eigenform F. This is the newform referred to in the enunciation.

We now deduce, conversely, that $C \cap M^-$ consists of imprimitive forms. For, if f were a primitive form in this space, then, for some $\alpha \neq 0, f - \alpha F$ would be an imprimitive form in C and so would lie in M^-; this would imply that $F \in M^-$, a contradiction.

We now show that

$$C \cap M(d)^+ = [0] \quad \text{when} \quad d > 1. \tag{9.4.25}$$

For, if not, then $C \cap M(d)^+$ contains a nontrivial eigenclass $C(d)$, as shown above. The argument given above, applied to $C(d)$ and $M(d)$ in place of C and M, now shows that $C(d)$ contains a newform F_d, which is a primitive eigenform for all the operators $T_m(N/d)$, where $(m, N/d) = 1$. Accordingly, $F - F_d$ is an imprimitive member of C and so lies in M^-. Since $C \cap M(d)^+ \subseteq M^-$, we have $F \in M^-$, a contradiction. Thus (9.4.25) holds. We deduce from (9.4.24) that

$$C \cap M(d) = C \cap M(d)^- \quad \text{when} \quad d > 1. \tag{9.4.26}$$

We now assume that $C \cap M^- \neq [0]$ and obtain a contradiction. As a consequence of this assumption it follows that there exists a maximum divisor d_0 of t_χ such that

$$M_0 := C \cap M(d_0)^- \neq [0].$$

We note that $d_0 < t_\chi$, since $M(t_\chi) = [0]$. Then, as we have shown, there exists a nontrivial eigenclass $C(d_0)$ in M_0 consisting of eigenforms for the operators $T_m(N/d_0)$, where $(m, N/d_0) = 1$.

Let f be a non-zero form in $C(d_0)$. Since $C(d_0) \subseteq M^-$, f is an imprimitive form in $M(d_0)$. By theorem 9.4.5 (with N and t replaced by N/d_0 and t/d_0, respectively), f is a sum of forms

$$f_d = g_d | J_d^{-1} \in M(dd_0) | J_d^{-1} \subseteq M(d_0),$$

where d runs through the squarefree divisors of t_χ/d_0 and $d > 1$. At least one of the forms g_d is non-zero. Moreover,

$$g_d \in C(d_0) \cap M(dd_0).$$

For, if $(m, N/d_0) = 1$, then

$$\begin{aligned} g_d | T_m(N/d_0) &= f_d | T_m^{(d)}(N/d_0) J_d = f_d | T_m(N/d_0) J_d \\ &= \lambda(m) f_d | J_d = \lambda(m) g_d. \end{aligned}$$

Since, by (9.4.26),

$$C(d_0) \cap M(dd_0) \subseteq C \cap M(dd_0) = C \cap M(dd_0)^-,$$

we deduce that

$$C \cap M(dd_0)^- \neq [0].$$

for some $d | t_\chi/d_0$ and $d > 1$. This contradicts the definition of d_0 and it follows that $C \cap M^- = [0]$. The theorem follows; for $M^- = [0]$ when $t_\chi | t_1^\infty$.

Since M^+ is spanned by eigenclasses in M^+, it follows that M^+ is spanned by the newforms in M. From this and theorem 9.4.4 we deduce

Theorem 9.4.7. *Let $M = \{N, k, \chi, t\}_0$. Then M^+ is spanned by the newforms in M. If F_d is a newform in $M(d)^+$, where $d | t_\chi$ and $d > 1$, then $F_d | J_\delta^{-1}$ is called an* oldform *in M for each $\delta | d$. The space M^- is spanned by the set of oldforms in M.*

Theorem 9.4.7 shows that a study of the spaces $\{N, k, \chi, t\}_0$ reduces to that of finding newforms. The Fourier coefficients of these forms have remarkable properties, as is shown in the next theorem.

Theorem 9.4.8. *Let F be a newform in $\{N, k, \chi, t\}_0$, with Fourier expansion given by (9.4.10). Then* (i)

$$F | T_n^{(t)} = \lambda(n) F(n) \quad (n \in \mathbb{Z}^+), \tag{9.4.27}$$

where the Fourier coefficients $\lambda(n)$ satisfy the relations

$$\lambda(m)\lambda(n) = \sum_{d|(m,n)} \chi(d)d^{k-1}\lambda(mn/d^2) \tag{9.4.28}$$

for all positive integers m and n. In particular,

$$\lambda(mn) = \lambda(m)\lambda(n) \quad \text{when} \quad (m,n)=1, \tag{9.4.29}$$

and

$$\lambda(p^{\nu+1}) = \lambda(p)\lambda(p^\nu) - \chi(p)p^{k-1}\lambda(p^{\nu-1}), \tag{9.4.30}$$

for each prime p and positive integer ν. Further,

$$\lambda(n) = \chi(n)\overline{\lambda(n)}, \quad \text{when} \quad (n,N)=1. \tag{9.4.31}$$

(ii) *Let q be a divisor of N satisfying*

$$N = qq_1, \quad (q, q_1t_1) = 1, \quad N_\chi | q_1 \tag{9.4.32}$$

and let H_q be the member of $\boldsymbol{H}_q(N, t_1)$ given by (8.4.11) with $N' = t_1$. Then

$$F|H_q = \lambda_q F, \tag{9.4.33}$$

where

$$\lambda_q^2 = \chi_q(-1)\chi_{q'}(q) = (-1)^k\chi_{q'}(-q) \tag{9.4.34}$$

in the notation of (8.4.13); here χ_q and $\chi_{q'}$ are characters modulis $q = (q, N)$ and $q_1 = (q', N)$ respectively, where $q' = q_1t_1$.

(iii) *Finally, let p be a prime dividing N and consider the following four mutually exclusive conditions on p:*

(*a*) $p|t_1$,
(*b*) $p \nmid t_1$, $p^2|t$ *and* $p|(N/N_\chi)$,
(*c*) $p \nmid t_1$, $p^2 \nmid t$ *and* $p|(N/N_\chi)$,
(*d*) $p \nmid t_1$, $p|N_\chi$

and

$$N_\chi | \tau t_1, \tag{9.4.35}$$

where τ is the largest factor of t prime to t_1.

Then $\lambda(p) = 0$ in cases (a) and (b). In case (c)

$$\lambda(p) = -p^{\frac{1}{2}k-1}\lambda_p, \tag{9.4.36}$$

so that

$$|\lambda(p)| = p^{\frac{1}{2}k-1}.$$

Finally, in case (d),

$$|\lambda(p)| = p^{\frac{1}{2}(k-1)}. \tag{9.4.37}$$

If t is a unitary divisor of N, then (9.4.35) is superfluous, and the four cases (a), (b), (c) and (d) cover all possibilities for a prime divisor p of N.

Proof. (i) The class of eigenforms to which F belongs consists, by theorem 9.4.6, of all scalar multiples of F. By considering the first Fourier coefficient of $F|T_n^{(t)}$, and by theorem 9.4.3(i), we obtain (9.4.27), and (9.4.28) follows from (9.2.24). From this (9.4.29, 30) follow immediately and (9.4.31) is a consequence of (9.4.6).

(ii) Now suppose that q and H_q satisfy the condition stated. From theorem 9.4.3(ii) we deduce (9.4.33) for some scalar λ_q. Now $H_q^2 = qS$, where $S \in \Gamma_0(N, t_1)$ and

$$\delta = q + \lambda' q_1 t_1.$$

Since

$$\chi(\delta) = \chi_q(\delta)\chi_{q'}(\delta) = \chi_q(-1)\chi_{q'}(q),$$

(9.4.34) follows from (9.4.33), since $\chi(-1) = (-1)^k$.

(iii) Now let p be a prime divisor of N. Since F has divisor t, it is clear that $\lambda(p) = 0$ in case (a). Suppose that $p \nmid t_1$ and that $p|(N/N_\chi)$. Then, by theorem 9.4.3(iii), we have

$$F|L_p^\chi = \nu(p)F,$$

for some scalar $\nu(p)$. But, by theorem 9.2.9,

$$F|L_p^\chi \in M(p)|J_p^{-1} \subseteq M^-,$$

where $M = \{N, k, \chi, t\}_0$, and so

$$F|L_p^\chi = 0.$$

It follows, from (9.2.45), that

$$F|T_p^{(t)} = -\varepsilon_p p^{\frac{1}{2}k-1}\bar{\chi}(J_p^{-1}H_p)F|H_p,$$

where ε_p is 0 or 1 according as p^2 does or does not divide N. Accordingly, we have

$$\lambda(p) = 0, \quad \text{when} \quad p^2|N,$$

which is case (b), and

$$\lambda(p) = -p^{\frac{1}{2}k-1}\lambda_p, \quad \text{when} \quad p^2 \nmid N,$$

which is case (c). Here λ_p is as defined previously (with p replacing q); note that the conditions (9.4.32) are satisfied.

Finally, suppose that p satisfies the conditions of case (d). We apply theorem 9.2.10 to the function F with $q=\tau$. If N' is the conductor of the character $\chi_{q'}=\varepsilon$, then (9.4.35) states that N' divides t_1; thus ε is, in fact, a character modulo t_1 and so the operator $D_t^{(\varepsilon)}$ of theorem 8.2.2 is defined.

By theorems 8.4.1 and 9.2.7, $F|H_q \in \{N, k, \chi^*, t\}_0$, where by (8.4.16),

$$\chi^* = \bar{\chi}_q \varepsilon.$$

By the results of §8.6, we have

$$F|H_qK \in \{N, k, \chi_q\bar{\varepsilon}, t\}_0$$

and therefore, by theorem 8.2.2,

$$G := F|H_qKD_t^{(\varepsilon)} \in M.$$

Now $F|H_qK$ is not the zero form and has divisor t; it follows that $G \neq 0$ and so, for some non-zero $\sigma \in \mathbb{C}$, $G = \sigma F$.

We therefore have, in the notation of theorem 9.2.10,

$$F|T_n^{(t)}H_qT_n^{(t)} = p^{k-1}\varepsilon(p)F|T_l^{(t)}H_qT_l^{(t)},$$

where $q = \tau$, as previously stated. Hence

$$\lambda(n)F|H_qT_n^{(t)} = p^{k-1}\varepsilon(p)\lambda(l)F|H_qT_l^{(t)}.$$

Now operate on both sides by the conjugate linear map $KD_t^{(\varepsilon)}$, to give

$$\overline{\lambda(n)}[F|H_qT_n^{(t)}KD_t^{(\varepsilon)}] = p^{k-1}\overline{\varepsilon(p)\lambda(l)}[F|H_qT_l^{(t)}KD_t^{(\varepsilon)}]. \tag{9.4.38}$$

By theorem 9.2.6 and (9.2.35),

$$\begin{aligned} F|H_qT_n^{(t)}KD_t^{(\varepsilon)} &= F|H_qKT_n^{(t)}D_t^{(\varepsilon)} = \overline{\varepsilon(n)}[F|H_qKD_t^{(\varepsilon)}T_n^{(t)}] \\ &= \overline{\varepsilon(n)}[\sigma F|T_n^{(t)}] = \sigma\overline{\varepsilon(n)\lambda(n)}. \end{aligned}$$

A similar result holds for the expression in square brackets on the right of (9.4.38) and we obtain

$$\sigma\overline{\varepsilon(n)}|\lambda(n)|^2 = \sigma p^{k-1}\overline{\varepsilon(n)}|\lambda(l)|^2.$$

Since $\sigma \neq 0$ and $\lambda(n) = \lambda(p)\lambda(l)$, we deduce (9.4.37).

In connexion with the last remark in the enunciation we recall what was stated at the end of §8.3.

Our last theorem concerns the groups Γ_n studied in theorem 8.3.1. As at the beginning of the present section, we study the space $M = \{\Gamma, k, u\}_0 \cap M_0$, where M_0, Γ and u satisfy the conditions (i), (ii) and (iii) stated there. For any $l \in \mathbb{Z}_N$ we put

$$M_l = \{\Gamma_l, k, u_l\}_0 \cap M_0, \tag{9.4.39}$$

where Γ_l and u_l are as in theorem 8.3.1, so that $M_1 = M$. We note that $M_l = M_m$ whenever $l \equiv m \pmod N$ and $(l, N) = 1$; also $M_{lm} = M_l$ for all $m \in \mathbb{Z}_\Gamma$.

Moreover, for any $m \in \mathbb{Z}_N$ and $S \in \Gamma(1)$,

$$R_m^{-1} S R_m \equiv S_{m^2} \pmod N,$$

by (8.3.4). Since $R_m \in \Gamma_0^0(N)$ it follows that $m^2 \in \mathbb{Z}_\Gamma$ for all $m \in \mathbb{Z}_N$, so that Γ_l, u_l and M_l remain unaltered when l is replaced by lm^2, where $(lm, N) = 1$. We deduce that $M_l = M_m$ whenever $lm \in \mathbb{Z}_\Gamma$.

By theorem 9.2.3,

$$M_l | T_m \subseteq M_{lm} \tag{9.4.40}$$

for all $m \in \mathbb{Z}_N^+$. Further, since the operators T_m and T_n commute, for m and n in $\mathbb{Z}_N^+$, it follows that, if f is an eigenform in M_l, then $f | T_m$ is a possibly zero eigenform in M_{lm} with the same eigenvalues $\lambda(n)$ for $n \in \mathbb{Z}_\Gamma^+$.

Hence, for each $l \in \mathbb{Z}_N$, M_l is a vector sum of eigenclasses $C_l^i (1 \le i \le q)$ for the operators $T_n (n \in \mathbb{Z}_\Gamma^+)$. For fixed i the different eigenclasses C_l^i are vector spaces of eigenforms in M_l having the same eigenvalues $\lambda_i(n)$ $(n \in \mathbb{Z}_\Gamma^+)$ for each l. It is possible that for some i and l, C_l^i may reduce to the zero class [0] but we may assume that, for each i there is some l such that $\dim C_l^i > 0$.

Theorem 9.4.9. *Let C_l and C_m be classes of eigenforms in the spaces M_l and M_m defined by (9.4.39) and having the same eigenvalues $\lambda(n)$ for $n \in \mathbb{Z}_\Gamma^+$, where $\Gamma_0^0(N) \subseteq \Gamma \subseteq \Gamma(1)$. Let r and s be positive integers such that rlm and $slm \in \mathbb{Z}_\Gamma$. Then* (i) *if*

$$\lambda(r, s) := \sum_{d|(r,s)} \chi(d) d^{k-1} \lambda(rs/d^2) \ne 0, \tag{9.4.41}$$

then $\dim C_l = \dim C_m$ *and, moreover,*

$$C_l | T_r = C_m.$$

Conversely (ii) *if* $\dim C_l \neq \dim C_m$, *then* $\lambda(r, s) = 0$ *for all positive integers r and s such that rlm and slm* $\in \mathbb{Z}_\Gamma$; *in particular, if* $(r, s) = 1$, *then* $\lambda(rs) = 0$ *and, if p is a prime such that plm* $\in \mathbb{Z}_\Gamma$, *then*

$$\lambda(p^2) = -p^{k-1}\chi(p). \tag{9.4.42}$$

Proof. (i) If $\lambda(r, s) \neq 0$ and $f \in C_l$, then, by (9.2.5),

$$\begin{aligned}(f|T_r)|T_s &= \sum_{d|(r,s)} d^{k-1}(f|R_d)|T_{rs/d^2}\\ &= \sum_{d|(r,s)} \chi(d)d^{k-1}\lambda(rs/d^2)f,\end{aligned}$$

since $rs/d^2 \in \mathbb{Z}_\Gamma^+$. Accordingly,

$$(C_l|T_r)|T_s = C_l$$

and we deduce that

$$\dim C_l \geq \dim(C_l|T_r) \geq \dim(C_l|T_rT_s) = \dim C_l,$$

so that

$$\dim(C_l|T_r) = \dim C_l.$$

But $C_l|T_r \subseteq C_{lr} = C_m$, so that

$$\dim C_l \leq \dim C_m.$$

The reverse inequality is proved similarly and this shows that $\dim C_l = \dim C_m$; it follows that $C_l|T_r = C_{lr} = C_m$.

(ii) If $\dim C_l \neq \dim C_m$, it follows that $\lambda(r, s) = 0$ for all r and s satisfying the conditions, and this completes the proof of the theorem.

9.5. Historical and other remarks. The origin of the theory of Hecke operators is to be found in the paper of Mordell (1920) in which he proved the multiplicative properties conjectured by Ramanujan for the coefficients of the modular discriminant. By their definition these operators involve transformations of order n and the study of such transformations may be traced back through the work of such writers as Hurwitz (1932, 1933) to the latter half of the nineteenth century and earlier; see, in particular, the letter from Dedekind (1877) to Borchardt.

Mordell applied his method also to modular forms belonging to subgroups of the modular group. Having accomplished what he set out to do he retained little further interest in the subject, and it was left to Hecke (1937), and later on to Petersson (1939–1940), to develop the elegant and powerful theory that now exists.

More recently the theory has been further generalized beyond the classical limits discussed in this book and various alternative methods of presenting the theory have been devised; see, for example, Shimura (1959, 1971), Ogg (1969*b*) and Serre (1970). The theory takes its simplest form when applied to cusp forms belonging to the full modular group; see, for example, Petersson (1969) or Gunning (1962). If we had been content with this application, the preceding analysis could have been considerably curtailed.

For cusp forms belonging to a group of level N the situation is, of necessity, more complicated, at any rate if operators T_n for which $(n, N)>1$ are introduced. The theory has been developed here by using properties of double coset modules, rather than in terms of correspondences or lattices. This permits the cases $(n, N)=1$ and $(n, N)>1$ to be taken together, as in §9.1, although ultimately it is necessary to introduce the special Hecke operators $T_n^{(t)}$ for functions of each divisor t of N. This might appear to to involve a different family of operators for each divisor; it is clear, however, that this can be avoided and that a single operator $T(n)$ may be defined on $M(N, k)$, that is identical with $T_n^{(t)}$ on each space $M_t(N, k)$. It may be remarked at this point that, if one is not concerned with forms of divisor t, then the theory for the group $\Gamma_0(M, N)$ scarcely differs from that for $\Gamma_0(MN)$, so that one can confine one's attention to groups of the form $\Gamma_0(N)$.

The greater part of §9.2 goes back to Hecke (1937) or generalizes work of Atkin and Lehner (1970). When $(n, N)=1$, the operators T_n are normal and it is possible to find bases of cusp forms, that are eigenforms for these operators and therefore have Fourier coefficients with multiplicative properties. Until Atkin and Lehner introduced their theory of newforms, it was not possible to remove the restriction $(n, N)=1$ expect in special cases. In §9.4 the Atkin–Lehner theory is generalized to apply to forms of arbitrary divisor and character. The key theorem is theorem 9.4.5; this generalizes and, it is hoped, makes theorem 1 of Atkin and Lehner (1970) more perspicuous. Theorem 9.4.8 summarizes the properties of newforms and includes fairly explicit information about the coefficient $\lambda(p)$, when p is a prime dividing N. This generalizes work of Atkin and Lehner, and of Ogg (1969*a*).

Since §§9.1–9.4 were written there has appeared the work of Li (1975), who has developed a similar theory of newforms on $\Gamma_0(M, N)$ for arbitrary character χ, but without consideration of divisors. If F is a newform in $\{\Gamma_0(M, N), k, \chi\}_0$ in Li's theory, then there are no other linearly independent forms in this space having the same eigenvalues for the operators T_n with $(n, MN) = 1$; in contrast to this see the concluding remarks in §10.3 below. Professor Lehner has informed me that Margaret Millington (née Ashworth) was working on an extension of the Atkin–Lehner theory of newforms at the time of her death in 1972. Indications of earlier generalizations of newforms for arbitrary character can be found in Miyake (1971) and Deligne (1973).

It is of interest to inquire whether two different newforms (of the same or different levels) can have, apart from a finite number of exceptions, the same eigenvalues $\lambda(p)$ for all operators T_p. That this cannot happen (for forms of codivisor 1 and principal character) has been shown by Atkin and Lehner; see also Miyake (1971).

That the Hecke operators commute with each other can be shown in various ways; see Hecke (1937, Satz 10) and Rankin (1962*a*, theorem 6) or (1964, theorem 6.3.3).

For the modular group the eigenvalues $\lambda(n)$ $(n \in \mathbb{Z}^+)$ are algebraic integers; see Rankin and Rushforth (1954). A similar result for certain spaces of theta functions of higher level is stated in Hecke (1940, Satz 41), but no proof is offered.

A necessary and sufficient condition, expressed in terms of trace operators, for a cusp form to belong to the subspace spanned by newforms has been stated by Serre (1973) and generalized by Li (1975).

In §§9.2–9.4 we have considered not only forms belonging to the groups $\Gamma(N)$ and $\Gamma_0(N, t_1)$ but also more general groups of level N. This gives rise to eigenforms with partially multiplicative properties; for a more detailed study with some applications, see Rankin (1964, 1967*a*) and also §10.4 below.

We have in this chapter been concerned with forms of integral weight k. A theory of Hecke operators can be developed for arbitrary real weight; see Wohlfahrt (1957) and van Lint (1957). However, except in the cases of integral and half-integral weight, there are difficulties that have not been resolved. For the case of half-integral weight see van Lint (1957), Kløve (1970), and, more particularly, Shimura (1973).

The work of Atkin and Swinnerton-Dyer (1971) indicates that results of great interest await discovery and proof in the theory of Hecke operators acting on modular forms belonging to noncongruence groups.

10: Applications

10.1. Dirichlet series. In this section we associate Dirichlet series with entire modular forms. We shall see that, when the modular form is a newform, the series has an Euler product of a particularly simple form.

Theorem 10.1.1. *Let $f \in \{\Gamma, k, v\}$, where $k>0$, and suppose that*

$$f(z)=\sum_{r+\kappa\geq 0} a_r e^{2\pi i(r+\kappa)z/n} \tag{10.1.1}$$

and

$$f_V(z)=f(z)|V=\sum_{r+\kappa'\geq 0} b_r e^{2\pi i(r+\kappa')z/n'}, \tag{10.1.2}$$

where

$$n=n(\infty,\Gamma), \quad n'=n(V\infty,\Gamma)$$

and

$$\kappa=\kappa(\infty,\Gamma,v), \quad \kappa'=\kappa(V\infty,\Gamma,v).$$

Put

$$\delta=\begin{cases}1 & \text{if } \kappa=0,\\ 0 & \text{if } \kappa>0.\end{cases} \quad \delta'=\begin{cases}1 & \text{if } \kappa'=0,\\ 0 & \text{if } \kappa'>0.\end{cases}$$

Write

$$\phi(s)=\sum_{r+\kappa>0}\frac{a_r}{(r+\kappa)^s}, \quad \phi_V(s)=\sum_{r+\kappa'>0}\frac{b_r}{(r+\kappa')^s} \tag{10.1.3}$$

and

$$\Phi(s)=(2\pi/n)^{-s}\Gamma(s)\phi(s), \tag{10.1.4}$$

$$\Phi_V(s)=(2\pi/n')^{-s}\Gamma(s)\phi_V(s). \tag{10.1.5}$$

The Dirichlet series for $\phi(s)$ and $\phi_V(s)$ are absolutely convergent for sufficiently large $\sigma = \text{Re } s$. Moreover, the function Ψ defined by

$$\Psi(s) = \Phi(s) + \frac{\delta a_0}{s} + \frac{\delta' i^k b_0}{k-s}$$

can be continued analytically as an entire function over the whole s-plane. In particular, ϕ is an entire function if $\delta' b_0 = 0$; this is the case whenever f is a cusp form.

The functions Φ and Φ_V satisfy the functional equation

$$\Phi(s) = i^k \Phi_V(k-s). \tag{10.1.6}$$

In particular, if $V \in \Gamma$, then

$$\Phi(s) = i^k v(V) \Phi(k-s). \tag{10.1.7}$$

Proof. Let the series for $\phi(s)$ and $\phi_V(s)$ be absolutely convergent for $\sigma > \sigma_0$. When f is a cusp form we can, by theorem 4.5.2(iv), take $\sigma_0 = \frac{1}{2}(k+1)$. If f is not a cusp form, we can, by theorem 4.5.3, take $\sigma_0 = k+1$; it is possible to improve this estimate to $\sigma_0 = k$ when $k > 2$, as indicated in the remarks following theorem 5.3.3, and also in certain other cases.

Now take any s with $\sigma > \sigma_0$. Since

$$\left\{\frac{2\pi(r+\kappa)}{n}\right\}^{-s} \Gamma(s) = \int_0^\infty u^{s-1} e^{-2\pi(r+\kappa)u/n} \, du$$

whenever $r + \kappa > 0$, we have, on multiplying by a_r and summing

$$(2\pi/n)^{-s} \Gamma(s) \phi(s) = \sum_{r+\kappa>0} a_r \int_0^\infty u^{s-1} e^{-2\pi(r+\kappa)u/n} \, du.$$

Because of absolute convergence we may invert the order of summation and integration and so obtain

$$\begin{aligned}
\Phi(s) &= \int_0^\infty u^{s-1}\{f(iu) - \delta a_0\} \, du \\
&= \int_1^\infty u^{s-1}\{f(iu) - \delta a_0\} \, du - \frac{\delta a_0}{s} + \int_1^\infty u^{-s-1} f(i/u) \, du \\
&= \int_1^\infty u^{s-1}\{f(iu) - \delta a_0\} \, du - \frac{\delta a_0}{s} + i^k \int_1^\infty u^{k-s-1} f_V(iu) \, du
\end{aligned}$$

$$= \int_1^\infty u^{s-1}\{f(iu)-\delta a_0\}\,du + i^k \int_1^\infty u^{k-s-1}\{f_V(iu)-\delta' b_0\}\,du$$

$$-\frac{\delta a_0}{s} - \frac{i^k \delta' b_0}{k-s}, \qquad (10.1.8)$$

where i^k means $e^{\frac{1}{2}\pi k i}$.

We now observe that, for large u,

$$f(iu)-\delta a_0 = O(e^{-2\pi c u}), \quad f_V(iu)-\delta' b_0 = O(e^{-2\pi c u})$$

for some constant $c>0$ and it follows that both integrals can be continued as entire functions over the whole complex s-plane. The statements about Ψ and ϕ follow, when we observe that the possible pole of ϕ at 0 is cancelled out by the gamma function.

Finally, since

$$f_V(z)|V = z^{-k}(-1/z)^{-k}f(z) = e^{-ik\pi}f(z),$$

we deduce from (10.1.8) that (10.1.6) holds. If $V \in \Gamma$, then $f_V = v(V)f$ and (10.1.7) follows.

If we apply the theorem with $f = \vartheta_3$, we have $k = \frac{1}{2}$, and $v(V) = e^{-\frac{1}{4}\pi i}$, by (7.1.15). Moreover, $\phi(s) = 2\zeta(2s)$, so that (10.1.7) becomes the well-known functional equation for the Riemann zeta-function, in the form

$$\Phi(s) = \Phi(\tfrac{1}{2}-s),$$

where

$$\Phi(s) = 2(2\pi)^{-s}\Gamma(s)\zeta(2s),$$

and we deduce also that $\zeta(s)$ is holomorphic except for a simple pole of residue 1 at 1. Note also that $\sigma_0 = k = \frac{1}{2}$ in this case.

Theorem 10.1.2. *Let f be an eigenform belonging to the space $\{N, k, \chi, t\}_0$, where $N = tt_1$, χ is a character modulo N and k is a positive integer satisfying (8.1.1). Let*

$$f(z) = \sum_{r=1}^{\infty} A_r e^{2\pi i r z/t_1} \qquad (10.1.9)$$

and suppose that $f|T_n = \lambda(n)f$ for all positive n prime to N. Then, if $\phi(s)$ is the associated Dirichlet series,

$$\phi(s) = \sum_{r=1}^{\infty} \frac{A_r}{r^s} \quad (\sigma = \operatorname{Re} s > \tfrac{1}{2}(k+1)), \qquad (10.1.10)$$

$\phi(s)$ has an Euler product of the form

$$\phi(s)=\phi_N(s)\prod_{p\nmid N}\{1-\lambda(p)p^{-s}+\chi(p)p^{k-1-2s}\}^{-1}. \tag{10.1.11}$$

The infinite product over the primes p not dividing N is absolutely convergent for $\sigma>\frac{1}{2}(k+1)$ and $\phi_N(s)$ is a Dirichlet series of the form $\sum A_r r^{-s}$ $(\sigma>\frac{1}{2}(k+1))$, the summation being restricted to positive integers r such that $r|N^\infty$.

Proof. Our notation is the same as that of theorem 9.4.2, except that we write $\lambda(n)$ for $\lambda_f(n)$. We assume, of course, that f is not the zero form. From (9.4.7) it follows that, for $\sigma>\frac{1}{2}(k+1)$,

$$\phi(s)=\phi_N(s)\psi(s), \tag{10.1.12}$$

where ϕ_N is as defined in the enunciation and

$$\psi(s)=\sum_{\substack{n=1\\(n,N)=1}}^{\infty}\lambda(n)n^{-s}.$$

Both series on the right of (10.1.12) are clearly absolutely convergent. For, if A_m is the first non-zero coefficient and $(n, N)=1$, then $m|N^\infty$, so that

$$\lambda(n)A_m=A_{mn}. \tag{10.1.13}$$

For any integer $q\geq 2$ and fixed s with $\sigma>\frac{1}{2}(k+1)$ put

$$\Pi_q(s)=\prod_{\substack{p\leq q\\p\nmid N}}\sum_{\nu=0}^{\infty}\lambda(p^\nu)p^{-\nu s}.$$

This is a product of a finite number of absolutely convergent series, so that

$$\Pi_q(s)=:\sum_{n=1}^{\infty}\lambda'(n)n^{-s},$$

where this series is again absolutely convergent. Now (9.4.11) shows that

$$\lambda(mn)=\lambda(m)\lambda(n)$$

whenever $(m, n)=(m, N)=(n, N)=1$. It follows that $\lambda'(n)$ is either equal to $\lambda(n)$ or zero; in any case

$$\lambda'(n)=\lambda(n)\quad\text{for}\quad n\leq q, (n, N)=1.$$

Accordingly,

$$\psi(s)-\Pi_q(s)=\sum_{\substack{n=q+1\\(n,N)=1}}^{\infty}\{\lambda(n)-\lambda'(n)\}n^{-s}.$$

As $q\to\infty$ the right-hand side tends to zero, so that

$$\psi(s)=\lim_{q\to\infty}\Pi_q(s)=\prod_{p\nmid N}\psi_p(s),$$

where

$$\psi_p(s)=\sum_{\nu=0}^{\infty}\lambda(p^\nu)p^{-\nu s}.$$

Now the coefficients of this Dirichlet series satisfy (9.4.30), as follows from (9.4.11) by taking $n=p^\nu$ $(\nu\geq 1)$, $r=p$. It is an easy consequence of this that

$$\psi_p(s)=\{1-\lambda(p)p^{-s}+\chi(p)p^{k-1-2s}\}^{-1}. \qquad (10.1.14)$$

Theorem 10.1.3. *Let f be a newform in $\{N, k, \chi, t\}_0$. Then, in the notation of theorem 10.1.2, $A_n=\lambda(n)$, the eigenvalue corresponding to the operator $T_n^{(t)}$ $(n\in\mathbb{Z}^+)$, and the associated Dirichlet series has the Euler product*

$$\phi(s)=\prod_p\{1-\lambda(p)p^{-s}+\chi(p)p^{k-1-2s}\}^{-1}\quad(\sigma>\tfrac{1}{2}(k+1)).$$
$$(10.1.15)$$

More generally, if $f\in\{N, k, \chi, t\}_0$ and is an eigenform of all the operators $T_n^{(t)}$ for $n\in\mathbb{Z}^+$, with eigenvalues $\lambda(n)$, then

$$\phi(s)=A_1\prod_p\psi_p(s),$$

in the notation of (10.1.14).

Proof. This is immediate from theorem 10.1.2 and theorem 9.4.8. Because of the multiplicative properties of all the coefficients $\lambda(n)$,

$$\phi_N(s)=\prod_{p|N}(1-\lambda(p)p^{-s})^{-1},$$

since, by (9.4.30), $\lambda(p^\nu)=\lambda^\nu(p)$ $(\nu\geq 0)$; for $\chi(p)=0$ when $p|N$. For some primes p the 'Euler factors' $\psi_p(s)$ will, of course, reduce to 1; this happens in cases (iii) (*a*), (*b*) of theorem 9.4.8.

The last part of the theorem follows similarly, since, by a remark following theorem 9.4.2, $A_n = \lambda(n)A_1$ for all $n \geq 1$.

10.2. Eigenforms for the full modular group. We apply the results of chapter 9 and §10.1 to the simplest case, when $N=1$ and the space is $\{\Gamma(1), k\}_0$, where k is an even integer. We confine our attention to the cases when $k=12$ or $k \geq 16$, since otherwise the dimension δ_k of the space is zero; see theorem 6.1.2. Here, of course, the character χ is the trivial one that takes the value 1 on $\mathbb{Z}$, and $t = t_1 = 1$.

Then every non-zero eigenform is a multiple of a newform. The space $\{\Gamma(1), k\}_0$ has an orthogonal basis of δ_k newforms. If f is any such newform, we may write

$$f(z) = \sum_{n=1}^{\infty} \lambda(n)\, e^{2\pi i n z}, \tag{10.2.1}$$

where $f|T_n = \lambda(n)f$ for all $n \geq 1$. The coefficients have the properties that

$$\lambda(mn) = \lambda(m)\lambda(n) \quad \text{for} \quad (m, n) = 1 \tag{10.2.2}$$

and

$$\lambda(p^{r+1}) = \lambda(p)\lambda(p^r) - p^{k-1}\lambda(p^{r-1}), \tag{10.2.3}$$

where m, n and r are positive integers and p is a prime.

If the associated Dirichlet series is

$$\phi(s) = \sum_{n=1}^{\infty} \lambda(n) n^{-s}, \tag{10.2.4}$$

then these properties imply, as in §10.1, that $\phi(s)$ has the Euler product

$$\phi(s) = \prod_p \{1 - \lambda(p)p^{-s} + p^{k-1-2s}\}^{-1}. \tag{10.2.5}$$

The series and product in (10.2.4, 5) are absolutely convergent for $\sigma = \operatorname{Re} s > \frac{1}{2}(k+1)$.

The cases when $k = 12$, 16, 18, 20, 22 and 26 are of particular interest, since then $\delta_k = 1$ and so the functions F_k defined by (6.3.2) are eigenforms whose coefficients $\gamma_k(n)$ satisfy the relations (10.2.2, 3). We have already used the associated Dirichlet series in §6.3 to evaluate the constants c_k occurring in the associated

Poincaré series; see (6.3.12). Since $\Delta | T_2 = \tau(2)\Delta$, we have

$$2^{11}\Delta(2z) + 24\Delta(z) + \tfrac{1}{2}\{\Delta(\tfrac{1}{2}z) + \Delta(\tfrac{1}{2}z + \tfrac{1}{2})\} = 0. \qquad (10.2.6)$$

We shall require this formula in §10.3.

We now consider briefly the six values of k for which $\delta_k = 2$, namely

$$k = 24, 28, 30, 32, 34 \text{ or } 38.$$

We define $F_k(z)$ as in (6.3.2) and write

$$H_4(z) = E_{k-24}(z)\,\Delta^2(z) = \sum_{n=2}^{\infty} h_k(n)\, e^{2\pi i n z}, \qquad (10.2.7)$$

so that F_k and H_k span the space $\{\Gamma(1), k\}_0$. Accordingly, if f is a newform, we must have

$$f = F_k + \mu H_k, \qquad (10.2.8)$$

and we assume that $f(z)$ is defined by (10.2.1). Then, since

$$\lambda(4) - \lambda^2(2) + 2^{k-1} = 0,$$

it follows that μ is a root of the quadratic equation

$$\gamma_k(4) + \mu h_k(4) - \{\gamma_k(2) + \mu h_k(2)\}^2 + 2^{k-1} = 0. \qquad (10.2.9)$$

The two roots of this equation give the two eigenforms we are seeking.

When $k = 24$, we find that

$$\mu = 12\left\{\frac{27017}{691} \pm \sqrt{144169}\right\},$$

for example.

In this and also in the general case it can be shown that numbers $\lambda(n)$ are algebraic integers. In fact, if r is the number of primes not exceeding δ_k, it can be shown that the degree over the rationals of the algebraic field containing the eigenvalues $\lambda(n)$ $(n \geq 1)$ does not exceed δ_k^r; see Rankin and Rushforth (1954).

Finally we note that, for each even integer $k \geq 4$, the Eisenstein series E_k is an eigenform; by (9.3.21),

$$E_k | T_n = \sigma_{k-1}(n) E_k \quad (n \in \mathbb{Z}^+). \qquad (10.2.10)$$

10.3. Eigenforms for $\Gamma_V(2)$ and $\Gamma(2)$. In this and the next section we apply the general theorems of the last chapter to a number of particular cases. We shall confine our attention to groups of level 2 and 4 and to functions of small weight in order to reduce the numerical calculations to manageable proportions. We shall also be particularly interested in eigenforms arising in the study of the representation of a number as a sum of squares.

We use the notations of §7.1. As in that section, we write $k=\frac{1}{2}s$, but we restrict our attention to the cases when s is a positive even integer. Write

$$D_k=\{\Gamma_V(2), k, v_s\}_0, \quad D_k^*=\{\Gamma_V(2), k, v_s^*\}_0 \qquad (10.3.1)$$

and put

$$\delta_k=\dim D_k, \quad \delta_k^*=\dim D_k^*. \qquad (10.3.2)$$

Although this notation conflicts with that used in chapter 6 and §10.2 for δ_k, this should not cause any confusion. By theorem 7.1.7, the spaces D_k and D_k^* are orthogonal.

By theorem 7.1.7,

$$C_k:=D_k\oplus D_k^*=\{\Gamma(2), k, v_s\}_0, \qquad (10.3.3)$$

so that

$$\gamma_k:=\dim C_k=\delta_k+\delta_k^*. \qquad (10.3.4)$$

It follows from theorems 7.1.4, 6, 7 that $\delta_k=\delta_k^*=0$ for $k\le 4$ and that, for $k>4$,

$$\delta_k=-1-[-\tfrac{1}{4}k], \quad \delta_k^*=-1-[\tfrac{1}{2}-\tfrac{1}{4}k]$$

and

$$\gamma_k=-2-[-\tfrac{1}{2}k]=[\tfrac{1}{2}(k-1)]-1. \qquad (10.3.5)$$

We accordingly obtain table 13 for $5\le k\le 12$.

The MS v_s coincides with the MS v defined in §8.1 on $\Gamma(2)$ when k is even, and on $\bar{\Gamma}(4)$ when k is odd; in both cases v_s takes the value 1 on the group $\Gamma^*(2)$; see (1.5.4).

Table 13. Dimensions of the spaces D_k, D_k^* and C_k

k	5	6	7	8	9	10	11	12
δ_k	1	1	1	1	2	2	2	2
δ_k^*	0	0	1	1	1	1	2	2
γ_k	1	1	2	2	3	3	4	4

We define the functions ψ_k $(k \geq 5)$ as in (7.4.26) and write

$$\psi_k^* = \tfrac{1}{16}\vartheta_3^{2k-12}\vartheta_2^4\vartheta_4^4(\vartheta_4^4 - \vartheta_2^4) \quad (k \geq 7). \qquad (10.3.6)$$

By theorem 7.1.4, ψ_k generates D_k for $5 \leq k \leq 8$, while, by theorem 7.1.6, ψ_k^* generates D_k^* for $7 \leq k \leq 10$. We note that in all these cusp forms the coefficient of $q = e^{\pi i z}$ is 1. The notations we use for the Fourier coefficients of ψ_5, ψ_6, ψ_7, ψ_8 and ψ_9^* are due to Glaisher; see (7.4.27) and (10.4.29) below.

We now apply theorem 8.3.1 to the group $\Gamma = \Gamma_V(2)$ with $N = 2$ or 4 according as k is even or odd. For each odd n the group Γ_n is identical with Γ but the multiplier systems u_n do not necessarily all coincide. Table 14 gives u_n for $u = v_s$ or v_s^* for different values of n and k. We deduce that, for odd positive n,

Table 14. The multiplier system u_n

	$k = \frac{1}{2}s$ even		$k = \frac{1}{2}s$ odd	
	$n \equiv 1 \pmod 4$	$n \equiv -1 \pmod 4$	$n \equiv 1 \pmod 4$	$n \equiv -1 \pmod 4$
$u = v_s$	v_s	v_s	v_s	v_s^*
$u = v_s^*$	v_s^*	v_s^*	v_s^*	v_s

$$D_k|T_n \subseteq D_k \quad \text{when } n \equiv 1 \pmod 4 \text{ or for even } k.$$

$$D_k|T_n \subseteq D_k^* \quad \text{when } n \equiv -1 \pmod 4 \text{ and for odd } k.$$

Similar results hold for $D_k^*|T_n$.

For small values of k we shall look for newforms in the spaces

$$M_t = \{N, k, \chi, t\}_0, \qquad (10.3.7)$$

where $t|N$, χ is a character modulo N with $\chi(-1) = (-1)^k$ and $N = 2$ or 4 according as k is even or odd. If k is even, $t = 1$ or 2 and χ is the unique (principal) character modulo 2. If k is odd, $t = 1, 2,$ or 4 and χ is given, as in (7.3.1), by

$$\chi(n) = \sin \tfrac{1}{2}n\pi \quad (n \in \mathbb{Z}).$$

It follows from this that, if (for $M = M_t$) the space $M(d)$ is defined by (9.4.12) for $d|t_\chi$, then we can have $d > 1$ only when k is even and $t = 2$, in which case

$$M(d) = \{\Gamma(1), k\}_0$$

for $d=2$, and so has dimension zero unless $k=12$ or $k\geq 16$. Accordingly, if $k\leq 12$, then $M_t^+=M_t$ except when $k=12$ and $t=2$; it follows that, except in this case, every eigenform in M_t is a constant multiple of a newform.

For the remainder of this section we restrict our attention to the cases $k=6$, 8, 10 or 12, so that $N=2$. We use table 13 and theorems 7.1.4, 6 to list bases for D_k and D_k^*. From these we deduce bases for the spaces M_t; for this purpose we write

$$C_k = C_k^1 \oplus C_k^2,$$

where $C_k^1 = M_1$ and $C_k^2 = M_2$, for the value of k considered. These two spaces are orthogonal by theorem 8.1.3.

It can be shown that $\{\Gamma_U(2), k, u_s\}_0 = C_k^1$ or C_k^2 and that $\{\Gamma_U(2), k, u_s^*\}_0 = C_k^2$ or C_k^1, according as $k\equiv 2$ or 0 (mod 4). Here u_s is the MS defined in (7.1.17, 18) and

$$u_s^*(S) = v_s^*(P^{-1}SP) \tag{10.3.8}$$

for $S\in\Gamma_U(2)$; see (7.5.1).

k = 6. We have

$$C_6 = D_6 = C_6^1,$$

and the space is generated by the function ψ_6 defined, as in (7.4.26, 33), by

$$\psi_6(z) = \Delta^{\frac{1}{4}}(z) = \sum_{n=1}^{\infty} \Omega(n) q^n,$$

where, as in §7.1, $q=e^{\pi i z}$. Note that $\Omega(1)=1$ and that $\Omega(n)=0$ for even n. It follows that ψ_6 is the sole newform in C_6.

k = 8. The spaces D_8 and D_8^* are generated by the forms ψ_8 and ψ_8^* as we have seen. These two functions are therefore eigenforms for the operators T_n with odd n. Write

$$F_8 = \tfrac{1}{4}(3\psi_8 + \psi_8^*), \quad F_8^* = \tfrac{1}{32}(\psi_8 - \psi_8^*). \tag{10.3.9}$$

It is easily verified from theorem 7.1.2 that $F_8|U = -F_8$ and $F_8^*|U = F_8^*$. In fact, $F_8 = \psi_8|D_1$ and $F_8^* = \frac{1}{8}\psi_8|D_2$. Accordingly,

$$F_8 \in C_8^1 \quad \text{and} \quad F_8^* \in C_8^2.$$

These two functions are therefore newforms, since their first Fourier coefficients have the value 1. Since all four functions are eigenfunctions for the operators T_n, it follows that they all have the same eigenvalues $(-1)^{n-1}\Theta(n)$ (n odd). It is easily verified that

$$\psi_8^*(z) = \sum_{n=1}^{\infty} \{2+(-1)^n\}\Theta(n)q^n, \tag{10.3.10}$$

$$F_8(z) = \sum_{\substack{n=1 \\ n \text{ odd}}}^{\infty} \Theta(n)q^n, \tag{10.3.11}$$

and

$$F_8^*(z) = \sum_{n=1}^{\infty} \Theta(n)q^{2n}. \tag{10.3.12}$$

We can apply theorem 9.4.8(ii) and (iii)(c) to F_8^*. Write

$$H = \begin{pmatrix} 0 & -1 \\ 2 & 0 \end{pmatrix}. \tag{10.3.13}$$

Then, by (9.4.33),

$$F_8^*|H = \lambda_2 F_8^*,$$

where $\lambda_2^2 = 1$ and, by (9.4.36),

$$\Theta(2) = -2^3\lambda_2.$$

The first equation can be verified with the help of theorems 7.1.2, 8 and yields the value $\lambda_2 = 1$. Accordingly, $\Theta(2) = -8$, as is easily checked from the expansion of ψ_8.

***k* = 10.** As a basis for D_{10} we may take

$$F_{10} = \Delta^{\frac{1}{2}}(\vartheta_3^8 - \vartheta_2^4\vartheta_4^4), \quad G_{10} = \Delta^{\frac{1}{2}}(\vartheta_3^8 + 2\vartheta_2^4\vartheta_4^4). \tag{10.3.14}$$

Moreover D_{10}^* is generated by ψ_{10}^* which is therefore an eigenform for the operators T_n (n odd). Write

$$f_{10} = \tfrac{1}{4}(G_{10} + 3\psi_{10}^*), \quad g_{10} = \tfrac{1}{64}(G_{10} - \psi_{10}^*). \tag{10.3.15}$$

It is easily checked from theorems 7.1.2, 8 that

$$g_{10}(z) = -\psi_{10}^*(2z+1) \tag{10.3.16}$$

and that

$$F_{10} \in C_{10}^1, \quad f_{10} \in C_{10}^1, \quad g_{10} \in C_{10}^2.$$

Since C_{10} is spanned by F_{10}, f_{10} and g_{10}, it follows that g_{10} is the sole newform in C_{10}^2. Moreover, since $G_{10} \notin C_{10}^1$,

$$\dim D_{10} \cap C_{10}^1 = 1$$

and so F_{10} is a newform in C_{10}^1.

We still have to find the remaining newform in C_{10}^1. This is f_{10}. To show this it is enough to show that f_{10} is orthogonal to F_{10}. But F_{10} is orthogonal to ψ_{10}^* and g_{10} by theorems 7.1.7 and 8.1.3. It follows from (10.3.15) that F_{10} is orthogonal to G_{10} and so to f_{10}.

Accordingly, F_{10}, f_{10} and g_{10} are the three newforms that span C_{10}. Write

$$F_{10}(z) = \sum_{n=1}^{\infty} a(n)q^n, \quad g_{10}(z) = \sum_{n=1}^{\infty} b(n)q^{2n}, \quad (10.3.17)$$

so that, by (10.3.16),

$$\psi_{10}^*(z) = \sum_{n=1}^{\infty} (-1)^{n-1} b(n)q^n. \quad (10.3.18)$$

It follows from (9.2.28, 30) and the multiplicative properties of $b(n)$ that ψ_{10}^* is an eigenform with the same eigenvalues $b(n)$ (n odd) as g_{10}; alternatively, we may use the fact that $g_{10} = -\frac{1}{16}\psi_{10}^*|D_2$. It follows from (10.3.15) that G_{10} and the newform f_{10} have the same eigenvalues as g_{10}. It can be shown that

$$G_{10}(z) = \sum_{n=1}^{\infty} \{2+(-1)^n\} b(n)q^n. \quad (10.3.19)$$

The analysis of §7.4 can be applied to express $r_{20}(n)$ as a linear combination of $\beta_{10}(n)$, $a(n)$ and $\{2+(-1)^n\}b(n)$.

As indicated by theorem 9.4.8(ii), (iii), we can show that

$$g_{10}|H_2 = -g_{10}$$

and that $b(2) = 16$.

$\boldsymbol{k = 12.}$ As a basis for D_{12} we may take Δ and Δ_1, where

$$\Delta = 2^{-8}\vartheta_2^8\vartheta_3^8\vartheta_4^8, \quad \Delta_1 = \psi_{12} = 2^{-4}\vartheta_2^4\vartheta_3^{16}\vartheta_4^4. \quad (10.3.20)$$

As a basis for D_{12}^* we may take

$$\Delta^* = \Delta^{\frac{1}{2}} E_6 = 2^{-4}\vartheta_2^4\vartheta_3^4\vartheta_4^4(\vartheta_3^8 + \tfrac{1}{2}\vartheta_2^4\vartheta_4^4)(\vartheta_4^4 - \vartheta_2^4), \quad (10.3.21)$$

and

$$\Delta_1^* = \psi_8^* E_4 = 2^{-4}\vartheta_2^4\vartheta_3^4\vartheta_4^4(\vartheta_3^8 - \vartheta_2^4\vartheta_4^4)(\vartheta_4^4 - \vartheta_2^4). \quad (10.3.22)$$

By theorem 7.1.8 we can show that

$$\Delta_1^*(z) = \Delta(\tfrac{1}{2}z) - 2^{12}\Delta(2z) \quad (10.3.23)$$

and that, as already stated in (7.4.36),

$$\Delta_1(z) = -\Delta(\tfrac{1}{2}z + \tfrac{1}{2}). \quad (10.3.24)$$

Moreover, C_{12}^2 is generated by Δ and Δ_2, where

$$\Delta_2(z) = \Delta(2z). \quad (10.3.25)$$

It may be deduced from (10.2.6) that

$$\Delta_2 = 2^{-13}\{\Delta_1 - \Delta_1^* - 48\Delta\}. \quad (10.3.26)$$

Also C_{12}^1 is generated by Δ^* and Δ_3, where

$$\Delta_3(z) = \tfrac{1}{2}\{\Delta(\tfrac{1}{2}z) - \Delta(\tfrac{1}{2}z + \tfrac{1}{2})\} = \sum_{\substack{n=1 \\ n \text{ odd}}}^{\infty} \tau(n)q^n, \quad (10.3.27)$$

and it may be verified from (10.3.23, 24) that

$$\Delta_3 = \tfrac{1}{2}\Delta_1^* + \tfrac{1}{2}\Delta_1 + 2^{11}\Delta. \quad (10.3.28)$$

Since $C_{12}^1 \cap D_{12}^*$ consists of multiples of Δ^* and is invariant under T_n (n odd), it follows that Δ^* is a newform. Moreover, by (10.3.28) and the multiplicative properties of $\tau(n)$, $\Delta_3|T_n = \tau(n)\Delta_3$ (n odd), so that Δ_3 is also a newform. The same argument Δ, Δ_1, Δ_1^* and Δ_2 are eigenforms for T_n (n odd) with the same eigenvalues $\tau(n)$, although they are not newforms.

In fact, $C_{12}^2 = (C_{12}^2)^-$ and is generated by the oldforms Δ and $\Delta|J_2^{-1}$. We have therefore found all the newforms in C_{12}, namely Δ^* and Δ_3.

The formula for $r_{24}(n)$ given in (7.4.37) is now easily derived by using the basis given in (10.3.20) for D_{12}.

We note in conclusion that the foregoing analysis has shown that newforms of different divisors may have the same eigenvalues, and that a newform and an oldform of a different divisor may have the same eigenvalues.

10.4. Eigenforms of level 4. We consider forms belonging to the spaces

$$M_t=\{4, k, \chi, t\}_0, \tag{10.4.1}$$

where t divides 4, k is an integer and $\chi(-1)=(-1)^k$. To begin with we allow k to be any positive integer, but later restrict our attention to odd k.

Write $\tau=1$ or 2 according as t is even or odd. Then it is clear that

$$M_t=\{\Gamma_0(4, \tau), k, v\}_0,$$

where

$$v(U^\tau)=-1, -1 \text{ or } 1$$

according as

$$t=1, 2 \text{ or } 4,$$

respectively.

Write B_t, for $t|4$, for the following families of functions:

$$B_4\text{: } c_r\vartheta_2^4(\vartheta_3^{2k-4-2r}\vartheta_4^{2r}-\vartheta_4^{2k-4-2r}\vartheta_3^{2r}) \quad (0<r<\tfrac{1}{2}k-1), \tag{10.4.2}$$

$$B_2\text{: } \tfrac{1}{32}\vartheta_2^4(\vartheta_3^{2k-4-2r}\vartheta_4^{2r}+\vartheta_4^{2k-4-2r}\vartheta_3^{2r}) \quad (0<r\le\tfrac{1}{2}k-1), \tag{10.4.3}$$

$$B_1\text{: } \tfrac{1}{4}\vartheta_2^2\vartheta_3^{2r}\vartheta_4^{2k-2-2r} \quad (0<r\le k-2). \tag{10.4.4}$$

Here

$$c_r=2^{-8}/(\tfrac{1}{2}k-1-r),$$

and the constant multipliers have been chosen to make the first non-vanishing Fourier coefficient unity. It is clear that the forms in each of the three families are linearly independent, and that, for $k\ge3$,

$$|B_4|=[\tfrac{1}{2}(k-1)]-1, \quad |B_2|=[\tfrac{1}{2}k]-1, \quad |B_1|=k-2,$$

so that

$$|B_4|+|B_2|+|B_1|=2k-5.$$

The group $\Gamma_0(4, \tau)$ is generated by W^4 and U^τ and $v(W^4)=1$. It follows from this and theorem 7.1.2 that $B_t\subseteq M_t$.

Since

$$\{4, k\}_0=M_4\oplus M_2\oplus M_1, \tag{10.4.5}$$

and since $\dim\{4, k_0\} \leq 2k-5$, by theorem 4.2.1 (for $\xi = 0$ and $\lambda = \hat{\lambda}(4) = 6$ by (2.4.17)), it follows that, for each t, B_t is a basis for M_t and therefore

$$\dim M_t = |B_t|.$$

From this we deduce the information given in table 15.

Table 15. Dimensions of the space M_t

k	3	4	5	6	7	8	9	10	11	12
$\dim M_4$	0	0	1	1	2	2	3	3	4	4
$\dim M_2$	0	1	1	2	2	3	3	4	4	5
$\dim M_1$	1	2	3	4	5	6	7	8	9	10
$\dim\{4, k\}_0$	1	3	5	7	9	11	13	15	17	19

It may be noted that all these dimensions may be obtained as upper bounds from theorem 4.2.1; the upper bounds are exact because the genus is zero.

Theorem 8.3.5 tells us that $M_2|J_2 \subseteq M_1$ and we note that, in fact, $\dim M_2|J_2 < \dim M_1$ for $k \geq 3$. Observe also that

$$C_k = \{2, k, v_s\}_0 = D_k \oplus D_k^* \subseteq M^{(2)}(4, k, \chi) = M_2 \oplus M_4. \qquad (10.4.6)$$

Here χ is principal or not according as k is even or odd.

Further, by theorem 8.3.2, since C_k is associated with the group $\bar{\Gamma}(2) = \Gamma_0(2, 2)$,

$$C_k|J_2^{-1} = M_4. \qquad (10.4.7)$$

As a verification we note that each space has dimension

$$\gamma_k = [\tfrac{1}{2}(k-1)] - 1.$$

Moreover, if $f \in C_k$ and

$$f|T_n = \lambda(n)f$$

for some odd n, then

$$(f|J_2^{-1})|T_n = (f|J_2^{-1})|T_n^{(2)} = (f|T_n)|J_2^{-1}$$
$$= \lambda(n)f|J_2^{-1}.$$

Accordingly, $f|J_2^{-1}$ is also an eigenform for T_n.

If, for any $f \in M_t$, $f|T_n = \lambda(n)f$, for some odd $n > 0$, then we deduce from (9.4.6) that the eigenvalue $\lambda(n)$ is real when k is even. If k is odd, then $\lambda(n)$ is real when $n \equiv 1 \pmod 4$, and is either zero or purely imaginary when $n \equiv 3 \pmod 4$.

The only even value of k that we shall consider is $k = 4$. In this case M_2 is generated by the sole member of B_2, namely

$$f_4 = \tfrac{1}{16}\vartheta_2^4\vartheta_3^2\vartheta_4^2, \tag{10.4.8}$$

which is therefore a newform. Write, as did Glaisher (1906),

$$f_4(z) = \sum_{n=1}^{\infty} P(n)q^n. \tag{10.4.9}$$

Then $P(n) = 0$ for even n and

$$f_4|T_n^{(2)} = P(n)f_4 \quad (n \geq 1).$$

Moreover, by theorems 7.1.2, 8, the two functions in B_1 satisfy the relations

$$f_4(\tfrac{1}{2}z) = \tfrac{1}{4}\vartheta_2^2(z)\vartheta_3^2(z)\vartheta_4^4(z) = \tfrac{1}{4}f_4(z)|J_2,$$

$$i^{-1}f_4(\tfrac{1}{2}z + \tfrac{1}{2}) = \tfrac{1}{4}\vartheta_2^2(z)\vartheta_3^4(z)\vartheta_4^2(z).$$

The expansions of these functions are, therefore,

$$\sum_{n=1}^{\infty} P(n)q^{\frac{1}{2}n} \quad \text{and} \quad \sum_{n=1}^{\infty} (-1)^{\frac{1}{2}(n-1)}P(n)q^{\frac{1}{2}n}.$$

From the multiplicative properties of $P(n)$ it follows that these are therefore newforms in M_1 with eigenvalues $P(n)$ and $(-1)^{\frac{1}{2}(n-1)}P(n)$ respectively, for odd n. In particular, we note that two different newforms of the same divisor can have infinitely many eigenvalues in common.

From now on we shall assume that k is an odd integer, so that χ is the non-principal character modulo 4. The spaces D_k and D_k^* have smaller dimensions than M_4 for $k \geq 7$ and it is accordingly easier to find eigenforms in these spaces than in M_4. We show how one can construct newforms in M_4 from eigenforms in D_k and D_k^*.

Let f be a normalized eigenform in D_k for the operators T_n with $n \equiv 1 \pmod 4$, and write

$$f(z) = \sum_{n=1}^{\infty} c(n)q^n. \tag{10.4.10}$$

Then, if $f|T_n = \lambda(n)f$ for $n \equiv 1 \pmod 4$, we have

$$c(n) = \lambda(n) \quad \text{for} \quad n \equiv 1 \pmod 4.$$

We assume that $c(n)$ is real for all $n \geq 1$. This is not a great restriction, since $\frac{1}{2}(f + f|K)$ is an eigenform with real coefficients and the same real eigenvalues $\lambda(n)$, where $n \equiv 1 \pmod 4$. Moreover, from the fact that D_k has, by theorem 7.1.4, a basis with real coefficients, it can be shown that $c(n)$ is necessarily real, when the class of equivalent forms in D_k to which f belongs has dimension 1.

Suppose also that, for some fixed $l \equiv 3 \pmod 4$,

$$c(l) \neq 0,$$

and write

$$f|T_l = c(l)g, \tag{10.4.11}$$

so that $g \in D_k^*$ and has first Fourier coefficient 1. The function g is therefore a normalized eigenform for the operators T_n with $n \equiv 1 \pmod 4$ and has the same eigenvalues $\lambda(n)$. Write

$$g(z) = \sum_{n=1}^{\infty} c^*(n)q^n, \tag{10.4.12}$$

so that

$$c(n) = c^*(n) = \lambda(n) \quad \text{for} \quad n \equiv 1 \pmod 4. \tag{10.4.13}$$

Note that

$$c(l)g|T_l = (f|T_l)|T_l = \sum_{d|l} \chi(d)d^{k-1}f|T_{l^2/d^2}$$

$$= \sum_{d|l} \chi(d)d^{k-1}\lambda(l^2/d^2)f,$$

since $l^2/d^2 \equiv 1 \pmod 4$. By considering the coefficient of q we deduce that

$$c(l)c^*(l) = \sum_{d|l} \chi(d)d^{k-1}\lambda(l^2/d^2).$$

Hence

$$(f|T_l)|T_l = c(l)c^*(l)f$$

so that

$$g|T_l = c^*(l)f.$$

Accordingly, by (9.3.37),

$$\begin{aligned} c^2(l)(g,g) &= (c(l)g, c(l)g) = (f|T_l, f|T_l) \\ &= \chi(l)(f, f|T_lT_l) \\ &= -c(l)c^*(l)(f,f). \end{aligned}$$

Since both (f,f) and (g,g) are positive, it follows that

$$\eta := -\frac{c^*(l)}{c(l)} > 0. \tag{10.4.14}$$

Now take any $n \equiv 3 \pmod 4$. Then

$$c(l)g|T_n = (f|T_l)|T_n = \left\{\sum_{d|(l,n)} \chi(d)d^{k-1}\lambda(ln/d^2)\right\}f,$$

so that

$$c(l)c^*(n) = \sum_{d|(l,n)} \chi(d)d^{k-1}\lambda(ln/d^2).$$

Accordingly,

$$c^*(l)f|T_n = (g|T_l)|T_n = c(l)c^*(n)g$$

which gives

$$c^*(l)c(n) = c(l)c^*(n).$$

We deduce that

$$f|T_n = c(n)g \qquad \text{for } n \equiv 3 \pmod 4, \tag{10.4.15}$$

$$g|T_n = c^*(n)f \qquad \text{for } n \equiv 3 \pmod 4 \tag{10.4.16}$$

and that

$$c^*(n) = -\eta c(n) \qquad \text{for } n \equiv 3 \pmod 4. \tag{10.4.17}$$

Now suppose that m and n are any two positive coprime odd integers. If one of these integers, say n, satisfies $n \equiv 1 \pmod 4$, then from the relation $f|T_n = c(n)f$ we get, on taking the coefficient of q^m,

$$c(mn) = c(m)c(n).$$

On the other hand, if $m \equiv n \equiv 3 \pmod 4$, then from (10.4.15) we get

$$c(mn) = c^*(m)c(n) = -\eta c(m)c(n).$$

Thus we have, for $(m, n) = 1$, and odd mn,

$$c(mn) = \begin{cases} c(m)c(n) & \text{if } m \equiv 1 \text{ or } n \equiv 1 \pmod 4, \\ -\eta c(m)c(n) & \text{if } m \equiv n \equiv -1 \pmod 4. \end{cases} \tag{10.4.18}$$

Similar relations hold for c replaced by c^* and η by $1/\eta$.

From the functions f and g we construct forms h and $h^K = h|K$ that are normalized eigenforms in C_k for all T_n (n odd). Put

$$h = \frac{\sqrt{(-\eta)}f + g}{\sqrt{(-\eta)} + 1}, \quad h^K = \frac{-\sqrt{(-\eta)}f + g}{-\sqrt{(-\eta)} + 1}. \tag{10.4.19}$$

Then it follows from (10.4.15, 16, 17) that

$$h|T_n = \lambda(n)h, \quad h^K|T_n = \overline{\lambda(n)}h \quad (n \text{ odd}), \tag{10.4.20}$$

where

$$\lambda(n) := \sqrt{(-\eta)}c(n) \quad \text{for } n \equiv 3 \pmod 4. \tag{10.4.21}$$

This is purely imaginary, when not zero, as we have deduced earlier.

It follows immediately that the forms

$$2^{\frac{1}{4}k}h|J_2^{-1} \quad \text{and} \quad 2^{\frac{1}{4}k}h^K|J_2^{-1}$$

are normalized eigenforms in M_4 for the operators T_n (n odd). Also the forms

$$h|D_2 \quad \text{and} \quad h^K|D_2$$

are normalized eigenforms in M_2 for all the operators $T_n^{(2)}$ ($n \in \mathbb{Z}^+$).

In the preceding discussion we began with a form f in D_k and from it constructed the forms g in D_k^* and h in C_k. We could equally well have begun with a form g in D_k^* and then constructed f and h.

We now apply the preceding analysis to a number of particular cases. We confine our attention to odd k in the range $3 \le k \le 11$.

k = 3. In this case the only non-vanishing basis form which arises is

$$\Delta^{\frac{1}{4}} = \tfrac{1}{2}\vartheta_2^2\vartheta_3^2\vartheta_4^2, \tag{10.4.22}$$

which is therefore a newform in M_1. In Glaisher's notation,

$$\Delta^{\frac{1}{4}}(z) = \sum_{n=1}^{\infty} \lambda(n)q^{\frac{1}{4}n},$$

and clearly $\lambda(n) = 0$ for $n \equiv 3 \pmod 4$.

$k = 5$. Here M_2, M_4, D_5 have dimension 1 and D_5^* has dimension zero. The form ψ_5 is a normalized eigenform in D_5 for the operators T_n (n odd); note that $\psi_5|T_n = 0$ when $n \equiv 3 \pmod 4$, since this form lies in D_5^*. Theorem 9.4.9(ii) shows that, when p is a prime,

$$\chi_4(p^2) = p^4 \quad \text{if } p \equiv 3 \pmod 4. \tag{10.4.23}$$

In fact it can be shown that the nth Fourier coefficient $\chi_4(n)$ of ψ_5 satisfies

$$\chi_4(n) = \tfrac{1}{4} \sum_{a^2+b^2=n} (a+ib)^4, \tag{10.4.24}$$

from which (10.4.23) follows.

Write $\hat{\psi}_5(z) = \psi_5(2z)$. Then, by theorem 7.1.8,

$$\hat{\psi}_5 = \tfrac{1}{128}\vartheta_2^4(\vartheta_3^4\vartheta_4^2 - \vartheta_3^2\vartheta_4^4) \in B_4, \tag{10.4.25}$$

and so $\hat{\psi}_5$ is the sole newform in M_4. Note also that, by (10.4.24), $\chi_4(2) = 4$ in agreement with theorem 9.4.8(d).

Moreover, if $\hat{f}_5(z) = f_5(2z)$, where

$$f_5 = \tfrac{1}{4}\vartheta_2^2\vartheta_3^4\vartheta_4^4 \in B_1, \tag{10.4.26}$$

then, by theorem 7.1.8,

$$\hat{f}_5 = \psi_5|D_2 = \tfrac{1}{32}\vartheta_2^4\vartheta_3^2\vartheta_4^2(\vartheta_3^2 + \vartheta_4^2) \in B_2. \tag{10.4.27}$$

Accordingly, $\hat{f}_5$ is the sole newform in M_2 and, like ψ_5 and $\hat{\psi}_5$, it has the eigenvalues $\chi_4(n)$ (n odd). Further f_5, being a multiple of $\hat{f}_5|J_2$, is an eigenform in M_1 with the same eigenvalues and so is one of the three newforms in M_1.

It can be shown that the two remaining newforms in M_1 are linear combinations of the two remaining forms in B_1, namely

$$g_5 = -\Delta^{\frac{1}{4}}(\rho\vartheta_3^4 + \rho^2\vartheta_4^4) \tag{10.4.28}$$

and

$$g_5|K = -\Delta^{\frac{1}{4}}(\rho^2\vartheta_3^4 + \rho\vartheta_4^4), \tag{10.4.29}$$

where, as usual, $\rho = e^{2\pi i/3}$. The coefficients of $q^{\frac{1}{4}n}$ (n odd) in these two newforms are $Q(n)$ for $n \equiv 1 \pmod 4$ and $\pm\sqrt{(3)}iQ(n)$ for $n \equiv 3 \pmod 4$, where $Q(n)$ is Glaisher's partially multiplicative function. It follows that, for $(m, n) = 1$, (10.4.18) holds with c replaced by Q and η by 3.

***k* = 7.** Here D_7 and D_7^* have dimension 1, while M_4 and M_2 have dimension 2; from now on we shall not consider M_1. The spaces D_7 and D_7^* are generated by the forms ψ_7 and ψ_7^*, respectively. These are therefore eigenforms for the operators T_n with $n \equiv 1 \pmod 4$ and are both normalized.

Moreover, since $W(3)$, the coefficient of q^3 in ψ_7, is -8 we can apply the analysis developed earlier in this section with $f = \psi_7$ and $l = 3$. It follows that

$$\psi_7|T_3 = W(3)\psi_7^*.$$

Since $W^*(3) = 120$, we have, by (10.4.14),

$$\eta = 15.$$

Accordingly, the function W satisfies the relations (10.4.18), with c replaced by W and η by 15, as conjectured by Glaisher. See Glaisher (1907*b*) and Rankin (1967*a*).

From ψ_7 and ψ_7^* we form two functions h_7 and h_7^K, say, as in (10.14.19). The two forms

$$2^{7/2}h_7|J_2^{-1} \quad \text{and} \quad 2^{7/2}h_7^K|J_2^{-1}$$

are therefore the newforms in M_4, while

$$h_7|D_2 \quad \text{and} \quad h_7^K|D_2$$

are the newforms in M_2. The coefficient of q^4 in $h_7(2z)$ is $\lambda(2) = 2(1+\sqrt{(-15)})$, so that $|\lambda(2)| = 8$, as expected from theorem 9.4.8(*d*).

***k* = 9.** The space D_9^* is generated by ψ_9^* and we write

$$\psi_9^*(z) = \sum_{n=1}^{\infty} G(n)q^n, \tag{10.4.30}$$

with Glaisher. Hence $\psi_9^*|T_n = G(n)\psi_9^*$ for $n \equiv 1 \pmod 4$. Since $G(3) = -80$, we put

$$-80f_9 := \psi_9^*|T_3,$$

so that f_9 is an eigenform in D_9. By elementary calculations we find that the coefficient of q^3 in f_9 is $624/5$. Accordingly, by (10.4.14) with $l = 3$,

$$\eta = \tfrac{39}{25}$$

and so the relations (10.4.18) hold with this value of η and c replaced by G, as conjectured by Glaisher (1970*c*).

Glaisher tabulated $G(n)$ for $n \leq 50$ and from the first few coefficients we can deduce that

$$f_9 = \tfrac{1}{80}\vartheta_2^4\vartheta_3^2\vartheta_4^4(5\vartheta_3^8 - 2\vartheta_2^4\vartheta_4^4).$$

From (10.4.19) with f and g replaced by ψ_9^* and f_9, respectively, and with the value obtained for η, we find eigenforms h_9 and h_9^K in C_9 for T_n (n odd). As in the case when $k=7$, we immediately deduce pairs of newforms in the spaces M_4 and M_2 having eigenvalues $G(n)$ for $n \equiv 1 \pmod 4$.

Since $G(9) = -3423 \neq 3^8$, it follows from theorem 9.4.9 that the eigenclasses in D_9 and D_9^* having the eigenvalues $G(n)$ for $n \equiv 1 \pmod 4$ have the same dimension. Since $\delta_9 = 2$, $\delta_9^* = 1$, it follows that the remaining eigenform g in D_9 must be such that

$$g|T_n = 0 \quad \text{for } n \equiv 3 \pmod 4.$$

Now we must have

$$g = \alpha g_9 + \beta f_9,$$

where

$$g_9 = \tfrac{1}{16}\vartheta_2^4\vartheta_3^2\vartheta_4^4(4\vartheta_3^8 + \vartheta_2^4\vartheta_4^4) = \sum_{n=1}^{\infty} \chi_8(n)q^n,$$

in Glaisher's notation, and α and β are constants to be determined.

Since $g|T_3 = 0$ and $\chi_8(3) = 0$, we deduce that $\beta = 0$ and so we can normalize by taking $\alpha = 1$. Thus $g = g_9$ and we deduce that $\chi_8(n) = 0$ whenever $n \equiv 3 \pmod 4$. In fact this follows since, as Glaisher showed,

$$\chi_8(n) = \tfrac{1}{4}\sum_{a^2+b^2=n} (a+ib)^8.$$

From g_9 we get newforms in M_1 in M_2, with the same eigenvalues, by operating on g_9 by J_2^{-1} and D_2.

***k* = *11*.** In this case $\delta_{11} = \delta_{11}^* = 2$ and it is no longer possible to find eigenforms in D_{11} and D_{11}^* with rational coefficients.

Write

$$F = \psi_{11}, \quad G = \tfrac{1}{256}\vartheta_2^8\vartheta_3^6\vartheta_4^8.$$

Then these two forms are in D_{11} and a normalized eigenform (which certainly must exist) in D_{11} must therefore be of the form

$$f = F + 4\mu G,$$

for some scalar μ. We use the notation (10.4.10). The relation

$$c(10) = c(2)c(5)$$

shows that μ satisfies the quadratic equation

$$\mu^2 + 13\mu - 84 = 0$$

and therefore belongs to the field $\mathbb{Q}(\sqrt{505})$. This gives two eigenforms in D_{11}. We find that

$$c(3) = 8(21 - 2\mu) \neq 0$$

and so obtain, by the method described, the corresponding eigenforms in D_{11}^* which yield, from (10.4.14), the value

$$\eta = \frac{9}{31^2}(16\mu + 469);$$

this is positive, in each case, as is expected.

The formulae (10.4.19) yield two pairs of functions in C_{11} that are eigenforms for all T_n (n odd); from these, by the methods already used, we can obtain the four newforms in each of the spaces M_4 and M_2. The coefficients of these forms lie in the field $\mathbb{Q}(\mu, i\sqrt{\eta})$. It can then be verified that, for each of the four newforms in M_4,

$$|\lambda(2)|^2 = 2^{10},$$

as proved in theorem 9.4.8(d).

10.5. Final remarks. Theorem 10.1 is based on the account given by Berndt (1970) (see his theorem 2.3, in particular) of the work of Hecke (1936, 1938) on the determination of Dirichlet series by their functional equations.

For the full modular group the Dirichlet series ϕ associated with an eigenform has the Euler product (10.2.5). The work of Deligne referred to in §4.6 proves that the eigenvalues $\lambda(p)$ satisfy the inequality

$$|\lambda(p)| < 2p^{\frac{1}{2}(k-1)} \quad (p \text{ prime}) \tag{10.5.1}$$

conjectured by Ramanujan and Petersson, and from this one can easily deduce the estimate (4.6.4) for the general coefficient $a_n = \lambda(n)$. Conversely, given (4.6.4), one can deduce (10.5.1); see Rankin (1973a).

Deligne and Serre have shown that the same inequality (10.5.1) holds for the eigenvalues of any newform. In the particular cases where p divides the level N we have, of course, the more precise information given by theorem 9.4.8(iii).

Many of the results obtained in §§10.3, 4 can be found in Rankin (1962*b*, 1964, 1967*a*). Multiplicative properties of arithmetical functions defined in terms of theta-functions can be proved, in some cases, by purely elementary methods. See, for example, Glaisher (1906) and other papers by that author in the old *Quarterly Journal of Mathematics.* These properties can be derived most easily and naturally when the functions are expressed in terms of sums over integral Gaussian numbers, quaternions or Cayley numbers; see Rankin (1946) and also the dissertation of Turton (1956).

Turton's work makes use of quaternions to prove the multiplicative properties of the coefficients of the function

$$\tfrac{1}{32}\vartheta_2^4\vartheta_3^3\vartheta_4^3(\vartheta_3^4+\vartheta_4^4),$$

which we have not considered in §§10.3, 4. This function is then used to derive formulae, in the missing cases not considered by Glaisher (1970*b*), for the number of representations of a positive integer as a sum of fourteen squares, an odd number of which are odd.

Finally, the analytic properties of the Dirichlet series associated with modular forms are of interest; this is the case particularly for the Dirichlet series associated with newforms. There is an unresolved conjecture of Weil (1967) relating to such Dirichlet series and elliptic curves; see also Li (1975). Further, many of the problems that arise with the Riemann zeta-function have their analogues for the Dirichlet series associated with newforms; see, for example, Rankin (1939*a*) and Moreno (1972, 1974).

Bibliography

In reference to periodicals the following abbreviations are used:

AA	= *Acta Arith.*	*JRAM*	= *J. Reine Angew. Math.*
AJM	= *Amer. J. Math.*	*MA*	= *Math. Ann.*
AM	= *Ann. of Math.*	*MMJ*	= *Michigan Math. J.*
DMJ	= *Duke Math. J.*	*MZ*	= *Math. Z.*
GMJ	= *Glasgow Math. J.*	*PAMS*	= *Proc. American Math. Soc.*
HA	= *Abh. Math. Sem. Univ. Hamburg*	*PCPS*	= *Proc. Cambridge Philos. Soc.*
IJM	= *Illinois J. Math.*	*PLMS*	= *Proc. London Math. Soc.*
JLMS	= *J. London Math. Soc.*	*QJM*	= *Quart. J. Math.*

Atkin, A. O. L. (1969), Note on a paper of Rankin. *Bull. London Math. Soc.* **1**, 191–2.

Atkin, A. O. L. and Lehner, J. (1970). Hecke operators on $\Gamma_0(m)$. *MA* **185**, 134–60.

Atkin, A. O. L. and Swinnerton-Dyer, H. P. F. (1971). Modular forms on non-congruence groups. *Combinatorics* (*Proc. Sympos. Pure Math., vol. 19*) pp. 1–25. Amer. Math. Soc., Providence, R.I.

Ayoub, R. (1963). *An introduction to the analytic theory of numbers.* Amer. Math. Soc., Providence, R.I.

Berberian, S. K. (1961). *Introduction to Hilbert space.* Oxford University Press.

Berndt, B. C. (1970). *Hecke's theory of modular forms and Dirichlet series.* Lecture notes, University of Illinois, Urbana.

Bol, G. (1949). Invarianten linearer Differentialgleichungen. *HA* **16**, 1–28.

Copson, E. T. (1935). *An introduction to the theory of functions of a complex variable.* Oxford University Press.

Dedekind, R. (1877). Schreiben an Herrn Borchardt über die Theorie der elliptischen Modulfunktionen. *JRAM* **83**, 265–92.

Deligne, P. (1969). Formes modulaires et représentations l-adiques. *Sém. Bourbaki* **21**, no. 335, 139–72.

Deligne, P. (1973). Formes modulaires et représentations de GL(2). *Lecture Notes in Mathematics* **349**, 55–105. Springer.

Deligne, P. (1974). La conjecture de Weil. *Inst. Hautes Études Sci. Publ. Math.* **53**, 273–307.

Deligne, P. and Serre, J.-P. (1974). Formes modulaires de poids 1. *Ann. Sci. École Norm. Sup.* **7**, 507–30.

Dickson, L. E. (1934). *History of the theory of numbers.* Stechert, New York.

Eichler, M. (1954). Quaternäre quadratische Formen und die Riemannsche Vermutung für die Kongruenz-zetafunktionen. *Arch. Math.* **5**, 355–66.
Eichler, M. (1974). *Quadratische Formen und orthogonale Gruppen.* Springer.
Estermann, T. (1930). Vereinfachter Beweis eines Satzes von Kloosterman. *HA* **7**, 82–98.
Ford, L. R. (1929). *Automorphic functions.* McGraw-Hill.
Frasch, H. (1933). Die Erzeugenden der Hauptkongruenzgruppen für Primzahlstufen. *MA* **108**, 229–52.
Fricke, R. (1887). Über die Substitutionsgruppen, welche zu den Legendre'schen Integralmodul $k^2(\omega)$ gezogenen Wurzeln gehören. *MA* **28**, 99–188.
Fricke, R. and Klein, F. (1926). *Vorlesungen über die Theorie der automorphen Funktionen*, 2 vols. Teubner.
Fueter, R. (1924, 1927). *Vorlesungen über die singulären Moduln und die komplexe Multiplikation der elliptischen Funktionen*, 2 vols. Teubner.
Gantmacher, F. R. (1960). *Matrix theory*, vol. 1. Chelsea.
Glaisher, J. W. L. (1906). The arithmetical functions $P(m)$, $Q(m)$, $\Omega(m)$. *QJM* **37**, 36–48.
Glaisher, J. W. L. (1907*a*). On the representations of a number as the sum of two, four, six, eight, ten and twelve squares. *QJM* **38**, 1–62.
Glaisher, J. W. L. (1907*b*). On the representations of a number as the sum of fourteen and sixteen squares. *QJM* **38**, 178–236.
Glaisher, J. W. L. (1907*c*). On the representations of a number as the sum of eighteen squares. *QJM* **38**, 289–351.
Gunning, R. C. (1955). General factors of automorphy. *Proc. Nat. Acad. Sci. U.S.A.* **41**, 496–8.
Gunning, R. C. (1962). *Lectures on modular forms.* Princeton University Press.
Hall, M. (1959). *The theory of groups.* Macmillan.
Hardy, G. H. (1940). *Ramanujan.* Cambridge University Press.
Hardy, G. H. and Wright, E. M. (1938). *An introduction to the theory of numbers.* Oxford University Press.
Hecke, E. (1924). Analytische Funktionen und algebraische Zahlen. Zweiter Teil. *HA* **3**, 213–36.
Hecke, E. (1925). Darstellung von Klassenzahlen als Perioden von Integralen 3. Gattung aus dem Gebiet der elliptischen Modulfunktionen. *HA* **4**, 211–23.
Hecke, E. (1927). Theorie der Eisensteinschen Reihen höhere Stufe und ihre Anwendung auf Funktionalgleichung und Arithmetik. *HA* **5**, 199–224.
Hecke, E. (1936). Über die Bestimmung Dirichletscher Reihen durch ihre Funktionalgleichung. *MA* **112**, 664–99.
Hecke, E. (1937). Über Modulfunktionen und die Dirichletschen Reihen mit Eulerscher Produktentwicklung I. *MA* **114**, 1–28; II. *MA* **114**, 316–51.
Hecke, E. (1938). *Dirichlet series, modular functions and quadratic forms.* Institute for Advanced Study, Princeton (planographed lecture notes).
Hecke, E. (1940). Analytische Arithmetik der positiven quadratischen Formen. *Danske Vid. Selsk. Mat.-Fys. Medd.* **17**, no. 12, 134 pp.
Hecke, E. (1959). *Mathematische Werke.* Vandenhoeck & Ruprecht.

Hua, L. K. and Reiner, I. (1951). Automorphisms of the unimodular group. *Trans. Amer. Math. Soc.* **71**, 331–48.

Hurwitz, A. (1932, 1933). *Mathematische Werke*, 2 vols. Birkhäuser.

Igusa, J.-I. (1959). Kroneckerian model of fields of elliptic modular functions. *AJM* **81**, 561–77.

Joris, H. (1975). An Ω-result for the coefficients of cusp forms. *Mathematika* **22**, 12–19.

Klein, F. and Fricke, R. (1890, 1892). *Vorlesungen über die Theorie der elliptischen Modulfunktionen*, 2 vols. Teubner.

Kløve, T. (1970). Recurrence formulae for the coefficients of modular forms. *Math. Scand.* **26**, 221–32.

Knopp, M. I. (1970). *Modular functions in analytic number theory*. Markham.

Knopp, M. I. and Lehner, J. (1962). On complementary automorphic forms and supplementary Fourier series. *IJM* **6**, 98–106.

Knopp, M. I., Lehner, J. and Newman, M. (1965). A bounded automorphic form of dimension zero is constant. *DMJ* **32**, 457–60.

Kogan, L. A. (1969). Liouville formulae and parabolic forms that are generated by generalized binary theta-series. *Litovsk Mat. Sb.* **9**, 519–33.

Kurosh, A. G. (1960). *The theory of groups*, vol. 2, second edition. Chelsea.

Lehmer, D. H. (1942). Ramanujan's function $\tau(n)$, *AJM* **64**, 488–502.

Lehmer, D. H. (1947). The vanishing of Ramanujan's function $\tau(n)$. *DMJ* **14**, 429–33.

Lehmer, D. H. (1959). Some functions of Ramanujan. *Math. Student* **27**, 105–16.

Lehner, J. (1959). On modular forms of negative dimension. *MMJ* **6**, 71–88.

Lehner, J. (1963). On the generation of discontinuous groups. *Pacific J. Math.* **13**, 169–70.

Lehner, J. (1964). *Discontinuous groups and automorphic functions*. American Mathematical Society.

Lehner, J. (1966). *A short course in automorphic functions*. Holt, Rinehart & Winston.

Lehner, J. (1968). On the multipliers of the Dedekind modular function. *J. Res. Nat. Bur. Standards Sect. B.* **72*B***, 253–61.

Lehner, J. (1969). *Lectures on modular forms*. National Bureau of Standards, Applied Mathematics Series 61.

Leutbecher, A. (1970). Die Automorphiefaktoren und die Dedekindschen Summen. *GMJ* **11**, 41–57.

LeVeque, W. J. (1974). *Reviews in number theory*, vol. 2. American Mathematical Society.

Li, W. C. W. (1975). Newforms and functional equations. *MA* **212**, 285–315.

Lyndon, R. (1973). Two notes on Rankin's book on the modular group. *J. Austral. Math. Soc.* **16**, 454–7.

Maas, H. (1964). *Lectures on modular functions of one complex variable*. Tata Institute of Fundamental Research, Lectures in Mathematics, no. 29.

Macbeath, A. M. (1961). *Discontinuous groups and birational transformations*. Dundee Summer School Lectures.

MacDuffee, C. C. (1946). *The theory of matrices*. Chelsea.

McQuillan, D. L. (1965). Classification of normal congruence subgroups of the modular group. *AJM* **87**, 285–96.
Mason, A. W. (1969). Lattice subgroups of free congruence groups. *GMJ* **10**, 106–15.
Miyake, T. (1971). On automorphic forms on GL_2 and Hecke operators. *AM* **94**, 174–89.
Mordell, L. J. (1920). On Ramanujan's empirical expansions of modular functions. *PCPS* **19**, 117–24.
Mordell, L. J. (1928). Poisson's summation formula and the Riemann zeta-function. *JLMS* **4**, 285–91.
Mordell, L. J. (1932). On a sum analogous to Gauss's sum. *QJM* (2)**3**, 161–7.
Moreno, C. J. (1972). Prime number theorems for the coefficients of modular forms. *Bull. Amer. Math. Soc.* **78**, 796–8.
Moreno, C. J. (1974). A necessary and sufficient condition for the Riemann hypothesis for Ramanujan's zeta function. *IJM* **18**, 107–14.
Newman, M. (1962). The structure of some subgroups of the modular group. *IJM* **6**, 480–7.
Newman, M. (1964). Free subgroups and normal subgroups of the modular group. *IJM* **8**, 262–5.
Newman, M. (1965). Normal subgroups of the modular group which are not congruence groups. *PAMS* **16**, 831–2.
Newman, M. (1967). Classification of normal subgroups of the modular group. *Trans. Amer. Math. Soc.* **126**, 267–77.
Newman, M. (1968). Maximal normal subgroups of the modular group. *PAMS* **19**, 1138–44.
Newman, M. (1972). *Integral matrices.* Academic Press.
Newman, M. and Smart, J. R. (1963). Notes on a subgroup of the modular group. *PAMS* **14**, 102–4.
Nielsen, J. (1948). Kommutatorgruppen for det frie produkt af cykliste grupper. *Mat. Tidskr. B*, pp. 49–56.
Ogg, A. P. (1969*a*). On the eigenvalues of Hecke operators. *MA* **179**, 101–8.
Ogg, A. P. (1969*b*). *Modular forms and Dirichlet series.* Benjamin.
Ogg, A. P. (1973). Survey of modular functions of one variable. *Lecture Notes in Mathematics* **320**, 1–35. Springer.
Petersson, H. (1930). Theorie der automorphen Formen beliebiger reeller Dimension und ihre Darstellung durch eine neue Art Poincaréscher Reihen. *MA* **103**, 369–436.
Petersson, H. (1931). Über die Entwicklungskoeffizienten der ganzen Modulformen und ihre Bedeutung für die Zahlentheorie. *HA* **8**, 215–42.
Petersson, H. (1932*a*). Über die Entwicklungskoeffizienten der automorphen Formen. *Acta Math.* **58**, 169–215.
Petersson, H. (1932*b*). Ein Fundamentalsatz aus der Theorie der ganzen automorphen Formen. *MA* **106**, 343–68.
Petersson, H. (1938*a*). Zur analytischen Theorie der Grenzkreisgruppen, I–IV. *MA* **115**, 23–67, 175–204, 518–72, 670–709; V. *MZ* **44**, 127–55.
Petersson, H. (1938*b*) Die linearen Relationen zwischen den ganzen Poincaréschen Reihen von reeller Dimension zur Modulgruppe. *HA* **12**, 415–72.

Petersson, H. (1939). Über eine Metrisierung der ganzen Modulformen. *Jber. Deutsch. Math. Verein.* **49**, 49–75.
Petersson, H. (1939–40). Konstruktion der sämtlichen Lösungen einer Riemannschen Funktionalgleichung durch Dirichlet-Reihen mit Eulerscher Produktentwicklung. I. *MA* **116**, 401–12; II. *MA* **116**, 39–64; III. *MA* **117**, 277–300.
Petersson, H. (1940). Über eine Metrisierung der automorphen Formen und die Theorie der Poincaréschen Reihen. *MA* **117**, 453–537.
Petersson, H. (1948). Automorphe Formen als metrische Invarianten II. *Math. Nachr.* **1**, 218–57.
Petersson, H. (1950). Konstruktion der Modulformen und die gewissen Grenzkreisgruppen gehörigen automorphen Formen von positiver reeller Dimension und die vollständige Bestimmung ihrer Fourierkoeffizienten. *S.-B. Heidelberger Akad. Wiss. Math.-Natur. Kl.* pp. 415–94.
Petersson, H. (1953). Über einen einfachen Typus von Untergruppen der Modulgruppe. *Arch. Math.* **4**, 308–15.
Petersson, H. (1956). Über Eisensteinsche Reihen und automorphe Formen von der Dimension −1. *Comment. Math. Helv.* **31**, 111–44.
Petersson, H. (1963). Über die Kongruenzgruppen der Stufe 4. *JRAM* **212**, 63–72.
Petersson, H. (1967*a*). Über Differentialoperatoren und die Kroneckersche Potenzdarstellung bei automorphen Formen. *JRAM* **231**, 163–91.
Petersson, H. (1967*b*). Über eine Spurbildung bei automorphen Formen. *MZ* **96**, 296–372.
Petersson, H. (1973). Über Heckesche Operatoren, Poincarésche Reihen und eine Siegelsche Konstruktion. *AA* **24**, 411–34.
Pick, G. (1887). Über gewisse ganzzahlige lineare Substitutionen, welche sich nicht durch algebraische Congruenzen erklären lassen. *MA* **28**, 119–24.
Rademacher, H. (1955). On the transformations of log $\eta(\tau)$. *J. Indian Math. Soc.* **19**, 25–30.
Rademacher, H. (1973). *Topics in analytic number theory*. Springer.
Rademacher, H. and Grosswald, E. *Dedekind sums*. Carus Mathematical Monographs no. 16. Mathematical Association of America.
Ramanujan, S. (1916). On certain arithmetical functions. *Trans. Cambridge Philos. Soc.* **22**, 159–84.
Rankin, F. K. C. and Swinnerton-Dyer, H. P. F. (1970). On the zeros of Eisenstein series. *Bull. London Math. Soc.* **2**, 169–70.
Rankin, R. A. (1939*a*). Contributions to the theory of Ramanujan's function $\tau(n)$ and similar arithmetical functions. I. The zeros of the function $\sum_{n=1}^{\infty} \tau(n)n^{-s}$ on the line $\Re s = 13/2$. *PCPS* **35**, 351–6.
Rankin, R. A. (1939*b*). Contributions etc. II. The order of the Fourier coefficients of integral modular forms. *PCPS* **35**, 357–72.
Rankin, R. A. (1940). Contributions etc. III. A note on the sum function of the Fourier coefficients of integral modular forms. *PCPS* **36**, 150–1.
Rankin, R. A. (1946). A certain class of multiplicative functions. *DMJ* **13**, 281–306.
Rankin, R. A. (1952). The scalar product of modular forms. *PLMS* (3)**2**, 198–217.

Rankin, R. A. (1954). On horocyclic groups. *PLMS* (3)**4**, 219–34.
Rankin, R. A. (1956). The construction of automorphic forms from the derivatives of a given form. *J. Indian Math. Soc.* **20**, 103–16.
Rankin, R. A. (1957). The construction of automorphic forms from the derivatives of given forms. *MMJ* **4**, 181–6.
Rankin, R. A. (1962*a*). Multiplicative functions and operators of Hecke type. *Acta Math. Acad. Sci. Hungar.* **13**, 81–9.
Rankin, R. A. (1962*b*). On the representation of a number as the sum of any number of squares, and in particular of twenty. *AA* **7**, 399–407.
Rankin, R. A. (1964). *Elliptic modular functions and forms.* Indiana University, Bloomington, Indiana (duplicated lecture notes).
Rankin, R. A. (1965). Sums of squares and cusp forms. *AJM* **87**, 857–60.
Rankin, R. A. (1967*a*). Hecke operators on congruence subgroups of the modular group. *MA* **168**, 40–58.
Rankin, R. A. (1967*b*). Lattice subgroups of free congruence groups. *Invent. Math.* **2**, 215–21.
Rankin, R. A. (1969). *The modular group and its subgroups.* Ramanujan Institute, Madras.
Rankin, R. A. (1973*a*). An Ω-result for the coefficients of cusp forms. *MA* **203**, 239–50.
Rankin, R. A. (1973*b*). Subgroups of the modular group defined by a single linear congruence. *AA* **24**, 313–23.
Rankin, R. A. and Rushforth, J. M. (1954). The coefficients of certain integral modular forms. *PCPS* **50**, 305–8.
Reiner, I. (1958). Normal subgroups of the unimodular group. *IJM* **2**, 142–4.
Resnikoff, H. L. (1966). On differential operators and automorphic forms. *Trans. Amer. Math. Soc.* **124**, 334–6.
Schoeneberg, B. (1974). *Elliptic modular functions.* Springer.
Schwandt, E. A. (1972). On the Fourier coefficients of the Poincaré series. *JLMS* (**2**)**5**, 584–8.
Selberg, A. (1939). Über die Fourierkoeffizienten elliptischer Modulformen negativer Dimension. *Neuvième Congrès des Mathematiciens Scandinaves. Helsingfors 1938*, pp. 320–2.
Selberg, A. (1940). Bemerkungen über eine Dirichletsche Reihe, die mit der Theorie der Modulformen nahe verbinden ist. *Arch. Math. Naturvid.* **43**, 47–50.
Selberg, A. (1965). On the estimation of Fourier coefficients of modular forms. *Proceedings of Symposia in Pure Mathematics*, vol. 8, pp. 1–15. American Mathematical Society.
Serre, J.-P. (1970). *Cours d'arithmétique.* Presses Universitaires de France.
Serre, J.-P. (1973). Formes modulaires et fonctions zêta p-adiques. *Lecture Notes in Mathematics* **350**, 192–265. Springer.
Shimura, G. (1959). Sur les integrales attachées aux formes automorphes *J. Math. Soc. Japan* **11**, 291–311.
Shimura, G. (1971). *Introduction to the arithmetic theory of automorphic functions.* Princeton University Press.
Shimura, G. (1973). On modular forms of half-integral weight. *AM* **97**, 440–81.

Springer, G. (1957). *Introduction to Riemann surfaces.* Addison-Wesley.

Stepanov, S. A. (1971). Estimation of Kloosterman sums. *Izv. Akad. Nauk SSSR Ser. Mat.* **35**, 308–23.

Tannery, J. and Molk, J. (1893–1902). *Éléments de la théorie des fonctions elliptiques*, vols. 1–4. Gauthier-Villars.

Turton, R. J. (1956). *An investigation into the construction of certain multiplicative functions.* University of Birmingham. Part of Ph.D. thesis (unpublished).

van Lint, J. H. (1957). *Hecke operators and Euler products.* Dissertation. University of Leiden.

van Lint, J. H. (1958). On the multiplier system of the Riemann–Dedekind function. *Nederl. Akad. Wetensch. Proc. Ser. A.* **61** = *Indag. Math.* **20**, 522–7.

Vivanti, G. (1906). *Elementi della teoria della funzioni periodiche e modulari.* Hoepli.

Vivanti, G. (1910). *Les fonctions polyédriques et modulaires* (translated by A. Cahen). Gauthier–Villars.

Walfisz, A. (1933). Über die Koeffizientensummen einiger Modulformen. *MA* **108**, 75–90.

Watson, G. N. (1922). *A treatise on the theory of Bessel functions.* Cambridge University Press.

Watson, G. N. (1949). A table of Ramanujan's function $\tau(n)$. *PLMS* (2)**51**, 1–13.

Weber, H. (1909). *Lehrbuch der Algebra*, vol. 3. Vieweg.

Weil, A. (1948). On some exponential sums. *Proc. Acad. Sci. U.S.A.* **34**, 204–7.

Weil, A. (1967). Über die Bestimmung Dirichletscher Reihen durch Funktionalgleichungen. *MA* **168**, 149–56.

Weil, A. (1971). Dirichlet series and automorphic forms. *Lecture Notes in Mathematics* **189**. Springer.

Whittaker, E. T. and Watson, G. N. (1927). *A course of modern analysis*, 4th edition. Cambridge University Press.

Williams, K. S. (1971). Note on the Kloosterman sum. *PAMS* **30**, 71–2.

Wilton, J. R. (1929). A note on Ramanujan's arithmetical function $\tau(n)$. *PCPS* **25**, 121–9.

Wohlfahrt, K. (1957). Über Operatoren Heckescher Art bei Modulformen reeller Dimension. *Math. Nachr.* **16**, 233–56.

Wohlfahrt, K. (1964). An extension of F. Klein's level concept. *IJM* **8**, 529–35.

Wohlfahrt, K. (1972). A direct proof of Leutbecher's lemma. *GMJ* **13**, 179–80.

Index of special symbols

Symbol	Meaning	Page
A	first entry in L	114
$A_r(n)$	Fourier coefficient	297
Aut G	automorphism group	13
$\mathbf{A}_n^a$	set of matrices	288
a	first entry in T	1
$a_r(n)$	Fourier coefficient	290
$a(r, m; L)$	Fourier coefficient	162
$a(r, m, L, S)$	Fourier coefficient	154
$a(r; C, D; N)$	Fourier coefficient	175
B	second entry in L	114
B_n	Bernoulli number	195
$\mathcal{B}_s$, $\mathcal{B}_s^*$	spaces of forms	224
$\mathbf{B}$, $\mathbf{B}_n^a$	right transversal	309
b	second entry in T	1
C	third entry in L	114
C_k	space of forms	344
C_k^1, C_k^2	spaces of forms	346
$C(r, m, n)$	Fourier coefficient	313
$C(r, m, n; \chi)$	Fourier coefficient	314
$\mathbb{C}$, $\bar{\mathbb{C}}$	complex numbers	1
c	third entry in T	1
c_k	Fourier coefficient	203
c_μ	module coefficient	283, 284
$c_\gamma(r)$	Ramanujan sum	167
$c(r, m)$, $c(r, m; \chi)$	Fourier coefficients	313
$c_k(r, m)$	Fourier coefficient	201
$c(L, M; X)$	module coefficient	279
$c(l, m; q)$	module coefficient	283
D	fourth entry in L	114
D_k, D_k^*	spaces of forms	344
D_t	linear operator	248
$D_t^{(\varepsilon)}$	linear operator	253

D'_τ	linear operator	257
$D(t)$	function	205
$D_T(z)$	modular function	211
$\mathcal{D}$	set of double cosets	275
$\mathbf{D}_n^a$	set of matrices	288
d	fourth entry in T	1
d_k	dimension of space	194
$d_k^{(r)}$	dimension of space	207
$d(q)$	divisor function	170
$\det T$	determinant of matrix T	1
$\dim M$	dimension of space M	102
$\operatorname{div} T$	divisor of matrix T	273
$E_0, E_0(z)$	constant Eisenstein series	198
E_k, E_k^*	Eisenstein series	175, 195
E_n	Euler number	233
$E_k(z), E_k^*(z)$	Eisenstein series	194
$E_2^+(z)$	Fourier series	196
$E_k(z; L)$	Eisenstein series	234
$E_k(z; C, D; N)$	Eisenstein series	174
$E_k^*(z; C, D; N)$	Eisenstein series	175
$\mathbb{E}, \mathbb{E}_2, \mathbb{E}_3$	sets of elliptic fixed points	45
$\mathbb{E}(\Gamma), \mathbb{E}_2(\Gamma), \mathbb{E}_3(\Gamma)$	sets of elliptic fixed points	46
e	2.71828 . . .	*passim*
e_2, e_3	numbers of elliptic fixed points	57
$e_2(\Gamma), e_3(\Gamma)$	numbers of elliptic fixed points	57
F_8, F_8^*	cusp forms	346
F_{10}	cusp form	347
F_q	Chebyshev polynomial	2
$F_k(z)$	cusp form	202
$F_k(z, k, \lambda)$	series	157
$\mathbb{F}$	fundamental region	47
$\mathbb{F}^*$	fundamental region	224
$\mathbb{F}_3$	fundamental region	61
$\mathbb{F}^{(1)}, \mathbb{F}^{(2)}$	triangles	51
$\mathbb{F}_I$	fundamental region	51
$\hat{\mathbb{F}}_I, \mathbb{F}_T, \hat{\mathbb{F}}_T$	fundamental regions	52
f_1, f_2, f_3	modular functions	228
f_4	cusp form	352
$f_5, \hat{f}_5$	cusp forms	356
f_9	cusp form	357–8
f_L	transformed modular form	89, 119

f^K	transformed modular form	270
f_L^*	function related to f_L	90
$f^{(\varepsilon)}$	modular form	253
$G(n), \hat{G}(n), \bar{G}(n)$	modulary groups	21
$G(n)$	Glaisher's function	357
$G_L(z, m) = G_L(z; m, \Gamma, k; v)$	Poincaré series	136
$G_k(z, m)$	Poncaré series	201
$G_j^*(z; N)$	difference of Poincaré series	189
$G_L(z; m, N)$	Poincaré series	188
$G_L(z; m, N; s)$	modified Poincaré series	183
g	genus	68, 104
g_2, g_3	Weierstrass invariants	212
g_5	cusp form	356
g_9	cusp form	358
g_{10}	cusp form	347
$g(\Gamma)$	genus of group Γ	68
$g(z, m), g(z, m, \chi)$	Poincaré series	313
$g_k(z, m)$	Poincaré series	201
H_q	matrix of order q	260, 261
$H(N, k)$	space of forms	245
$H(\Gamma(1), k)$	space of forms	194
$H(\Gamma, k, v)$	space of forms	94
$H(N, k, \chi)$	space of forms	246
$H^{(t)}(N, k, \chi)$	space of forms	250
$H(N, k, \chi, t)$	space of forms	253
$\mathbb{H}, \bar{\mathbb{H}}, \mathbb{H}'$	upper half planes	1
$\boldsymbol{H}_q(N, N')$	set of matrices	260
I	identity matrix	2
I_{k-1}	Bessel function	156
$I(\zeta, s), I_r(\zeta, s)$	series and integrals	185
Im	imaginary part	*passim*
Inn	inner automorphism group	8
J	matrix	14
$J, J(z)$	Klein's absolute invariant	199
J_1, J_2, J_3	modular functions	228
$J_{k-1}(z)$	Bessel function	156
J_n	matrix	113
$J^*(z)$	function	15
$j, j(z)$	modular function related to J	199

K	conjugate linear map	270
k	weight of modular form	70, 88
L	matrix	114
$L_q^\chi, L_q^\chi(N, N')$	linear operator	266
$L(s, \chi)$	Dirichlet L-series	171, 233
$\mathrm{LF}(2, \mathbb{Z})$	linear fractional group	8
$\mathbb{L}_U, \mathbb{L}_V$	families of pairs of sides	60
l_U, l_V	sides of fundamental region	52
$l(T)$	length of word T	13
$\mathrm{lev}\, \Gamma$	level of group Γ	19
M	inverse of matrix L	135
M	space of modular forms	104, 245, 292, 295, 319
M_t	subspace $M\|D_t$ of M	249, 292, 350
$M^{(t)}$	subspace of M	249
M^χ	subspace $M\|R^\chi$ of M	247, 292
M_t^χ	intersection of M_t and M^χ	252
M^-, M^+	orthogonal subspaces of M	321
$M(d)$	space of forms of level N/d	321
$M(d)^-, M(d)^+$	orthogonal subspaces of $M(d)$	322
$M(N, k)$	space of forms of level N	245
$M_t(N, k)$	subspace of $M(N, k)$	248
$M^{(t)}(N, k)$	subspace of $M(N, k)$	249
$M(\Gamma(l), k)$	space of forms of constant MS	194
$M(N, k, \chi)$	subspace of $M(N, k)$	246
$M^{(t)}(N, k, \chi)$	subspace of $M^{(t)}(N, k, \chi)$	250
$M(\Gamma, k, v)$	space of modular forms	91
$M'(\Gamma, k, v)$	set of unrestricted modular forms	88
$M(N, k, \chi, t)$	subspace of $M(N, k)$	253
$\mathcal{M}$	free left-module of double cosets	279
m_L	integer parameter associated with $L\rho$	78
N	level of group $\Gamma(N)$	*passim*
N_χ	conductor of character χ	258
n_2, n_3, n_∞	branch schema parameters	60
n_L	order of $L\infty$ (mod Γ)	19, 77
$n(z, \Gamma)$	order of z (mod Γ)	46
$\mathrm{ord}(f, \zeta)$	order of f at ζ	40
$\mathrm{ord}(f, \Gamma)$	total order of f (mod Γ)	95
$\mathrm{ord}(f, \zeta, \Gamma)$	order of f at ζ (mod Γ)	91

P, P^2	matrices	8
P_1	matrix	16
$P(n)$	Glaisher's function	352
$\mathbb{P}$	set of cusps	1
$\wp$	Weierstrass elliptic function	213
ph	principal phase (argument)	70
Q	matrix	18, 33
$Q(n)$	Glaisher's function	356
$\mathbb{Q}$	rational numbers	1
q	$e^{\pi i z}$	218
q_0, q_1, q_2, q_3	infinite products	226
R_d	matrix (mod N)	246
R^x	linear operator	246
$R_k(z)$	cusp form	241
Re	real part	*passim*
$\mathscr{R}$	finite set of weights	206
$\mathscr{R}$	right transversal	*passim*
$\mathbb{R}, \bar{\mathbb{R}}$	real numbers and compactification	1
$\boldsymbol{R}$	right transversal of $\Gamma(N)$ in $\Gamma(1)$	278
$r(T)$	index determining MS	83
$r_s(n)$	Fourier coefficient of ϑ_3^s	241
res(f, ζ)	residue of f at ζ	122
res(f, ζ, Γ)	residue of f at ζ (mod Γ)	122
S	matrix	1
S_n	matrix associated with S	254
$S(u, v; q)$	Kloosterman sum	166
SL(2, $\mathbb{Z}$)	special linear group	8
$\mathbb{S}$	strip in $\mathbb{H}$	148
$\mathbb{S}_k, \mathbb{S}_k(\delta)$	strip sets in $\mathbb{C}$	49
s_L	integer parameter associated with Li	79
T	matrix	1
T^*	various meanings	202, 210, 270, 273, 317
T_n	Hecke operator	289, 311
$T_n^{(t)}$	conjugate Hecke operator	295
$T_n(N), T_n(N/d)$	Hecke operators of levels $N, N/d$	321
Tr	trace operator	107
$\mathscr{T}$	transversal of double cosets	159
$\mathscr{T}_L, \mathscr{T}_L(\gamma)$	sets of matrices	160
$\boldsymbol{T}_n, \boldsymbol{T}_n^*$	transversals	113
$\boldsymbol{T}_n^0, \boldsymbol{T}_r^*, \boldsymbol{T}_n^+$	transversals	287

${}^{+}\boldsymbol{T}_n$	transversal	288
$\boldsymbol{T}_L$	transversal	275
$\boldsymbol{T}_L^+$	transversal	284
$\boldsymbol{T}_n^L$	transversal	116
t, t_1	divisor and codivisor	247
t_x	divisor of t	321
$\operatorname{tr} T$	trace of matrix T	2
U	matrix	8
u	multiplier system	219
u_n	MS associated with u	255, 345
u_s	MS	221
u_s^*	MS associated with u_s	241
V	matrix	8
V_1, V_2	matrices	16
$V(f, \Gamma)$	valence of f (mod Γ)	109
$\mathscr{V}_s$	space of forms of weight $\frac{1}{2}s$	224
$\mathbb{V}(\zeta, \mathbb{F})$	vertex set	55
$v, v(T)$	multiplier system	71, 219, 245
v^*	MS	270
$\tilde{v}$	MS	81, 180
v_0	MS	180
v_s	MS	221
v_s^*	MS associated with v_s	223
$v^{(0)}, v^{(1)}, v^{(2)}$	multiplier systems	182
$v^{(r)}$	MS	206
v^L	MS of transformed group	73
W	matrix	8
$W(n)$	Glaisher's function	242
$W^*(n)$	Fourier coefficient associated with $W(n)$	353, 357
$W(r, m; \gamma)$	generalized Kloosterman sum	161, 165
w	multiplier system	219
w_s	MS	221
w_s^*	MS associated with w_s	244
$w(S, T)$	function taking values 0, ±1	74
X	matrix	180
$\mathfrak{X}(\Gamma, \Delta)$	set of characters on Γ/Δ	104
$\mathbb{Z}, \mathbb{Z}^+$	sets of integers	1
$\mathbb{Z}_N, \mathbb{Z}_N^+$	subsets of $\mathbb{Z}, \mathbb{Z}^+$	256

$\mathbb{Z}_\Gamma$, $\mathbb{Z}_\Gamma^+$	subsets of $\mathbb{Z}$, $\mathbb{Z}^+$	256
$\mathbb{Z}_\Gamma(n)$, $\mathbb{Z}_\Gamma^+(n)$	subsets of $\mathbb{Z}$, $\mathbb{Z}^+$	256
$\boldsymbol{Z}_q$, $\boldsymbol{Z}_q(N, N')$	transversals	265
z	member of $\mathbb{C}$	*passim*
α	first entry in S	1
α_k	parameter in Eisenstein series E_k	194
$\alpha_k(n, L)$	Fourier coefficient of Eisenstein series	234
β	second entry in S	1
β_k	parameter occurring in c_k	204
$\beta_k(n)$	Fourier coefficient of Eisenstein series	240
Γ, $\hat{\Gamma}$	homogeneous and inhomogeneous groups	3
$\bar{\Gamma}$	group $\Gamma\Lambda$	4
Γ^*	subgroup of index 2 in Γ (when $-I \in \Gamma$)	80
Γ^*	group $J^{-1}\Gamma J$	270
Γ^+	semigroup in $\Gamma(1)$	84
Γ^2	subgroup of $\Gamma(1)$ of index 2	16
Γ^3, Γ^4	subgroups of $\Gamma(1)$ of index 3 and 4	15
$\hat{\Gamma}^2$, $\hat{\Gamma}^3$, $\hat{\Gamma}^4$	corresponding groups of mappings	18
$\Gamma^{2'}$, $\Gamma^{3'}$	commutator groups	36
$\hat{\Gamma}^m$	group of mappings	33
Γ_n	group associated with Γ	147, 255
Γ_z, $\hat{\Gamma}_z$	stabilizer groups	46
Γ_L	group $\Gamma^L \cap \Gamma$	116, 274
Γ_L^*	group $\Gamma^L \cap \Gamma(1)$	120
Γ^L	conjugate group $L^{-1}\Gamma L$	72
Γ_T	group $T^{-1}\Gamma(1)T \cap \Gamma(1)$	211
Γ_U	group generated by $\pm U$	8
$\hat{\Gamma}_U$	corresponding group of mappings	9
Γ_{U^k}	group generated by $\pm U^k$	9
$\Gamma(1)$	homogeneous modular group	7
$\hat{\Gamma}(1)$	inhomogeneous modular group	8
$\Gamma'(1)$	commutator subgroup of $\Gamma(1)$	16
$\hat{\Gamma}'(1)$	corresponding group of mappings	18
$\bar{\Gamma}'(1)$	group $\Lambda\Gamma'(1)$	17
$\hat{\Gamma}''(1)$	commutator subgroup of $\hat{\Gamma}'(1)$	34
$\Gamma_z(1)$, $\hat{\Gamma}_z(1)$	stabilizer subgroups	46
$\Gamma_L(1)$	group $L^{-1}\Gamma(1)L \cap \Gamma(1)$	117
$\Gamma(2)$	*see* $\Gamma(n)$	
$\Gamma^*(2)$	subgroup of index 2 in $\Gamma(2)$	30, 81
$\tilde{\Gamma}(2)$	subgroup of index 2 in $\Gamma(2)$	81, 180, 272
$\Gamma_P(2)$, $\Gamma_U(2)$, $\Gamma_W(2)$	subgroups of index 3 in $\Gamma(1)$	29

$\hat{\Gamma}_P(2)$	group of mappings	30
$\Gamma_V(2)$	subgroup of index 3 in $\Gamma(1)$	29, 33
$\Gamma_P(3), \Gamma_{P_1}(3)$	subgroups of index 4 in $\Gamma(1)$	31, 33
$\Gamma_U(3), \Gamma_W(3)$	subgroups of index 4 in $\Gamma(1)$	31
$\Gamma_V(3), \Gamma_{V_1}(3), \Gamma_{V_2}(3)$	subgroups of index 2 in Γ^3	31
$\Gamma_U(4), \Gamma_V(4), \Gamma_W(4)$	subgroups of index 6 in $\Gamma(1)$	33
$\Gamma(k)$	gamma function	149, 156
$\Gamma(n), \hat{\Gamma}(n), \bar{\Gamma}(n)$	principal congruence groups of level n	20
$\Gamma_0(n), \Gamma^0(n), \Gamma^0_0(n)$	groups of level n	25
$\hat{\Gamma}_0(n), \hat{\Gamma}^0(n), \hat{\Gamma}^0_0(n)$	corresponding groups of mappings	25
$\Gamma_t(N)$	group of level N	249
$\Gamma(N, l)$	group $\Gamma(N) \cap \Gamma^0(lN)$	274
$\Gamma_0(m, n), \hat{\Gamma}_0(m, n)$	groups $\Gamma_0(m) \cap \Gamma^0(n), \hat{\Gamma}_0(m) \cap \hat{\Gamma}^0(n)$	27
γ	third entry in S	1
γ_k	dimension of space C_k	344
$\gamma_k(n)$	Fourier coefficient of cusp form F_k	202
$\Delta, \Delta(z)$	modular discriminant function	196
Δ_1, Δ^*	cusp forms related to Δ	348
$\Delta_1^*, \Delta_2, \Delta_3$	cusp forms related to Δ	349
$\Delta(n), \hat{\Delta}(n)$	normal subgroups containing U^n	19, 27, 37
δ	fourth entry in matrix S	1
δ_k	dimension of space of cusp forms	194
δ_k, δ_k^*	dimensions of spaces D_k, D_k^*	344
$\delta_k^{(r)}$	dimension of space of cusp forms	207
δ_L	parameter	138
δ_L^*	parameter	177
δ_N^*	parameter	171
$\delta_k(n)$	Fourier coefficient	205
$\delta_k(\Gamma)$	dimension of space of forms	104
$\varepsilon_2, \varepsilon_3, \varepsilon_2(\Gamma), \varepsilon_3(\Gamma)$	numbers of orbits	57
$\varepsilon(n)$	character	293
$\varepsilon(\zeta), \varepsilon(\zeta, \nu)$	elliptic fixed point parameters	94
$\zeta(s)$	Riemann zeta function	135
η	Dedekind function	213
η	partially multiplicative function parameter	354
Θ	SL(2, $\mathbb{C}$)	1
$\hat{\Theta}$	LF(2, $\mathbb{C}$)	4
$\Theta(n)$	Glaisher's function	242

$\Theta_{\mu,\nu}(z)$	Hermite function	215
θ_n	group isomorphism	254
ϑ_T	parameter associated with T	2
$\vartheta_1', \vartheta_2, \vartheta_3, \vartheta_4$	theta functions	218
$\vartheta_1'(0\|z)$, etc.	theta functions	218
$\vartheta(z)$	function formed from theta functions	231
$\vartheta_\alpha(\nu\|\tau)$	generalized theta function	243
$\kappa_2, \kappa_3, \kappa_\infty$	fixed point parameters	102
κ_L	cusp parameter	77
Λ	group generated by $-I$	2
$\lambda, \lambda(\Gamma)$	number of incongruent cusps	57
$\lambda, \lambda(z)$	modular function of level 2	226
λ'	parameter related to cusp parameters	102
λ_q	eigenvalue of operator H_q	329
$\lambda(n)$	multiplicative function associated with $\Gamma(n)$	22
$\lambda(n)$	eigenvalue of Hecke operator T_n	320
$\lambda(n)$	Glaisher's function	355
$\hat{\lambda}(n)$	$\hat{\mu}(n)/n$	62
$\lambda_f(n)$	eigenvalue of Hecke operator T_n	318
$\mu, \mu(\Gamma)$	index of group	55, 144
$\mu(n), \bar{\mu}(n)$	index of $\Gamma(n), \bar{\Gamma}(n)$	21
$\mu(n)$	Möbius function	167
$\hat{\mu}(n)$	index of $\hat{\Gamma}(n)$	21, 26
$\mu_L(\Gamma)$	index of Γ_L in Γ	116
$\mu(T, z)$	$(T{:}\,z)^k$	71, 119
$\nu, \nu(T, z)$	automorphic factor	70
ν^L	AF of transformed group	72
Ξ	orthogonal group of matrices	74
$\xi, \xi(\Gamma, v)$	fixed point parameter	102
π	3.14159...	*passim*
$\pi(T)$	period of abelian integral	126
ρ	$e^{2\pi i/3}$	14
Σ	group of matrices	74
σ	real part of s	338
$\sigma_{k-1}(r)$	divisor function	177

$\sigma'_{k-1}(n), \sigma''_{k-1}(n), \sigma'''_{k-1}(n)$	divisor functions	233
$\sigma^*_{k-1}(n), \sigma^\dagger_{k-1}(n)$	divisor functions	233
$\sigma(S, T)$	function associated with matrices S, T	72
$\sigma(l, N)$	divisor function	285
$\sigma_{k-1}(r; C, D; N)$	divisor function	177
$\tau(n)$	Ramanujan's function	197
$\Phi_n(\tau, t)$	modular equation polynomial	209
$\phi(n)$	Euler's function	26
$\phi(s)$	Dirichlet series	339
$\phi_k(s)$	Dirichlet series	204
$\phi_k(z)$	cusp form	242
χ	character	*passim*
χ^*	character	262
χ_0	principal character	172
$\chi_q, \chi_{q'}$	characters	261
$\chi(T)$	multiplier system	246
$\chi(U)$	sixth root of unity	83
$\chi_4(n)$	Glaisher's function	242, 356
$\chi_8(n)$	Glaisher's function	358
$\hat{\psi}_5$	newform	356
ψ_k	cusp form	242
ψ^*_k	cusp form	345
$\psi(n), \hat{\psi}(n)$	index of $\Gamma_0(n)$, $\hat{\Gamma}_0(n)$	26
$\psi_N(l)$	index of $\Gamma(N, l)$ in $\Gamma(N)$	274
Ω	$SL(2, \mathbb{R})$	1
$\hat{\Omega}$	$LF(2, \mathbb{R})$	4
Ω^+, Ω^*_n	groups of nonsingular matrices	113
$\Omega_0, \Omega^*_0, \Omega_n$	sets of nonsingular matrices	273
$\Omega(n)$	Glaisher's function	242
ω	number of unit modulus	82
ω_k	poisitive parameter	202
$\omega(q)$	number of different prime factors of q	170
$\varnothing$	empty set	*passim*
$\oplus$	direct sum sign	224
$\|A\|$	cardinality of finite set A	5
$\|T\|$	determinant of matrix T	1

$|z|$ absolute value of z *passim*
$\|f\|$ norm of f 131, 148
(L) double coset 275
$[0]$ class consisting of zero function 320
$[L]$ set of double cosets 276
$[f]$ eigenclass containing eigenform f 320
$[x]$ integral part of x 94
$[z]$ orbit containing z 47
$[\![L]\!]$ set of double cosets 284
$\{x\}$ fractional part of x 103
$x := y, x =: y$ definitions 7
$f|T$ stroke operator 89, 104, 119
$f|T_n$ Hecke operator 289
$q|r$ q divides r 22
$n|N^\infty$ each prime p dividing n divides N 288
$q \nmid r$ q does not divide r *passim*
$T:z$ $cz+d$ 42, 119
$A \cdot B$ product with uniqueness condition 5
$(c, d), (c, d, n)$ highest common factors 22
$(f, g), (f, g; \Gamma)$ scalar product 146, 190
$G * H$ free product 13
$[S, T]$ commutator 16
$[c, d]$ second row of T 9
$[\Gamma_1 : \Gamma_2]$ index of Γ_2 in Γ_1 5
$\langle A, B \rangle$ group generated by A, B 7
$\{m, n\}$ least common multiple of m and n 23
$\{N, k\}, \{N, k\}_0$ spaces of forms 245
$\{N, k, \chi\}, \{N, k, \chi\}_0$ spaces of forms 246
$\{N, k, \chi\}^{(t)}, \{N, k, \chi\}_0^{(t)}$ spaces of forms 250
$\{N, k, \chi, t\}, \{N, k, \chi, t\}_0$ spaces of forms 253

Index of authors

Atkin, A. O. L. 272, 334, 336
Ayoub, R. 213, 258

Berberian, S. K. 148
Berndt, B. C. 359
Bol, G. 131
Borchardt, C. W. 333

Copson, E. T. 244

Dedekind, R. 333
Deligne, P. 132, 335, 360
Dickson, L. E. 244

Eichler, M. 132
Estermann, T. 169

Ford, L. R. 67
Frasch, H. 37
Fricke, R. 32–3, 37–8, 66, 272
Fueter, R. 214, 244

Gantmacher, F. R. 317
Glaisher, J. W. L. 242, 244, 352, 355–7, 360
Grosswald, E. 213
Gunning, R. C. 21, 68, 71, 104, 334

Hall, M. 24, 117
Hardy, G. H. 132, 167, 170, 244
Hecke, E. 127, 131, 141, 183, 191–2, 271–2, 334–5, 359
Hua, L. K. 13
Hurwitz, A. 272, 333

Igusa, J.-I. 132

Jacobi, C. G. J. 244
Joris, H. 133

Klein, F. 32–3, 37–8, 66, 272
Kløve, T. 335
Knopp, M. I. 130, 192, 213
Kogan, L. A. 244
Kurosh, A. G. 34

Lehmer, D. H. 205, 213
Lehner, J. 66–7, 86, 130–1, 192, 213, 272, 334–5
Leutbecher, A. 37, 87
LeVeque, W. J. 244
Li, W. C. W. 335, 360
Liouville, J. 244
Lyndon, R. 35

Macbeath, A. M. 67
MacDuffee, C. C. 114–15
McQuillan, D. L. 37
Mason, A. W. 35, 38
Millington, M. 335
Miyake, T. 335
Molk, J. 243–4
Mordell, L. J. 168, 184, 333
Moreno, C. J. 360

Newman, M. 34–5, 38, 130
Nielsen, J. 34–5

Ogg, A. P. 272, 334

Petersson, H. 33, 36, 80, 86, 107, 131, 141, 191–2, 214, 334, 359
Pick, G. 38

Rademacher, H. 86, 213
Ramanujan, S. 213, 244, 333, 359
Rankin, F. K. C. 213
Rankin, R. A. 33–4, 37–8, 67, 105, 131–3, 214, 244, 335, 343, 357–60
Reiner, I. 13, 38
Resnikoff, H. L. 131
Rushforth, J. M. 335, 343

Schoeneberg, B. 67–8
Schreier, O. 34–5
Schwandt, E. A. 191
Selberg, A. 132–3, 191
Serre, J.-P. 132, 334–5, 360
Shimura, G. 334–5
Springer, G. 126
Stepanov, S. A. 193
Swinnerton-Dyer, H. P. F. 231, 336

Tannery, J. 243–4
Turton, R. J. 360

van Lint, J. H. 86, 214, 335
Vivanti, G. 37

Walfisz, A. 131, 133
Watson, G. N. 156–7, 205
Weber, H. 33, 244
Weil, A. 132, 192, 360
Whittaker, E. T. 156
Williams, K. S. 193
Wilton, J. R. 131
Wohlfahrt, K. 33, 37, 87, 335
Wright, E. M. 167, 170

Subject index

abelian integral 126
absolutely uniformly convergent 134
adjoint operator 317
AF 70
algebraic curves 192
alternating group 37
associative law 74
automorphic factor 70, 135
automorphic form 192
automorphic function 191
automorphism 13, 254
awkward case 165

Bernoulli numbers 195, 233
Bessel functions 156
bilinear mapping 2
branch schema 60
branched covering surface 68
branchpoint 68

canonical form 261
canonical fundamental region 69
Cayley numbers 360
character 79, 232, 246, 258, 271
character of real type 262
class 320
coboundary 87
cocycle 87
codivisor 248
cohomology 87
commutator 16
commutator group 16
complementary forms 192
conductor 258
conformal mapping 40
congruence group 20, 25
congruent point 47
conjugate character 246
conjugate linear map *K* 270, 300, 314
conjugate Hecke operator 295
constant multiplier system 79
core 81
correspondence 334
cusp 45
cusp form 94
cusp parameter 78, 89
cusp width 47
cycloidal group 36

Dedekind eta function 213
differential 126
dimension (weight) 88
dimension of space 102
direct sum 106
Dirichlet region 67
Dirichlet series 171–2, 233, 337–42, 359, 360
discriminant (modular) 197–8, 211–12, 222
divisor 248, 271
divisor functions 177, 233–4, 285
domain 39
double coset 113, 273, 275, 291
double coset module 279, 334

eigenclass 320
eigenform 317
eigenvalue 317
eigenvector 313
Eisenstein series 136, 142, 175, 177, 191, 194, 213, 232–8, 239–40, 312, 343
elliptic fixed point 43
elliptic transformation 43, 45
entire modular form 94
equivalent point 47
eta function 213
Euler factor 341
Euler numbers 233
Euler product 341, 359
extended complex plane 1

fixed point 42
form of character χ 246
form of divisor t 248
four-group 13
Fourier coefficients 128–30, 155, 173–9, 194, 197, 213–14, 234–8, 247, 290, 297, 313–15, 316, 328
fractional part 103
fundamental region 50, 67
fundamental set 67

gamma function 156, 339
Gauss–Bonnet formula 67
Gaussian numbers 360
genus 68, 104
Gram–Schmidt orthogonalization 148

Grenzkreisgruppe 86, 191
groups of level 2 28, 215–44
groups of level 3 30

Hankel's formula 156
Hauptkongruenzgruppe 20
Hecke operator 132, 263, 289 *et seq.*
height of point 51
Hermite functions 215–18, 243
Hermite's normal form 114
Hilbert space 131, 144, 148, 317
holomorphic 39
homogeneous group 3
homogeneous modular group 8
horocyclic group 86, 191
hyperbolic area 66
hyperbolic distance 66
hyperbolic fixed point 43
hyperbolic length 66
hyperbolic straight line 66
hyperbolic transformation 43, 45

icosahedral group 37
identity transformation 45
imprimitive cusp form 317
inhomogeneous group 3
inhomogeneous modular group 8
inner automorphism 13
inner product 146, 191
integral part 94
invariants g_2, g_3 213
isometric circle 67

Klein's absolute invariant 199, 213, 227–30
Kloosterman sum 132, 161, 164–71, 175, 191–2

lattice 334
left-equivalent 113
left-module 279
left transversal 5
length of word 12, 13
level 19
linear fractional group 4
linear fractional mapping 2
local uniformizing variable 90
logarithmic winding point 68
LUV 90

meromorphic 39
Möbius function 167
Möbius mapping 2
modular discriminant 197
modular equation 118, 209–12, 230
modular form 88–91
modular forms of weight 2 122–8
modular function 108
modular function λ 226–32, 244
modular functions of level 2 226–32
modular group 8
Modulargruppe 21
modulary group 21, 37, 254
Modulgruppe 8
MS 71
multiplicative function 22, 204
multiplier 71
multiplier system 71–87 *passim*, 135, 180, 182, 206, 208–9, 214, 219, 221, 223, 238, 244–5, 255, 270, 345

newform 326
non-congruence group 38
norm 131, 148
normal operator 317
normal polygon 67
normalized eigenform 318
number of cusps 57

octahedral group 37
oldform 328
orbit 47, 68
order of function at cusp 91
order of point 46
order of pole 40
orthogonal basis 319
orthogonal spaces 155, 247
orthonormal basis 317

parabolic fixed point 43
parabolic transformation 43, 44, 45
partially multiplicative function 356
period 127
Picard's theorem 244
Poincaré series 134–93 *passim*, 201–2, 213, 238, 271, 307–15
Poisson summation formula 184, 243
pole 40
primitive character 258
primitive cusp form 316
primitive matrix 113, 273
primitive pair 21, 22
primitive transformation 113
principal character 254, 258, 345
principal congruence group 20
principal value of logarithm 70
projection operator 268
proper fundamental region 47, 67

quaternions 360

Ramanujan–Petersson conjecture 132–3, 359
Ramanujan's function 197, 213, 243

Ramanujan's sum 167
regular matrix 141
residue 122–3
Riemann mapping theorem 200
Riemann–Roch theorem 68, 104, 192
Riemann surface 48, 68, 126
Riemann zeta-function 135, 171, 194–5, 204, 339, 360
right-equivalent 113
right transversal 5
Rouché's theorem 40

scalar product 146
Schwarz's lemma 41
self-adjoint operator 247
semisymmetric multiplier 87
side 52, 55
Smith's normal form 115
special linear group 2
stabilizer 46
stroke operator 89, 119, 245
Stufe 19
sums of squares 238–43, 349
symmetric multiplier 87

tetrahedral group 37
theta functions 215–42 *passim*, 345, 347–50, 352, 355–8
total order 95
trace operator 107, 272
transform 89
transformation equation 118
transformation group of order n 116
transformation of order n 112
translation operator 247
transversal 5
triangle 51
trivial class 320

unitary divisor 260
unrestricted modular form 88
upper half-plane 1

valence 109
vectorial modular form 132
vertex 55
vertex angle 55
vertex set 55

Weierstrass elliptic function 213
Weierstrass point 192
weight 88
width of cusp 47, 89
word 12–13

zero function 88